Design of Blast-Resistant Buildings in Energy and Industrial Facilities

Other Titles of Interest

Seismic Evaluation and Design of Petrochemical and Other Industrial Facilities, Third Edition, by Task Committee on Seismic Evaluation and Design of Petrochemical Facilities (ASCE/Energy Division 2020). This report provides practical recommendations affecting the safety of new and existing petrochemical and other industrial facilities during and following an earthquake. (ISBN 978-0-7844-1548-1)

Anchorage Design for Petrochemical and Other Industrial Facilities, Second Edition, by Task Committee on Anchorage Design (ASCE/Energy Division 2022). This book intends to serve as a source for uniformity in the design, fabrication, and installation of anchorage in petrochemical and other industrial facilities. (ISBN 978-0-7844-1605-1)

Wind Load Design for Petrochemical and Other Industrial Facilities, Second Edition, by Task Committee on Wind-Induced Forces (ASCE/Energy Division 2020). This report represents the state-of-the-practice for wind load design at industrial facilities and is aimed at engineers familiar with design of industrial type structures. (ISBN 978-0-7844-1561-0)

Structural Design for Physical Security, by Task Committee on Structural Design for Physical Security; edited by Peggy Van Eepoel and Sharon M. Gallant (ASCE/Structural Engineering Institute 2021). This Manual of Practice provides guidance to structural engineers in the design of civil structures to resist the effects of terrorist bombings. (ISBN 978-0-7844-1549-8)

Lessons From the Oklahoma City Bombing: Defensive Design Techniques, by Eve E. Hinman and David J. Hammond (ASCE/ASCE Press 1997). Hinman and Hammond document the 1995 bombing of the Murrah Federal Office Building in Oklahoma City from the structural engineer's perspective, covering the weapon used, the specific damage sustained, and vulnerabilities inherent in the building's design. (ISBN 978-0-7844-0217-7)

Design of Blast-Resistant Buildings in Energy and Industrial Facilities

Third Edition

Sponsored by
Task Committee on Blast-Resistant Design of the Energy Division of the
American Society of Civil Engineers

Edited by
Luca Magenes, P.E.

ASCE

Published by the American Society of Civil Engineers

Published by American Society of Civil Engineers
1801 Alexander Bell Drive
Reston, Virginia 20191-4382
www.asce.org/bookstore | ascelibrary.org

Errata: **Errata, if any, can be found at https://dx.doi.org/10.1061/9780784485897.**

ISBN 978-0-7844-8589-7 (PDF)
ISBN 978-0-7844-1629-7 (Print)
ISBN 978-0-7844-8590-3 (ePub)

Manufactured in the United States of America.

29 28 27 26 25 1 2 3 4 5

Photo Credit: The photo appearing on the back cover of the book shows a deflagration test of a pre-engineered metal building. It is provided courtesy of Baker Engineering and Risk Consultants and the Explosion Research Cooperative.

Contents

Contributors

TASK COMMITTEE ON BLAST RESISTANT DESIGN

Luca Magenes, P.E., P.Eng, M.ASCE, *Chair*, Powell Industries
Matthew T. Edel, P.E., M.ASCE, *Co-chair*, Baker Engineering & Risk Consultants
Timothy B. Sisson, P.E., *Secretary*, US General Services Administration

MEMBERS

Charles Auverny Bennetot, *Total Energies*
Darrell D. Barker, P.E., M.ASCE
William L. Bounds, P.E., M.ASCE, *FLUOR*
Matthew J. Brightman, P.E., S.E., M.ASCE, *ExxonMobil*
Haritha Cherkatu Vasala, *KBR (Energo)*
Ozer Dereli, P.E., M.ASCE, *Wood*
Jason R. Florek, Ph.D., P.E., M.ASCE, *Stone Security Engineering*
Song F. Jan, Ph.D., P.E., M.ASCE
David Kerins, P.E., F.ASCE, F.ACI
Theodor Krauthammer, Ph.D., M.ASCE, *University of Florida*
Phillip Lange, A.M.ASCE, *RedGuard*
Guzhao Li, Ph.D., P.E., S.E., M.ASCE, *Simpson Gumpertz & Heger*
Michael J. Lowak, A.M.ASCE, *Baker Engineering and Risk Consultants*
Rudy Mulia, P.E., S.E., M.ASCE, *Chevron*
Ravi Mullapudi, Ph.D., P.E., S.E., M.ASCE, *Walter P Moore*
Franz Mutis, P.E., S.E., M.ASCE, *Synergeer Engineering*
Carlos A. Ortega, Ph.D., P.E., S.E., M.ASCE, *Simpson Gumpertz & Heger*
Charles Oswald, Ph.D., P.E., M.ASCE
Mehmet Cem Ozbey, Ph.D., Aff.M.ASCE
Tom Tanas Panayotidi, D.Eng.Sc, P.E., M.ASCE, *Columbia University*

Mike Rookstool, P.E., S.E., M.ASCE, *Chevron*

Ali Sari, Ph.D., P.E., M.ASCE, *Omega-Risk*

Mohammad Sharafbayani, Ph.D., P.E., M.ASCE

Seung Han Song, Ph.D., P.E., M.ASCE, *Bechtel Corporation (Retried)*

Paul B. Summers, P.E., S.E., F.ASCE, *Simpson Gumpertz & Heger*

James W. Wesevich, P.E., S.E., M.ASCE, *Jensen Hughes*

Yu Zhao, Ph.D., P.E., S.E.

Acknowledgments

The Task Committee for this publication update would like to thank the following individuals who independently reviewed the manuscript and provided valuable comments: Raymond Bennett, Ph.D, P.E., John Geigel, and Lauren Stewart, Ph.D. In particular, we would like to acknowledge John Geigel for his substantial contribution to this publication and its previous editions and for his lifelong contributions to the civil engineering field at large.

A special acknowledgment goes to Jason Florek, Ph.D., P.E., who lead the significant updates to the practical example chapters (Ch.10, 11, and 12). The committee would like to recognize the efforts of the previous two Task Committees, and the past committee Chair, William Bounds, who built the foundation for this publication update.

The authors are grateful for the support received from the ASCE Energy Division and Committee on Technical Advancement, which continue to sponsor the updates of this important industry publication.

Additional Contributors

Clay Efferson

Kieran Joseph Glynn, BP

Paulus Goudswaard

Sauren Guha-Majumdar, P.E.

Haijian Shi, P.E., *Pepco Holdings Inc.*

Preface

This report is the third edition of this publication. The previous version was titled *Design of Blast-Resistant Buildings in Petrochemical Facilities* and was published in 2010. The title of this report was changed for this edition to better represent the broad range of applications for which the report has been used and for its intended audience. In addition to the previously mentioned petrochemical industry, some additional industries to which this report holds applicability include, but are not limited to, power generation, hydrogen and ammonia production, battery manufacturing and utilization, pharmaceuticals, and other chemical processing industries.

This new edition includes updates that are in line with new developments in the practice of structural blast engineering in energy and industrial facilities including

- Expanded discussions on risk-based facility siting,
- New building types (e.g., timber and unreinforced masonry buildings),
- Loading on elevated buildings,
- New screening-level guidance,
- Mitigation using energy-absorbing systems,
- Updated response criteria for prestressed concrete and reinforced concrete, and
- Updated and expanded example chapters.

We hope that this updated report provides useful guidance to a wide community of structural engineers working on the design of structures and buildings to resist the effects of blasts, explosions, and impulsive/impactive loads.

Nomenclature

a	= Acceleration
a_H	= Horizontal acceleration
a_V	= Vertical acceleration
a_θ	= Rotational acceleration
A_1	= Bearing area
A_2	= Pedestal area
A_b	= Nominal bolt area
A_g	= Gross area
A_n	= Nominal area
A_{nv}	= Net section area
A_{ps}	= Area of prestressed reinforcing in the tension zone
A_s	= Area of steel
A_v	= Shear area of steel
A_w	= Weld area
b	= Member width
b_f	= Flange width
B_H	= Building height
B_L	= Building length
B_p	= Plate width
B_W	= Building width
BL	= Blast load
c	= Impulse factor for a decaying shock wave
c_1	= Distance from the center to the edge of the element
c_h	= Horizontal weld eccentricity
c_v	= Vertical weld eccentricity
C	= Viscous damping constant
C_d	= Drag coefficient
C_e	= Side wall reduction factor
C_F	= Loading reaction coefficient
C_r	= Reflection coefficient
C_R	= Resistance reaction coefficient
C_V	= Shear coefficient
d	= Depth
d_b	= Bolt diameter

d_h = Hole diameter
d_{na} = Depth to neutral axis
d_p = Depth to the center of prestressing steel
d_{sb} = Depth of a concrete stress block
DL = Dead load

E = Modulus of elasticity
E_c = Modulus of elasticity of concrete
E_m = Modulus of elasticity of masonry
E_s = Modulus of elasticity of steel

f = Frequency of vibration
f_c = Concrete stress
f'_c = Concrete strength
f_{cr} = Compression stress
f'_{dc} = Dynamic concrete strength
f'_{dm} = Dynamic masonry strength
f'_m = Masonry strength
f_{ps} = Calculated stress in prestressing steel at the design load
f_r = Modulus of rupture
f_{vh}, f_{v1}, f_{v2} = Shear stress
F = Blast force
$F_1 \ldots F_7$ = Blast force
F_{cr} = Flexural buckling stress
F_C = Compression force
F_{du} = Steel dynamic ultimate strength
F_{ds} = Dynamic design stress
F_{dy} = Steel dynamic yield strength
F_e = Equivalent force
F'_e = Elastic critical buckling stress
F_{EXX} = Weld electrode strength
F_H = Horizontal dynamic force
F_{nt} = Nominal bolt tension strength
F_{nv} = Nominal bolt shear strength
F_o = Peak blast force
F_{RL} = Reflection load
F_s = Static load
F_{SL} = Stagnation load
F_T = Tension force
F_u = Steel ultimate strength
F_V = Vertical dynamic force
F_w = Weld strength
F_y = Steel yield strength
F_θ = Rotational dynamic force

FS = Factor of safety

g = Acceleration of gravity
g_b = Bolt gauge
G = Shear modulus

h = Height
J_w = Polar moment of inertia

i = Time step number
I = Moment of inertia
I_a = Averaged moment of inertia
I_{bar} = Nondimensionalized impulse
I_{cr} = Cracked moment of inertia
I_g = Gross moment of inertia
I_o = Positive phase impulse
I_o^- = Negative phase impulse
I_w = Equivalent triangular impulse
I_x = Moment of inertia about the x-axis
I_y = Moment of inertia about the y-axis
ip = (subscript) In-plane deformations

k_v = Shear width-thickness coefficient
K = Stiffness
K_a = Coefficient of active soil pressure
K_e = Equivalent or effective stiffness
K_{ep} = Elastic-plastic stiffness
K_H = Horizontal soil stiffness
K_L = Load or stiffness transformation factor
K_{LM} = Load-mass transformation factor
K_M = Mass transformation factor
K_p = Coefficient of passive soil pressure
K_V = Vertical soil stiffness
K_x = Effective length factor for the x-axis
K_y = Effective length factor for the y-axis
K_θ = Rotational soil stiffness
KE = Kinetic energy
KEY = Response indicator

L = Length
L_1 = Length of the surface parallel to a traveling blast wave
L_c = Clear distance
L_h = Horizontal weld length
L_p = Limiting unbraced length for plastic design

L_s = Spacing
L_v = Vertical weld length
L_w = Blast wave length
L_x = Unbraced length in the x-axis
L_y = Unbraced length in the y-axis
LL = Live load

m_1 = Baseplate design dimension
M = Mass
M_1 = Foundation moment
M_c = Midspan moment
M_{cr} = Cracking moment
M_e = Equivalent mass
M_p = Plastic moment
M_{ps} = Moment capacity at the support
M_{pc} = Moment capacity at midspan
M_y = Yield moment
M_θ = Mass moment of inertia

n = Modular ratio
n' = Baseplate design dimension
n_1 = Baseplate design dimension
n_w = Horizontal weld center
N = Number of structural elements
N_p = Plate length
N_u = In-plane capacity of the element
N_u' = Ultimate applied in-plane load

OM = Overturning moment
op = (subscript) Out-of-plane deformations

P = Pressure
$P_1 \ldots P_5$ = Vertical static load on the foundation
P_2 = Vertical static load on the foundation
P_3 = Vertical static load on the foundation
P_a = Effective side-on overpressure
P_b = Rear face overpressure
P_c = Permissible compression
P_{bar} = Nondimensionalized pressure
P_{nc} = Nominal compression capacity
P_{nt} = Nominal tension capacity
P_o = Ambient atmospheric pressure
P_p = Available strength
P_r = Peak-reflected overpressure

P_s	= Stagnation pressure
P_{so}	= Peak side-on overpressure
P_{so}^-	= Negative peak side-on overpressure
P_t	= Permissible tension
P_u	= Factored axial load
q	= Bearing pressure
q_o	= Peak dynamic pressure
Q_u	= Shear capacity of the element
Q_u'	= Ultimate applied shear load
r_x	= Radius of gyration about the x-axis
r_y	= Radius of gyration about the y-axis
R	= Resistance
R_1	= Axial component of a battered pile
R_b	= Bending resistance
R_{b-}	= Negative bending resistance
R_e	= Equivalent resistance
R_{HORIZ}	= Lateral load component of a battered pile
R_m	= Maximum resistance
R_{MAX}	= Maximum axial pile load
R_{MIN}	= Minimum axial pile load
R_n	= Nominal resistance
R_s	= Shear resistance
R_u	= Ultimate resistance
R_{u-}	= Rebound resistance
R_y	= Yield resistance
RM	= Resisting moment
S	= Clearing distance
S_x	= Section modulus about the x-axis
SE	= Strain energy
t	= Time
t_a	= Time of arrival
t_c	= Clearing time
t_d	= Positive phase duration
t_d^-	= Negative phase duration
t_e	= Equivalent duration
t_f	= Flange thickness
t_m	= Time of maximum response
t_n	= Natural period
t_o	= Total positive phase duration
t_p	= Plate thickness

t_r = Rise time
t_w = Web thickness

U = Shock front velocity

v = Velocity
V = Dynamic reaction
$V_1..V_9$ = Lateral load on the foundation
V_f = Fixed end reaction
V_n = Nominal shear
V_p = Pinned end reaction
V_u = Ultimate shear

w_p = Reinforcement index
W = Width
X = Baseplate design factor
X_a = Allowable sidesway
X_e = Elastic sidesway
X_m = Maximum sidesway

y = Deflection
y_d = Dead load deflection
y_e = Yield deflection
y_{e-} = Rebound yield deflection
y_{ep} = Elastic-plastic deflection
y_H = Horizontal deflection
y_m = Maximum deflection
y_{m-} = Maximum rebound deflection
y_p = Plastic deflection
y_{st} = Static deflection
y_V = Vertical deflection
y_θ = Rotational deflection

Z = Calculation case indicator
Z_x = Plastic modulus about the x-axis

α = Angle of incidence
α_1 = Angle
α_2 = Angle

Δ_a = Allowable deformation (ductility ratio or support rotation)
Δ_d = Computed deformation (ductility ratio or support rotation)
Δ_f = Flexural deflection
Δ_s = Shear deflection

ε_u = Maximum strain
ε_c = Concrete strain
ε_s = Steel strain
ε_u = Ultimate concrete strain
ε_y = Steel yield strain

ϕ = Strength reduction factor
Φ = Curvature or deflected shape

γ = Soil density

λ = Roughness coefficient
λ_{bp} = Baseplate design factor
λ_c = Compression slenderness factor
λ_f = Flange slenderness factor
λ_p = Limiting plastic slenderness factor
λ_{ps} = Limiting seismic plastic slenderness factor
λ_v = Shear slenderness factor
λ_w = Web slenderness factor

μ = Ductility ratio
μ_a = Allowable ductility ratio
μ_d = Demand ductility ratio
μ_{d-} = Demand rebound ductility ratio
μ_f = Friction coefficient

ν = Poisson's ratio

θ = Rotation or hinge rotation
θ_a = Allowable support rotation
θ_d = Demand support rotation
θ_1 = Support rotation
θ_2 = Midspan rotation

ρ = Tension-reinforcing ratio

τ = Ratio of load duration to the natural period (t_d/t_n)

CHAPTER 1
Introduction

The focus of this report is on blast-resistant structural aspects of designing new buildings, as well as evaluating and modifying/strengthening existing buildings. The term "petrochemical facility", which was included in the original title of this publication, typically refers to a facility that refines or manufactures petroleum-based products. However, the guidelines related to such facilities can also be used for buildings located on any type of industrial facility that have hazards with the potential to generate blast overpressure from accidental events arising from handling hydrocarbons and other flammable or reactive materials, or from generating dust. These industrial facilities can include, for example, upstream oil and gas processing and storage facilities, refineries, liquefied natural gas (LNG) liquefaction and regasification facilities, chemical manufacturing plants, sugar factories, biofuel plants, and others. For these reasons, the title of this publication was changed to *Design of Blast-Resistant Buildings in Energy and Industrial Facilities*. The term petrochemical facility is still used in the body of this book and should be interpreted as inclusive of those energy and industrial facilities that have the potential for accidental explosions.

The methodology to design blast-resistant buildings uses the following steps:

1. Identify a range of potential accident scenarios that could result in blast overpressure.
2. Quantify the free field blast loads that could result at the building location.
3. Establish the design blast loads for the various surfaces of the building [walls, roofs, and sometimes floor (for elevated buildings)].
4. Set the structural performance requirements, including building functional performance, after a blast event as it relates to structural, architectural, and mechanical elements.
5. For existing buildings—assess the performance of the structure and other ancillary and architectural features, such as doors, windows, and heating, ventilation, and air-conditioning (HVAC) penetrations, and compare them to structural performance requirements.
6. For new buildings and upgrades to existing buildings—design the structure and other ancillary and architectural features to resist the design blast loads within the structural performance requirements.

Note that Steps 1 and 2 are outside of the purview of these guidelines. See Chapter 3, "Determination of Loads," for additional commentary.

Buildings in energy and industrial facilities can be intended to be occupied by personnel and/or intended to host equipment that controls the facility and its emergency systems. Blast-resistant design, or the structural strengthening of these buildings subjected to explosion hazards, is one of the measures an owner may employ to minimize the risk to people and facilities from the hazards of accidental explosions in a plant. Other mitigative or preventive measures, including siting (adequate spacing from potential explosion hazards) and hazard reduction (inventory and process controls, occupancy limitations, etc.), are not covered in this report.

1.1 BACKGROUND

Petrochemical facilities handle flammable and reactive materials, which can produce (and have) produced accidental explosions in the past. It is recognized that plant design and procedures (operations and maintenance) can reduce the likelihood of accidental explosions; however, if an accidental explosion does occur, severe consequences including safety (casualties and injuries to on-site and off-site populations) and financial loss (property damage, business interruption) could result.

A number of accidental explosions at petrochemical facilities in the past have resulted in fatalities to occupants of plant buildings. For example, the 1989 explosion in Pasadena, Texas, occurred during maintenance of a polyethylene unit and included the collapse of a control building. Losses included 23 fatalities and 120 injured. The 2005 explosion in Texas City, Texas, occurred during the restart of a refinery isomerization unit and resulted in the destruction of trailers used as temporary offices. Losses caused by this accident included 15 fatalities and 170 injured.

Other explosions that have occurred at industrial facilities have resulted in a significant number of fatalities and injuries from the severe damage or collapse of buildings. A sugar dust explosion that occurred in Port Wentworth, Georgia (2009), resulted in 14 fatalities and 39 injuries. More recent explosion accidents that have resulted in damage to buildings, injuries and fatalities include Amauy, Venezuela (2012); West, Texas (2013); Veracruz, Mexico (2016); and Cambria, Wisconsin (2017). Figures 1-1 to 1-4 show some of the extent of damage caused by some of these incidents.

During the Mexichem vapor cloud explosion (VCE) of April 20, 2016, in Coatzacoalcos, Veracruz, Mexico, there were 32 fatalities and more than 100 injured. Figure 1-3 shows a heavily damaged building in the process unit.

A review of these accidents points to the need to design occupied plant buildings to withstand explosion effects so that the building does not pose an added hazard to its occupants. This is an expression of "neutral risk" that has largely been replaced with more stringent protection performance criteria.

Figure 1-1. Port Wentworth, Georgia.
Source: Chemical Safety Board (2009).

Figure 1-2. West, Texas.
Source: US Chemical Safety Board (2013).

In addition to personnel safety, it has been common practice in the industry to consider blast resistance for buildings such as control centers, substations, and emergency response centers that may have limited or no personnel occupancy, but that house critical operations and emergency equipment required to aid in the safe shutdown and emergency response of the facility to an accidental explosion.

Figure 1-3. Veracruz, Mexico (2016).
Source: Thornton Tomasetti Database.

(a)

(b)

Figure 1-4. Cambria, Wisconsin (2017) (a) before and (b) after the accident.
Source: Chemical Safety Board (CSB.gov).

For buildings, blast loading can be the most damaging effect from an accidental explosion in a process plant. However, in addition to blast loading, such incidents can result in fires, toxic gas ingress, and projectiles that can cause damage to buildings and potentially cause harm to occupants and/or damage equipment within the building.

Historically, blast-resistant design technology in the petrochemical and other industries has evolved on two fronts, namely: (1) hazard analysis and explosion modeling, and (2) structural responses. The following describes the evolution in chronological order:

1. Equivalent static loads and conventional static design methods (Bradford and Culbertson 1967).
2. Blast loading derived using TNT equivalence and simplified dynamic design methods that take into account dynamic characteristics and ductility of structural components (Forbes 1982).
3. More complex and rational methods involving VCE models to characterize the blast loading (Pierorazio et al. 2005) and nonlinear multi-degree-of-freedom dynamic models to analyze the building structure.
4. High-fidelity physics-based numerical methods such as computational fluid dynamics (CFD), turbulent combustion models to simulate blast loading, and dynamic nonlinear finite-element (FE) models for structural response.

It should be noted that current practices within the industry include or refer to some (or all) of the approaches described previously.

1.2 PURPOSE AND SCOPE

This report is intended to provide guidelines on the various methods available for the structural analysis and design of blast-resistant buildings, both new and existing, at industrial facilities. It provides the basic considerations, principles, procedures, and details for structural analysis and design of buildings to resist blast overpressure effects. Evaluation and retrofit of existing buildings are also included. The guidelines do not address considerations regarding protection of buildings from potential thermal hazards and toxic gas or smoke ingress that may or may not accompany a VCE. However, some applicable industry references are listed subsequently, and further context is provided in Chapter 2, "General Considerations."

The report focuses primarily on "how to" design, evaluate, and retrofit buildings for blast resistance once the blast loading is defined for a postulated explosion scenario. Chapter 2, "General Considerations," discusses the basic philosophy and general considerations involved in establishing design requirements for blast resistance of buildings to resist the effects of accidental explosions in energy and industrial facilities. Chapter 3, "Determination of Loads," describes the types of explosions that may occur and the general characteristics of the resulting blast load but does not prescribe values for design. The chapter provides a brief review of the approaches used in the industry to quantify blast loads for design purposes and gives typical examples of such loads. In Chapter 4, "Types of Construction," the types of building construction appropriate for various levels of blast resistance are discussed. The dynamic ultimate strength design criteria, including the dynamic material properties and deformation limits applicable to blast-resistant design, are covered in Chapter 5, "Dynamic Material Strength and Response Criteria."

The methods and procedures for blast-resistant design can vary considerably in terms of complexity, accuracy, cost, and efficiency from a simple conventional static design approach to a transient nonlinear, multi-degree-of-freedom dynamic design methods to a combination of CFD and finite-element analysis. To assist engineers in striking a balance among these, Chapter 6, "Dynamic Analysis Methods," provides a discussion of the various blast-resistant analysis methods, while identifying the main features, advantages, and disadvantages of each method. Chapter 7, "Design Procedures," outlines recommended procedures and provides aids for the design of the typical components of reinforced concrete, reinforced masonry, and structural steel buildings. Chapter 7 also provides some typical structural details for doors and frames, wall penetrations, and connections for steel and reinforced concrete components. Blast protection considerations for nonstructural items such as interior details, windows, openings, and HVAC ductwork are covered in Chapter 8, "Ancillary and Architectural Considerations." Chapter 9, "Evaluation and Upgrade of Existing Buildings," provides some practical measures for upgrading masonry and metal buildings, which represent some of the most common types of building construction for plants in the energy industry. Design examples are provided in Chapter 10, "Shear Wall Building Design Example," Chapter 11, "Metal Building Design Example," and Chapter 12, "Masonry Retrofit Design Example," to illustrate the use of these procedures in the design of typical buildings for blast resistance.

1.3 RELATED INDUSTRY GUIDELINES, SPECIFICATIONS, AND CODES

The design practices used by some operating companies and contractors are based on a number of existing documents dealing with this subject, including

1. PIP STC01018 (PIP 2014), *Blast Resistant Building Design Criteria.*
2. ACI 370R-14 (ACI 2014), *Report for the Design of Concrete Structures for Blast Effects.*
3. CIA (2020), *Guidance for the Location and Design of Occupied Buildings on Chemical Manufacturing and Similar Major Hazard Sites.*
4. UFC 3-340-02 (US DoD 2014), *Structures to Resist the Effects of Accidental Explosions.*
5. ASCE 59-22 (ASCE 2022), *Blast Protection of Buildings.*

6. Steel Design Guide 26 (AISC 2013), *Design of Blast Resistant Structures.*
7. PDC TR 06-08 (USACE 2005), *Single Degree of Freedom Structural Response Limits for Antiterrorism Design.*
8. API RP 752 (API 2009), *Management of Hazards Associated with Location of Process Plant Permanent Buildings* (A new edition of this document was published in January 2024.)
9. API RP 753 (API 2007), *Management of Hazards Associated with Location of Process Plant Portable Buildings.*
10. API RP 756 (API 2014), *Management of Hazards Associated with Location of Process Plant Tents.*
11. API RP 2FB (API 2006), *Recommended Practice for the Design of Offshore Facilities against Fire and Blast Loading.*
12. CCPS (2012), *Guidelines for Evaluating Process Plant Buildings for External Explosions, Fires and Toxic Releases.* (A new edition of this document was published in January 2024.)

The PIP and CIA documents are similar and cover the siting, design, and/or construction of buildings in petrochemical plants for the simplified dynamic (elasto-plastic, single degree of freedom) design approach. The other documents cited previously are more comprehensive but are usually geared to design for high-yield explosives for military and munitions applications. However, the fundamentals and design principles covered in these documents are applicable to designs for other types of explosions.

In addition to the publications cited previously, the American Institute of Chemical Engineers, Center for Chemical Process Safety (CCPS) committee and the American Petroleum Institute (API) have addressed various aspects of blast protection technology relevant to this report. In particular, CCPS has developed *Guidelines for Evaluating the Characteristics of Vapor Cloud Explosions, Flash Fires, and BLEVEs* (CCPS Explosion Guidelines) (CCPS 1994) and *Guidelines for Evaluating Process Plant Buildings for External Explosions, Fires, and Toxic Releases* (CCPS Building Guidelines) (CCPS 2012).

1.4 BLAST-RESISTANT DESIGN PROCESS

The overall process involved in the evaluation and design of energy and industrial plant buildings for explosion hazards is illustrated in Figure 1-5, which shows 15 basic steps as follows:

1. Define Scope: Steps 1 and 2 are to define the owner's requirements and needs for the building.
2. Analyze Explosion Hazards: Steps 3 and 4 are to identify the explosion scenarios to be used to quantify the design blast overpressures (refer to Chapter 3, Section 3.4).

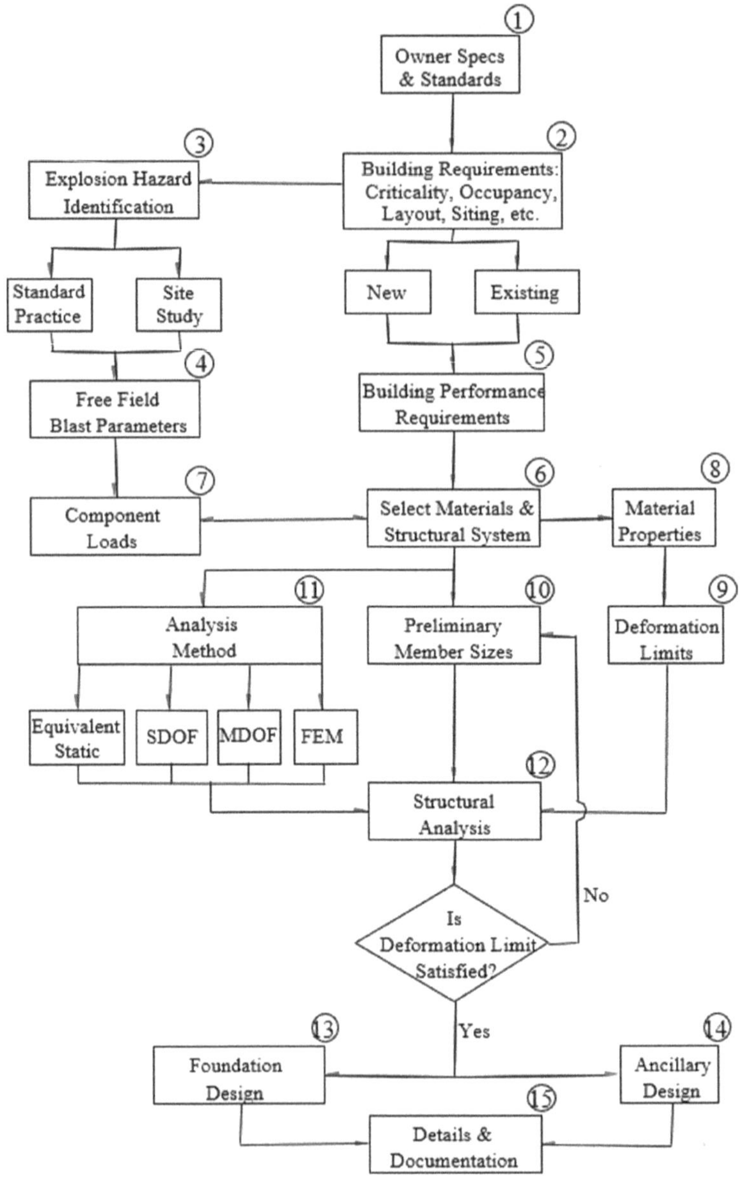

Figure 1-5. Blast-resistant design process.

3. Determine Performance Criteria: Step 5 is to determine how the building should perform during the explosion scenario (refer to Chapter 5).
4. Determine Blast Loads: Step 7 is to determine the applied blast loads for the various components that make up the building envelope (refer to Chapter 3).

5. Select Structural System and Material and Response Criteria: Steps 6, 8, and 9 are to choose the structural materials and systems for the building and the associated structural properties and response limits consistent with the performance requirements for the building (refer to Chapters 4 and 5).
6. Perform Structural Analysis and Component Design: Steps 10 to 12 are to select and perform the level of structural calculations appropriate for the particular situation (refer to Chapters 6 and 7).
7. Finalize and Detail Design: Steps 13 to 15 are to proportion and detail building components and document design (refer to Chapter 7).

It is expected that the owner will provide or direct the aforementioned Items 1 to 3 (Steps 1 to 5). *CCPS Building Guidelines, CCPS Explosion Guidelines,* and API RPs 752, 753, and 756 provide guidance on these steps. The design engineer's major responsibilities fall in Items 4 to 7 (Steps 6 to 15) of the process. These steps are the primary focus of this ASCE guideline.

References

ACI (American Concrete Institute). 2014. *Report for the design of concrete structures for blast effects,* ACI 370R-14. Farmington Hills, MI: ACI.

AISC (American Institute of Steel Construction). 2013. *Design of blast resistant structures,* Steel Design Guide 26. Chicago: AISC.

API (American Petroleum Institute). 2006. *Recommended practice for the design of offshore facilities against fire and blast loading,* API RP 2FB. Washington, DC: API.

API. 2007. *Management of hazards associated with location of process plant portable buildings,* API RP 753. Washington, DC: API.

API. 2009. *Management of hazards associated with location of process plant permanent buildings.* 3rd ed. API RP 752. Washington, DC: API.

API. 2014. *Management of hazards associated with location of process plant tents,* API RP 756. Washington, DC: API.

ASCE. 2022. *Blast protection of buildings,* ASCE 59-22. Reston, VA: ASCE.

Bradford, W. J. and T. L. Culbertson. 1967. "Design of control houses to withstand explosive forces." In *Proc., Loss Prevention Symposium.* 28–30. New York: American Institute of Chemical Engineers.

CCPS (Center for Chemical Process Safety). 1994. *Guidelines for evaluating the characteristics of vapor cloud explosions, flash fires, and BLEVEs.* New York: CCPS.

CCPS. 2012. *Guidelines for evaluating process plant buildings for external explosions, fires, and toxic releases.* 2nd ed. New York: CCPS.

CIA (Chemical Industries Association). 2020. *Guidance for the location and design of occupied buildings on chemical manufacturing and similar major hazard sites.* 4th ed. London: CIA.

Forbes, D. J. 1982. "Design of blast-resistant buildings in petroleum and chemical plants." In *Safety and accident prevention in chemical operations.* 2nd ed. 489–506. Hoboken, NJ: Wiley.

Pierorazio, A. J. J. K. Thomas, Q. A. Baker, and D. E. Ketchum. 2005. "An update to the Baker–Strehlow–Tang vapor cloud explosion prediction methodology flame speed table." *Process Saf. Prog.* 24 (1): 59—65. https://doi.org/10.1002/prs.10048.

PIP (Process Industry Practice). 2014. *Blast resistant building design criteria*, PIP STC01018. Austin, TX: PIP.

USACE (US Army Corps of Engineers). 2005. *Single degree of freedom structural response limits for antiterrorism design*, PDC TR 06-08. Washington, DC: USACE, Protective Design Center.

US DoD (US Department of Defense). 2014. *Structures to resist the effects of accidental explosions*, UFC 3-340-02 (formerly TM 5-1300). Arlington, VA: US DoD.

CHAPTER 2
General Considerations

2.1 INTRODUCTION

The need and requirements for blast resistance in plant buildings within energy and industrial facilities have evolved over recent years. Petrochemical processes have become more complex, plants have aged, plants have experienced multiple owners/operators, and plants have increased in size. Accidental explosions have resulted in significant damage to plant buildings and process units, in some cases resulting in personnel casualties and financial losses. Such events have heightened the concerns of the industry, plant management, and regulatory agencies about the issues of blast protection in plants having the potential for explosions. In general, these issues relate to plant safety and risk management to prevent or minimize the frequency of occurrence of such incidents and to siting, design, and construction practices for plant buildings to mitigate the effects on plant workers and operations.

This chapter covers the general considerations pertaining to the design of plant buildings to resist the effects of accidental explosions in petrochemical plants. First, the relevant regulatory requirements are briefly discussed. Next is a discussion of current industry practice and the objectives for providing blast resistance in plant buildings. In Section 2.4, some factors are discussed on how to identify the plant buildings that should be considered for blast resistance. The location of buildings relative to potential blast scenario locations [also known as facility siting by the Occupational Safety and Health Administration (OSHA)] plays a key role in blast protection of buildings in a plant. Often, the need for blast protection has to be balanced with functional or operational needs. These considerations are discussed in Section 2.5. Section 2.6 discusses blast design considerations for offshore facilities, and an overview of the performance of nonbuilding structures, equipment, and components is presented in Section 2.7.

2.2 OSHA REQUIREMENTS

The General Duty Clause of the Occupational Safety and Health Act of 1970 states that

> Each employer ... shall furnish to each of his employees, employment and a place of employment which are free from recognized hazards that are causing or are likely to cause death or serious physical harm to his employees;...

More specifically, Section (e)(3) of 29 CFR 1910.119 (OSHA 1992) states that process hazard analysis shall address facility siting. Occupational Safety and Health Act has recognized and pointed out the potential hazards associated with process control centers of typical conventionally designed construction. Appendix C 13 of 29 CFR 1910.119 states,

> The use of process control centers or similar process buildings in the process area as safe areas is discouraged. Recent catastrophes have shown that a large life loss has occurred in these structures because of where they have been sited and because they are not necessarily designed to withstand overpressures from shockwaves resulting from explosions in the process area.

2.3 OBJECTIVES OF BLAST-RESISTANT DESIGN

The primary objectives for providing blast-resistant design for buildings are to

1. Protect personnel.
2. Maintain power and control of process unit safety systems.
3. Reduce business interruption.

Blast-resistant design should provide a level of safety for people in the building that is no less than that for people outside the building in the event of an explosion. Evidence from past incidents has shown that many of the fatalities and serious injuries were caused by collapse of buildings onto the people inside the building. The personnel safety objective is to reduce the probability that the building itself becomes a hazard to its occupants in an explosion. Personnel protection from other hazards after an explosion may be required if the building is to provide a shelter-in-place function should fire and toxic hazards occur after an explosion. This would require that the blast resistance of the building be sufficient to maintain the building envelope integrity and the functionality of emergency systems required to ensure safety of personnel using the building as a shelter-in-place. (Note, however, that, although it is possible to design a building for fire and toxic exposure following a blast, it is often not cost-effective to design a building that is exposed to a blast to also serve as a shelter-in-place following a blast. There is very little design guidance on nonstructural systems and their ability to provide reliable shelter-in-place-level functionality following a blast event. Refer to Chapter 9 for some guidance.)

Preventing cascading events caused by loss of control of process units not involved in the initial event is another objective of blast-resistant design. An incident in one unit should not affect the continued safe operation or orderly shutdown of other units by, for example, damaging critical power and control equipment within a building or rendering the building untenable for operational personnel.

Preventing or minimizing business interruptions and/or financial losses is another objective of blast-resistant design. Buildings containing business information, emergency response equipment, critical or essential equipment, equipment that can operate other units, expensive and long lead-time equipment, or equipment, which if destroyed, would constitute significant interruption or financial loss to the owner, may warrant protection.

2.4 BUILDINGS REQUIRING BLAST-RESISTANT DESIGN

The decision regarding blast-resistant requirements is made by the owner, typically through standard practice or by following site-specific methodologies such as those described in *CCPS Building Guidelines* (CCPS 2012) or API RP 752 (API 2009) (a new edition of this was released in 2024), API RP 753 (API 2007) (a new edition of this was released in 2024), and API RP 756.

The design requirements to meet the objectives for an individual building are greatly influenced by the factors of distance from blast source, criticality of the function, and expected occupancy. For example, an occupied building sited far enough from a potential blast source may not need specific blast-resistant design features. Conversely, if a remote location is unavailable, or the proximity of the building to the unit is important, then the choice may be to provide a blast-resisting building. Buildings that are required to perform and remain operational post event (e.g., as a shelter-in-place or to control operations) may need to meet additional blast response criteria, as well as other building design criteria, to mitigate additional hazards that may follow the design blast event.

One should keep in mind that every building has some level of blast resistance based on its design for local conventional loads (i.e., wind, seismic, snow, dead and live loads). Therefore, the term *blast resistance* is not synonymous with a bunker design. Blast-resistant construction is sometimes referred to as *blast proof*. This is a misnomer because it is not realistic to provide an absolute level of blast protection. In other words, there is always some probability that failure may occur, for example, if the design basis event were exceeded or if a nonstructural component fails (such as a door left open). In addition, in practice, three minimum ingredients are required for blast-resistant design: knowledge of overpressure, duration (or impulse), and response or performance criteria. Note that several older sources of building response to blast loads provided tables in terms of overpressure only, and API RP 752 has clarified that blast loads must address both overpressure and duration or impulse.

In the blast-resistant design methods presented in this guideline, the blast loads are strength-level events. Plastic response of the structural material is accepted unlike typical static load design where the structural response is elastic (and where strengths less than elastic limit are used.) The target reliability is set by the blast load determination and not by the materials and/or response. There is limited information on blast events and associated building response. A small increase in congestion or confinement may result in a large increase in blast loads on the building. This approach is sensitive to any uncertainty in the blast loading, so it is critical that the specified loading matches the target reliability for the building.

More specifically, it is important to note that the structural blast engineer and the facility owner should be aware that, depending on the methodology chosen, there can be significant variations between the selected design blast load and the maximum blast load that could potentially affect a building. These variations are typically lower for design blast loads obtained from consequence-based siting studies that use the maximum credible events as the design blast load. However, the variations can be high for blast loading developed by quantitative risk-based approaches as these approaches typically select a design load that relates to a specified exceedance frequency and therefore the selected design load can be significantly lower than the maximum calculated blast loading. Moreover, there can also be variations owing to the process safety engineers' experience and approach, as well as to the different company practices and procedures for conducting facility siting studies. Additional guidance and background can be found in the *Guidance for the Location and Design of Occupied Buildings on Chemical Manufacturing and Similar Major Hazard Sites* (CIA 2020).

It is important that facility owners also be aware of the level of uncertainty associated with the prediction of blast loads and building response. The "blast damage" curve methods were primarily developed decades ago for general building types (e.g., pre-engineered metal buildings, load-bearing masonry wall buildings, wood-framed buildings.) for more typical building footprints and eave heights, on the basis of a combination of simplified single degree of freedom, high explosive testing, and some actual vapor cloud explosion events, where the blast loads were inferred from post-documented damage indicators. The original intention was that these blast damage curves would be used for screening-level results so that the more problematic buildings could be identified and addressed using a more advanced analysis of the specific building configuration, framing, and connections. The challenge is that these blast capacity screening curves may be either conservative or unconservative when considering nontypical framed buildings, building modifications and additions over the years, and the era of construction, which can include buildings that can well be older than 60 years. In general, these models are not intended for siting of new buildings or for the design of new or upgraded buildings. A specific structural evaluation of the actual building framing configuration and connections, as well as its material properties, can provide an estimate of blast capacity for a particular building.

When a building or installation is not sited far enough away from a blast source, the building is potentially exposed to damaging overpressures. A blast-resistant design is then recommended if any of the following applies:

1. The building is considered occupied by the owner's occupancy criteria (API RP 752). Even where evacuation is considered as a mitigation strategy, blast resistance should be considered for occupied buildings because complete evacuation may not be possible in the short response time because of the number of occupants or size and layout of the building.
2. The building or installation is expected to perform critical services, which may require personnel to remain inside during an accidental explosion to regain or maintain control or to safely shut down operating units. Buildings housing critical services may include power supply and controls for one or multiple units or equipment that controls a particularly high-risk unit. Critical services may also include emergency response equipment, such as fire-fighting vehicles, spill control equipment, and firewater pumping equipment.
3. The building is required to serve as a shelter-in-place where immediate evacuation is not recommended because of anticipated knock-on hazards such as fire and toxic releases. However, as noted in Section 2.3, it is particularly challenging (and it may not be cost-effective) to design a building exposed to a blast to also serve as a shelter-in-place following a blast, as there is very little design guidance on the ability of nonstructural systems to provide reliable shelter-in-place-level functionality following a blast event.

2.5 SITING CONSIDERATIONS

The siting of a typical plant building is unlikely to be based on a single factor. Hazards, exposures, future expansions, and spacing play a role in establishing the selected site.

Siting a plant building should consider the hazards in the adjacent and nearby processing operations and the possible results of an incident involving these hazards. The hazards from neighboring plants may also need to be considered.

As a minimum, blast-resistant buildings should be sited to meet the appropriate guidelines for fires such as those in insurance recommendations and company engineering practices.

Blast protection can be provided by adequate spacing from a potential hazard or by strengthening the building. Increased spacing should be the primary choice in providing blast protection. The guiding principles of API RP 752 should be followed, which are as follows:

1. Locate personnel away from process areas consistent with safe and effective operations.
2. Minimize the use of buildings intended for occupancy in close proximity to process areas.
3. Manage the occupancy of buildings in close proximity to process areas.
4. Design, construct, install, modify, and maintain buildings intended for occupancy to protect occupants against explosion, fire, and/or toxic material releases.
5. Manage the use of buildings intended for occupancy as an integral part of the design, construction, maintenance, and operation of a facility.

In general, buildings designed for conventional loads can be sited in areas where the overpressure is relatively low. The site owner is typically responsible for defining the blast acceptance criteria for building siting.

In the previous version of this document, a peak side-on overpressure of 1 psi and a peak side-on impulse of 30 psi·ms were given as general thresholds below which heavy structural damage or collapse of buildings was unlikely. This has been misunderstood on occasions by users, because for buildings exposed to accidental explosions of less than 1 psi, damage has ranged from none at all to complete replacement of the building, with life-threatening injuries possible but relatively uncommon. Because screening of buildings depends on building type and configuration, as well as each specific owner's tolerance to risk, no definitive screening criteria, below which performance can be guaranteed, are presented in this edition of the guidelines. Case-specific evaluations are recommended.

As noted, the aforementioned limits are provided for high-level guidance, and a more specific blast capacity for a given building can be determined on the basis of structural analysis and evaluation of the actual design and specific construction details that are present. See Chapter 9, "Evaluation and Upgrade of Existing Buildings," for additional guidance.

When siting buildings, one can consider the following: (1) Buildings should be oriented, if possible, such that the short side faces the most probable explosion source. (However, note that buildings are often exposed to multiple potential explosions from different directions and orientation, and therefore, having the short side face the blast may provide little improvement.) (2) Building roof framing should not be supported, if possible, on walls that face the most probable explosion source.

Other siting considerations that are not related to blast design include the following:

1. Buildings should not be sited downhill from potential release sources of heavier-than-air materials.
2. Buildings should not be sited in a prevailing downwind direction from potential release sources.
3. Buildings should not be sited in a location where flammable liquid could pool.

4. Buildings intended for shelter-in-place or safe haven often require a more rigorous design of nonstructural building components and systems to address the effects of fire and toxics in addition to blast loads, as applicable.

Siting of plant buildings, including determination of safety of existing buildings, should be performed by structural engineers experienced with blast-resistant analysis and design. Process safety professionals performing facility siting studies (often chemical, process safety, or technical safety engineers) have sometimes accepted certain building types as being able to safely resist certain levels of blast overpressure, and such determinations have been made without receiving building-specific input from structural engineers. This practice is discouraged. The reasons for this are clear—not all buildings of a certain construction type perform the same way in response to blast loading. The actual building geometry, structural configuration, load path, connections, detailing, and materials used need to be considered when establishing the blast-resisting capacity of a particular building.

2.6 OFFSHORE FACILITIES

Blast loading can lead to partial or total collapse of an offshore platform and result in loss of life and environmental pollution. Guidelines and recommended practice for satisfactory design of offshore structures against blast loading are described in API RP 2FB (API 2006), which contains an extensive list of references.

Many of the aspects of blast-resistant design discussed in this guideline relate equally either to onshore petrochemical facilities or to offshore facilities. Among these are dynamic material properties and analysis methods. The primary differences are in siting considerations and in details of the design overpressures and pressure waveforms.

On an offshore facility, available space is usually limited and costly so that significant mitigation of blast effects by distance is impractical unless a multiplatform bridge link solution is adopted. Consequently, control rooms, living quarters, escape routes, evacuation facilities (muster stations and lifeboats), critical structural supports, and safety-critical items, such as firewater lines and their supporting structures, must be designed explicitly to resist blast effects or be protected by blast walls and/or decks.

Relative to an onshore petrochemical facility, an offshore facility has a very compact footprint and can be multistory (i.e., processing areas may be adjacent to, above or below occupied areas). Potential explosion sources are therefore typically close to critical items that need to survive the explosion. Preliminary design is often based on nominal overpressures and impulses are based on typical or similar facilities. Detailed design usually relies on results of numerical models, such as computational fluid dynamics, which solve equations describing gas flow, turbulence, and combustion processes to determine specific blast pressure time histories on various surfaces of the building.

2.7 NONBUILDING STRUCTURES, EQUIPMENT, AND INFRASTRUCTURE

An overview of the historical performance record of nonbuilding structures, equipment, and infrastructure subjected to blast testing and accidental explosions is presented in this section. Much of the data reported in the various public documents is based on atomic bomb tests conducted by the United States during the 1950s. Glasstone and Dolan (1977) compiled a great deal of the performance data for a wide range of industrial structures, equipment, vehicles, and infrastructure subjected to blast loads. Much of these data is available in the TNO (1992).

In 1970, the US Department of the Interior's Office of Oil and Gas published a handbook (Stephens 1970) intended for use by the petroleum refining industry for the purpose of building additional protection into refineries and petrochemical facilities against all forms of disaster, including nuclear attacks. Stephens (1970) provides an overview of the performance of various nonbuilding structures, equipment, and infrastructure subjected to long duration (i.e., nuclear weapon) blast loading. Although it may be feasible to perform detailed structural analyses and damage evaluations of nonbuilding structures, equipment, and infrastructure, such efforts run the risk of becoming extremely complex to adequately address the many variables involved in such an undertaking. Utilizing screening data as presented in the aforementioned sources, with an appropriate level of engineering judgment and caution, can provide a basic understanding on the expected performance under blast loading conditions, but a detailed analysis can provide improved estimates of anticipated damage.

References

API (American Petroleum Institute). 2006. *Recommended practice for the design of offshore facilities against fire and blast loading*, API RP 2FB. Washington, DC: API.

API. 2007. *Management of hazards associated with location of process plant portable buildings*, API RP 753. Washington, DC: API.

API. 2009. *Management of hazards associated with location of process plant permanent buildings*, API RP 752. 3rd ed. Washington, DC: API.

CCPS (Center for Chemical Process Safety). 2012. *Guidelines for evaluating process plant buildings for external explosions, fires, and toxic releases*. 2nd ed. New York: CCPS.

CIA (Chemical Industries Association). 2020. *Guidance for the location and design of occupied buildings on chemical manufacturing and similar major hazard sites*. 4th ed. London: CIA.

Glasstone, S., and P. J. Dolan, eds. 1977. *The effects of nuclear weapons*. 3rd ed. Washington, DC: US Department of Defense/US Department of Energy.

OSHA (Occupational Safety and Health Administration). 1992. *Process safety management of highly hazardous chemicals*, 29 CFR 1910.119. Washington, DC: OSHA.

Stephens, M. M. 1970. *Minimizing damage to refineries from nuclear attack, natural and other disasters*, NTIS AD-773-048. Washington, DC: US Department of the Interior, Office of Oil and Gas.

TNO (The Netherlands Organization of Applied Scientific Research). 1992. *Green Book: Methods for the determination of possible damage to people and objects resulting from release of hazardous materials. Rep. CPR 16E*. The Hague, The Netherlands: Director-General of Labour.

CHAPTER 3

Determination of Loads

3.1 INTRODUCTION

The preceding chapters discussed the considerations involved in deciding the need for blast protection for buildings located in energy and industrial plants. Structural strengthening or design to resist the effects of accidental explosions was identified as one of the options available to achieve the appropriate level of blast protection. Blast-resistant design requires that the loads from such events be quantified and that the structural performance requirements be established for buildings subjected to these loads. Methods to determine the blast loading and structural performance limits are well established in UFC 3-340-02 (US DoD 2008) for buildings exposed to shock-based explosions from trinitrotoluene (TNT) or other high-yield explosives in military applications and munitions plants. However, blast loading and structural performance methods for the kinds of accidental explosions that have occurred in petrochemical plants are not as well established.

This chapter provides general information on the characteristics of shock and blast loads. A detailed discussion can be found in several publications, including Baker et al. (1996) and CCPS Explosion Guidelines (CCPS 1994). The chapter also discusses how explosions that occur in process plants are characterized to determine the blast loads for structural design. First, Section 3.2 discusses the types of explosions that may occur in petrochemical plants. Section 3.3 provides a description of the basic parameters that define a blast wave. Section 3.4 covers some of the methods currently in use in the industry and some blast overpressure values for accidental explosions used for design. Section 3.5 provides a simplified conservative method for determining blast loads applied to a typical building envelope. Section 3.6 provides an overview of computational fluid dynamic (CFD) methods for providing more accurate detailed spatial blast load variations across the building envelope, including the effects of clearing, drag, shielding, and channeling. These CFD methods can propagate either shock or blast waves and include negative phase loading.

3.2 TYPES OF EXPLOSIONS

Explosions in the petrochemical industry can be classified into four basic types: vapor cloud explosions (VCEs), pressure vessel explosions, condensed phase explosions, and dust explosions. Baker et al. (1996) and CCPS Explosion Guidelines (CCPS 1994) also provide information for characterizing some of these types of explosions.

3.2.1 Vapor Cloud Explosions

Four conditions are necessary for a VCE with damaging overpressures to occur (CCPS Explosion Guidelines).

First, there must be a release of a flammable material at suitable conditions of pressure or temperature. These include liquefied gases under pressure, ordinary flammable liquids (especially at elevated pressures and/or temperatures), and flammable gasses. When a flammable liquid spills, some or all of it will vaporize and/or form an aerosol. This dispersion is called a*vapor cloud.*

Second, ignition must be delayed long enough for a vapor cloud of sufficient size to form. Maximum flammable cloud size can be reached in 30 to 60 s; thus, the ignition delay is not long. If ignition occurs relatively instantly, a fire, fireball, or flash fire, but not a VCE, would more likely occur.

Third, the fuel–air ratio of a sufficient amount of the vapor cloud must be within the flammable range. The more uniform the fuel–air mixture, near the stoichiometric fuel-air ratio, the stronger the explosion.

Finally, there must be a flame acceleration mechanism, such as congested and confined areas, within the flammable portion of the vapor cloud. The overpressures produced by a VCE are determined by the speed of flame propagation through the cloud. Objects in the flame pathway (such as congested areas of piping, process equipment, and staggered or reoccurring objects) enhance vapor and flame turbulence. This turbulence results in a much faster flame speed which, in turn, can produce significant overpressures. Confinement that limits flame expansion, such as solid decks in multilevel process structures, also increases flame speed. Both congestion and confinement lead to greater overpressure. Without flame acceleration, a fire, fireball, or flash fire can result but not as a VCE.

Thus, the center of a VCE is not necessarily where the flammable material is released, the point of ignition, or the center of the vapor cloud. Rather, the center of a VCE is usually an area of congestion/confinement within the vapor cloud. If there are multiple areas of congestion or confinement within the flammable portion of a vapor cloud, multiple explosions can occur as the flame front propagates through each congested/confined area, and in some cases can result in more significant overpressures generated as the flame front passes through co-located areas of congestion and confinement.

In many VCE cases, the release of flammable material that forms the cloud continues to burn after the blast event has occurred. The fire can continue after the blast event until the inventory of flammable material is completely released or until the fire is extinguished. The objective of this book is not to discuss the protection of buildings to fire hazards; several guides and recommended practice documents have been published to provide guidance to assess and design building protection against hydrocarbon fires, such as the ones that can follow VCEs (API 2013, CCPS 2003, NFPA 2017, NIST 2010, UL 2018).

3.2.2 Pressure Vessel Explosions

In petrochemical plants, vessel explosions may occur as one of several subtypes:

1. Deflagrations and Detonations of Pure Gases Not Mixed with Oxidants: Acetylene is an example of a gas that would undergo a self-sustaining decomposition that releases energy. Acetylene can burn with the oxygen in the air as either a deflagration or a detonation. However, acetylene alone, with no oxygen, can also deflagrate or detonate.
2. Combustion Deflagrations and Detonations in Enclosures: These can be fueled by gaseous, liquid, or dust particle fuels (refer to Section 3.2.4). If an enclosure is too weak to sustain the pressure resulting from the combustion, it will explode.
3. Runaway Exothermic Chemical Reactions: Many industrial chemical reactions are exothermic, (i.e., they release energy). Certain reactions can go into accelerated (runaway) conditions if the released energy is not removed fast enough. If a containment vessel has insufficient venting capabilities, considerable pressure can build up. If this pressure exceeds the pressure capabilities of the vessel, it will explode.
4. Simple Overpressure of Equipment with Nonreactive Gaseous Contents: These are also called *mechanical explosions* Rupture of pressure vessels caused by overpressure may occur if human error or ancillary equipment failures allow too high an internal pressure to accumulate.
5. Physical Vapor Explosions: Physical vapor explosions occur when two streams of widely differing temperatures mix suddenly, such that the cooler liquid flashes rapidly to vapor and generates a pressure beyond the pressure capability of the container. The container thus explodes. Foundries may experience such explosions if molten metal is accidentally poured into a moist mold or water into hot oil.
6. Boiling Liquid Expanding Vapor Explosions (BLEVEs): This occurs when a large amount of pressurized liquid is suddenly vented to the atmosphere as the result of a containment vessel rupture. The rupture may be from a number of causes, but often it is from excessive heating by external fire that contacts the vessel walls above the liquid level. In this case, the vessel is not pressured above its rated pressure but is weakened by the heat. Much of the liquid flash vaporizes, and much of the remainder is broken up into aerosol droplets. The vapor aerosol mixture is typically ignited as the material is suddenly vented to the atmosphere. The combustion rate is limited to the

rate at which air can mix into the fuel. In terms relative to the speed of flames, the rate of mixing with air is relatively slow. A huge, billowing, highly radiant fireball results, and a pressure wave may also occur.

It is important to note that in some cases a pressure vessel explosion can result in a secondary explosion event fueled by the material inside of the vessel. These potential events can add to the overpressure effects to buildings.

3.2.3 Condensed Phase Explosions

Condensed phase materials are those in the liquid or solid phase, in contrast to the gaseous phase. The classic example of condensed phase materials that can detonate is high explosives. Some materials found in petrochemical plants have properties that cause them to explode under upset process conditions. These materials contain oxidizers and thus do not require the same conditions that VCEs require to produce damaging overpressure.

3.2.4 Dust Explosions

Many solid particulates handled in various industrial processes are combustible. Fine combustible particles, under the correct circumstance, can present an explosion hazard. Dust explosions are a phenomenon of ignition and flame propagation between the solid particles when suspended in air, which may be classified as being either primary or secondary in nature. Primary dust explosions typically occur within individual pieces of equipment inside process plants or similar enclosures (e.g., dust collector, cyclone, hammer mill) and are usually controlled by pressure relief (e.g., deflagration venting, chemical suppression). Secondary dust explosions are the result of dust accumulations outside of equipment within the facility, which are disturbed, suspended, and ignited by the primary explosion, resulting in a more dangerous uncontrolled explosion inside the workplace. Major facility destructions caused by dust explosions are normally the result of secondary dust explosions. Control of fugitive dust emissions and housekeeping outside of equipment are therefore a focus of all combustible dust–related standards [e.g., NFPA 61 (NFPA 2020), NFPA 68 (NFPA 2018), FM Datasheet 7-76 (FM Global 2020)] and are the key to reducing the potential for a catastrophic event.

It is noted that a dust suspended in air, combined with concentrations of flammable gas, even well below the lower flammable limit of the gas or dust, can contribute to a more severe explosion than that of the dust or gas alone. Such mixtures are called *hybrid mixtures* and must be handled with extreme care. Consultation with a subject matter expert is recommended when such hazards are present.

3.3 BLAST WAVE PARAMETERS

For blast-resistant design, the most significant feature of an explosion is the sudden release of energy to the atmosphere, which results in a pressure transient, or blast wave. This section provides basic phenomenology behind the magnitude and shape of a blast pressure-time history that can be helpful when understanding blast load histories applied along a building envelope. A blast wave propagates outward in all directions from the source at sonic or supersonic speed. The magnitude and shape of the blast wave depends on the nature of the energy release and on the distance from the explosion epicenter. The characteristic shapes of blast waves are shown in Figure 3-1.

The two types of blast waves are (1) Shock wave: This has a sudden, almost instantaneous rise in pressure above ambient atmospheric conditions to a peak free field (side-on or incident) overpressure. The peak side-on overpressure gradually returns to ambient condition with some highly damped pressure oscillations. This results in a negative pressure wave following the positive phase of the blast wave. Many company practices require that the blast loading be represented as a shock wave to invoke a more severe structural response. (2) Pressure wave: This has a gradual pressure rise to the peak side-on overpressure followed by a gradual pressure decay and a negative phase similar to that for a shock wave.

Shock waves in the near and far fields usually result from condensed phase detonations or from an extremely energetic VCE. Most vapor cloud deflagrations will give rise to pressure waves in the near field, which may propagate as a shock wave, or "shock-up," in the far field.

The negative phase of a shock or pressure wave typically has a lower peak pressure, more gradual buildup in pressure, and longer decay than the positive phase waveform, and consequently has been historically ignored in blast-resistant design. However, there may be instances when a consideration of the negative phase can have a significant effect on structural response. For situations where negative phase blast loading may be important, the reader is referred to UFC 3-340-02 (US DoD 2008) and Geng et al. (2015) for the characterization and treatment of this loading.

In Figure 3-1, the time over which the blast wave overpressure lasts is referred to as the "positive phase duration," or simply "duration." The area under the pressure–time curve is the impulse of the blast wave. Consequently, the positive phase impulse, I_o, is defined as follows:

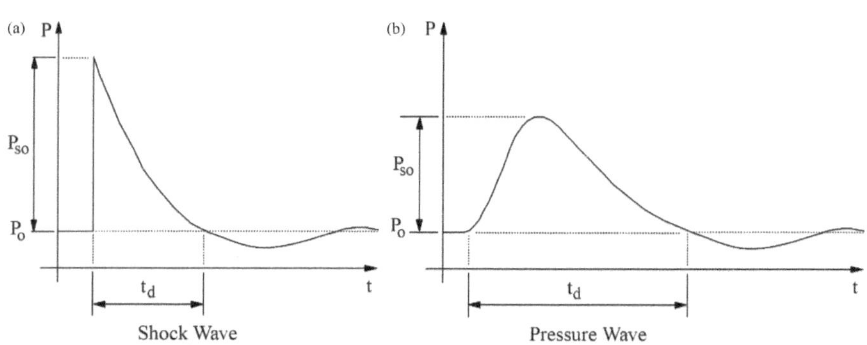

Figure 3-1. Characteristic shapes of blast waves.

$$
\begin{aligned}
I_o &= \int_0^{td} P(t)dt \\
&= 0.5P_{so}t_d \quad \text{for a simplified triangular shock wave} \\
&= 0.64P_{so}t_d \quad \text{for a half-sine wave} \\
&= cP_{so}t_d \quad \text{for an exponentially decaying shock wave}
\end{aligned}
\tag{3-1}
$$

where

$P(t)$ = Overpressure function with respect to time;

P_{so} = Peak, or incident, side-on overpressure;

t_d = Duration of the positive phase; and

c = Usually a value between 0.2 and 0.5 depending on P_{so}.

3.3.1 Blast Wave Parameters for Blast Loading

For blast-resistant design of buildings, the principal parameters of the blast wave required to define the blast loading for a building's components are as follows: peak side-on positive overpressure, P_{so}, positive phase duration, t_d, and the corresponding positive impulse, I_o; and peak side-on negative pressure (suction), P_{so}^-, negative phase duration, t_d^-, and the associated negative impulse, I_o^-.

Note that the positive and negative impulses are equal to the area underneath the positive phase pressure-time history and negative pressure-time history.

The blast wave attenuates as it propagates outward from the explosion epicenter. Consequently, the values of peak overpressure and impulse decrease with distance, whereas the duration tends to increase. Values for these blast wave parameters can be determined from published data in the form of scaled values (overpressure, impulse, or duration) as a function of scaled distance. UFC 3-340-02 (US DoD 2008) provides data on high-energy condensed phase explosives, whereas Baker (1983), Tang and Baker (1999), TNO (1985), and *CCPS Explosion Guidelines* provide values for VCE according to their respective models. These sources do not provide data on the negative phase of the blast wave from a VCE. The values of blast overpressure and duration appropriate for petrochemical design are discussed in Sections 3.4 and 3.6.

In addition to peak overpressure, duration, and impulse, other blast wave parameters that may enter into the determination of the blast loads for a structure include

- Peak reflected pressure, P_r;
- Peak dynamic (blast wind) pressure, q_o;
- Shock front velocity, U; and
- Blast wave length, L_w.

Usually, these secondary parameters can be determined from the primary blast wave parameters as discussed subsequently.

3.3.2 Peak Reflected Pressure, P_r

When the free field blast wave from an explosion strikes a surface, it is reflected. The effect of this blast wave reflection is that the surface will experience a pressure greater than the incident side-on value. The magnitude of the reflected pressure is usually determined as an amplifying ratio of the incident pressure:

$$P_r = C_r P_{so} \tag{3-2}$$

where C_r is the reflection coefficient.

The reflection coefficient depends on the peak overpressure, the angle of incidence of the wave front relative to the reflecting surface, and on the type of blast wave. The curves in Figure 3-2 shows reflection coefficients for shock waves and pressure waves, for propagation directions varying from 0 degree d (wave front direction perpendicular to surface) to 90 degrees (wave front direction parallel to surface), and for peak overpressures up to about five times atmospheric pressure. It should be noted that full reflection can be experienced for incidence angles as high as 60 to 75 degrees for P_{so} values typical of many VCEs.

For shock waves, the reflection coefficient appears to be higher for certain angles compared to the wave front hitting the surface at right angles. These higher values arise from theoretical calculations and tests that show a spike in pressure having very little associated impulse. The values shown in the dashed lines can be used for the reflection coefficient determination, which are consistent with their associated reflected impulse (TNO Green Book 1992).

CFD methods account for reflections and angle-of-incident spatially over the entire building envelope relative to the VCE source location. For peak overpressures up to 20 psi (140 kPa), which represents the expected range for most accidental VCEs, *Newmark* provides a simple formula for the blast wave reflection coefficient at normal, 0 degree, incidence as follows:

$$\begin{aligned} C_r = P_r / P_{so} &\approx (2 + 0.05P_{so})(P_{so} \text{ in psi}) \\ &\approx (2 + 0.0073P_{so})(P_{so} \text{ in kPa}) \end{aligned} \tag{3-3}$$

The duration of the reflected pressure depends on the dimensions of the reflecting surface, up to a maximum time approximately equal to the positive phase duration of the incident blast wave. This upper limit corresponds to the total reflection of the entire blast wave without any diffraction around the edges of the reflecting surface. Further details of the duration are provided in Section 3.5.1.

3.3.3 Dynamic (Blast Wind) Pressure, q_o

Dynamic pressure is a blast effect caused by air movement as the blast wave propagates through the atmosphere. The velocity of the air particles, and hence the wind pressure, depends on the peak overpressure of the blast wave. Baker (1983) and UFC 3-340-02 (US DoD 2008) provide data to compute this blast effect for shock waves. In the low overpressure range with normal atmospheric

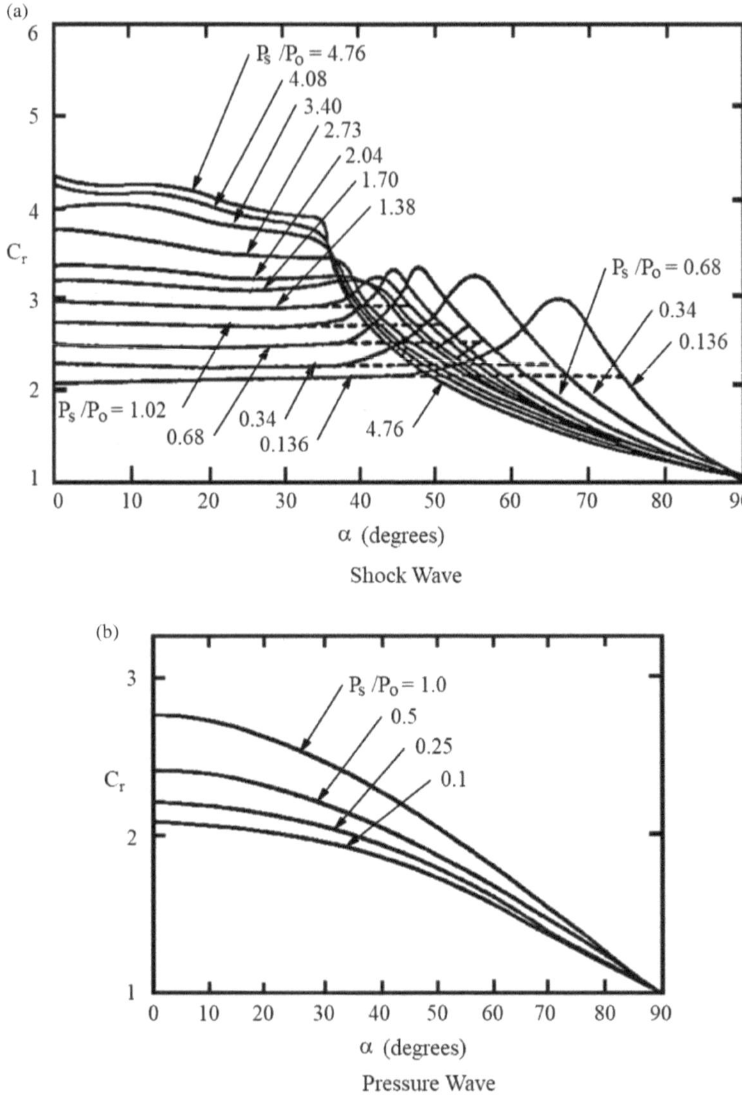

Figure 3-2. Blast wave reflection coefficient vs. angle of incidence.
Source: TNO Green Book.

conditions, the peak dynamic pressure for shock waves can be calculated using the following empirical formula from *Newmark*:

$$q_o = 2.5P_{so}^2 / (7P_o + P_{so}) \approx 0.022P_{so}^2 \quad \text{(psi)}$$
$$\approx 0.0032P_{so}^2 \quad \text{(kPa)} \qquad (3\text{-}4)$$

where P_o is the ambient atmospheric pressure.

For VCE events, Equation (3-4) is applicable in the far field (i.e., $R/R_o > 10$, where R is the distance from the explosion origin and R_o is the hemispherical explosion source radius). Simplified methods for cases in the near- and midfield have been developed, as published in ASME PVP (Geng et al., 2018).

The net dynamic pressure on a structure is the product of the dynamic pressure and a drag coefficient, C_d. The drag coefficient depends on the shape and orientation of the obstructing surface. For a rectangular building, the drag coefficient may be taken as +1.0 for the front wall and −0.4 for the side walls, rear wall, and roof.

The dynamic pressure exerts the dominant blast effect on open frame structures, framed structures with frangible cladding, and on small structures or components such as poles and stacks. The dynamic pressure also influences, but to a lesser extent, the net blast loads on the walls and roof of an enclosed building, as discussed in Section 3.5. Computational fluid dynamic methods include the effects of drag.

3.3.4 Shock Front Velocity, *U*

In the free field, the blast wave from an explosion travels at or above the acoustic speed for the propagating medium. UFC 3-340-02 (US DoD 2008) provides plots of shock front velocity versus scaled distance for high-energy TNT explosives. No similar plots are available for pressure wave propagation. Computational fluid dynamic methods compute the proper shock and pressure wave propagation velocity over distance for any magnitude of overpressure. With that said, for design purposes, it can be assumed that a pressure wave travels at the same velocity as a shock wave. In the low-pressure range, and for normal atmospheric conditions, the shock/pressure front velocity in air can be approximated using the following relationship from *Newmark*:

$$\begin{aligned} U &\approx 1{,}130(1+0.058P_{so})^{0.5} \quad \text{(ft/s)} \\ &\approx 345(1+0.0083P_{so})^{0.5} \quad \text{(m/s)} \end{aligned} \tag{3-5}$$

3.3.5 Blast Wave Length, L_w

The propagating blast wave at any instant in time extends over a limited radial distance as the shock/pressure front travels outward from the explosion. The pressure is the largest at the front and trails off to ambient over a distance, L_w, the blast wave length. Values of L_w for high-energy explosives can be obtained from UFC 3-340-02 (US DoD 2008). For relatively low pressures and long durations, the length of the blast wave can be approximated by

$$L_w \approx Ut_d \tag{3-6}$$

3.3.6 Idealized Blast Wave Parameters

To simplify the blast loads, the generalized blast wave profiles shown in Figure 3-1 can be idealized, or linearized, as illustrated in Figure 3-3 for a shock wave and

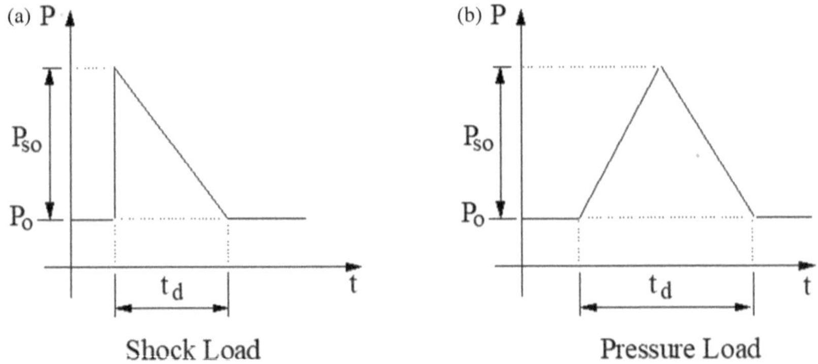

Figure 3-3. Idealized shock and pressure loads.

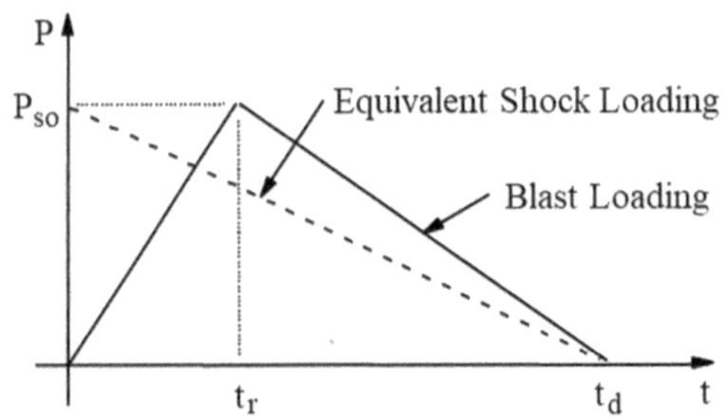

Figure 3-4. Idealized equivalent pressure load.

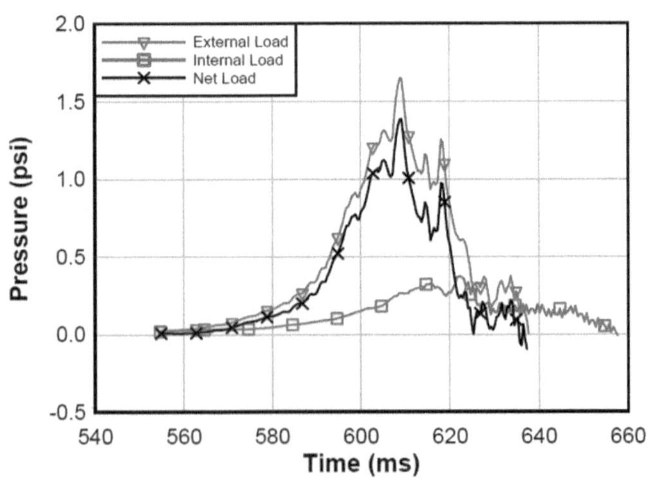

Figure 3-5. Representative VCE pressure gauge readings from VCE testing.
Source: Anderson et al. (2017).

pressure wave. Furthermore, to use certain design charts and formulas in UFC 3-340-02 (US DoD 2008), a pressure wave can be further conservatively simplified by using an equivalent shock loading, which has the same peak overpressure and impulse. This simplification is shown in Figure 3-4. The blast loads on the various parts of a building based on these simplified blast wave parameters are discussed in Section 3.5. Figure 3-5 shows a time history plot of a VCE load shape measured from a pressure gauge. This figure is included to illustrate how measured VCE overpressure waves differ from simplified triangular waves.

3.4 DETERMINATION OF VAPOR CLOUD DESIGN OVERPRESSURES

Although there is a wide range of explosion types, VCEs are a primary concern in the petrochemical industry. Because there are no codes or industry standards providing specific blast overpressures that should be used, the design blast loads are usually supplied by the facility owner or developed by specialty consultants applying commonly accepted methods to identify blast design loads. Considering the wide variety of processes, it is easy to understand why these overpressures will be different from one owner or specialty consultant to the next, even for different locations within a single facility. Some owners have several hazard levels that are used to classify different plant areas. These hazard levels are based on the material handled and the process used.

The actual design overpressures may be stated to the design engineer in two ways: (1) The simplest is a set blanket statement such as "All buildings shall be designed for a peak reflected overpressure of *X* psi (kPa), a peak side-on overpressure of *Y* psi (kPa), and duration of *Z* milliseconds." (2) A further refinement is to specify overpressures and durations based on the distance between the structure and a potential source. The distances may be given in stepped blocks or a continuous function. The building engineer would then determine design loads on the basis of the appropriate distance. Computational fluid dynamic codes have the ability to mimic provided overpressures and durations at stated ranges and can be used to interpolate blast loads across building envelopes at any range.

The basis for the aforementioned design criteria may have been developed from a site-specific study, from commonly used criteria, or from historical data. This method does not take into consideration the effects of shielding and channeling effects that buildings and process equipment may have on these predetermined blast loads as a function of distance.

A site-specific study is the most comprehensive approach. Site-specific studies to identify and quantify explosion hazards are usually conducted by the owner's process safety specialist or by specialty consultants. Several steps need to be taken, each of which may be done in a variety of ways. The steps are outlined subsequently with some of the available methods. More detailed information is available in CCPS Building Guidelines (CCPS, 1996) and API RP 752 (API 2009).

1. Define the Release(s): This step may be based on a worst-possible case depending on the maximum amount of material within a process loop, or a worst-probable (credible) case selected from a hazards review. This step may also be based on all possible release scenarios that are each associated with a probability of occurrence.
2. Formation of an Explosive Cloud: This step is often done using two computer models. The first is a source emissions model that calculates what happens at the interface between the contained material and the atmosphere into which it is being released. The second is a dispersion model that calculates how the released material disperses and mixes with the air.
3. Amount of Energy Contributing to the Explosion: This may be based on a fraction of the total amount of material available or by determining the mass of the cloud that is within the flammable limits. It may be further refined by looking at the level of confinement within the area of the cloud.
4. Calculation of Blast Overpressure Parameters: Three major methods are in use today. One is the TNT Equivalency Method that gives inaccurate results for VCEs. The other two methods are the Strehlow curves from Baker (1983) [or modified BST curves from Tang and Baker (1999) and Baker (1983)] and the Multi-Energy Method from TNO (1985). Both provide a family of curves on the basis of flame speed or explosion strength. These curves are used to select dimensionless parameters, which are then unscaled to determine the actual overpressures.

Overpressures may be conservatively determined at the point of the structure closest to the source and then applied to the entire structure. If the structure is large, the average overpressure on the surface or the overpressure at the centroid of the surface may be used if the pressure gradient on the upstream face is not significantly different from that on the downstream face. Normally, a building should be designed considering the potential blast wave from each VCE source direction. Computational fluid dynamic methods can provide the effects of directionality and the actual applied blast loading at any location over the building envelope for any building size and shape.

Accidental explosion scenarios in petrochemical facilities can produce blast loads of varying severity. The consequences associated with these blast loads may range from negligible building damage and personnel injuries to catastrophic building damage with significant levels of injuries and/or fatalities. Defining design blast loads may be challenging when the number of possible explosion scenarios is large or when the magnitude of the blast loads differs significantly. A risk-based approach may be used to consider all credible events in developing the design blast loads to meet prescribed owner criteria. A variety of different risk-based methodologies are used in the industry. An example of a risk-based approach is described in the Chemical Industry Association (CIA 2020) document.

Blast overpressures are specific to companies, processes, and sites, and it is therefore impractical to quantify a uniform minimum or maximum blast overpressure. A survey of the blast-resistant design practices of some operating companies and contractors within the industry shows that blast-resistant design is considered for buildings 50 to 1,200 ft (15 to 365 m) from VCE hazards. However, most industry standards cover buildings in the 100 to 400 ft (30 to 120 m) range. The blast loading specified varies considerably depending on the plant type, spacing, and model used to quantify the explosion. Overall, the specified blast loads used for design have side-on overpressures ranging from 1.5 to 15 psi (10 to 103 kPa) with positive phase duration ranging from 20 to 200 ms. These loads are for buildings spaced from 100 to 200 ft (30 to 60 m) from an explosion source. In general, the greater the spacing, the lesser the overpressure and impulse, but the longer the duration of the blast loading.

Historical data from industrial explosions are hard to accurately quantify as these can be approximated only by back-calculating from observed deformations of structures. Blast overpressures from VCEs are especially difficult to quantify because they tend to be directional, come from multiple sources, and vary with site conditions. In addition, less information is available than for the effects from high explosives. In one company's review of five recent VCE incidents, measured at a range of 200 to 1,000 ft (60 to 300 m), peak reflected pressures ranging from 2 psi (14 kPa) with a 35 ms duration to 12 psi (83 kPa) with a 33 ms duration occurred. These pressures correspond to side-on overpressures ranging from 1 psi (7 kPa) to 5.5 psi (38 kPa). An extensive list of this type of explosion data is included in Lenoir and Davenport (1993). This review represents a very limited set of data having short durations compared to longer durations that are more common in VCE events.

3.5 BUILDING BLAST LOADING

To design a blast-resistant building, first the design engineer has to determine loads on the building as a whole and on each individual structural component such as wall, roof, and frame from the free field blast overpressure usually provided by the facility owner or specialty consultant. To establish these loads, the design engineer should understand the interaction of the propagating blast wave with the building if the loads are provided as overpressures and not provided by CFD methods. The following provides important background in understanding the applied blast loads over the building envelope so as to adjust overpressure loads to applied blast loads and to better understand CFD-provided blast loads.

When a blast wave strikes a building, the building is loaded by the overpressure and drag forces of the blast wave. The interaction between the blast wave and a structure is quite complex, as shown schematically in Figure 3-6. For the purposes of design, the resulting blast loading can be simplified, as illustrated in Figure 3-7, on the basis of the idealized shock wave discussed in Section 3.3.6. The blast shock wave in Figure 3-6 is shown traveling horizontally left to right. However, depending on the location of potential explosion hazards relative to the

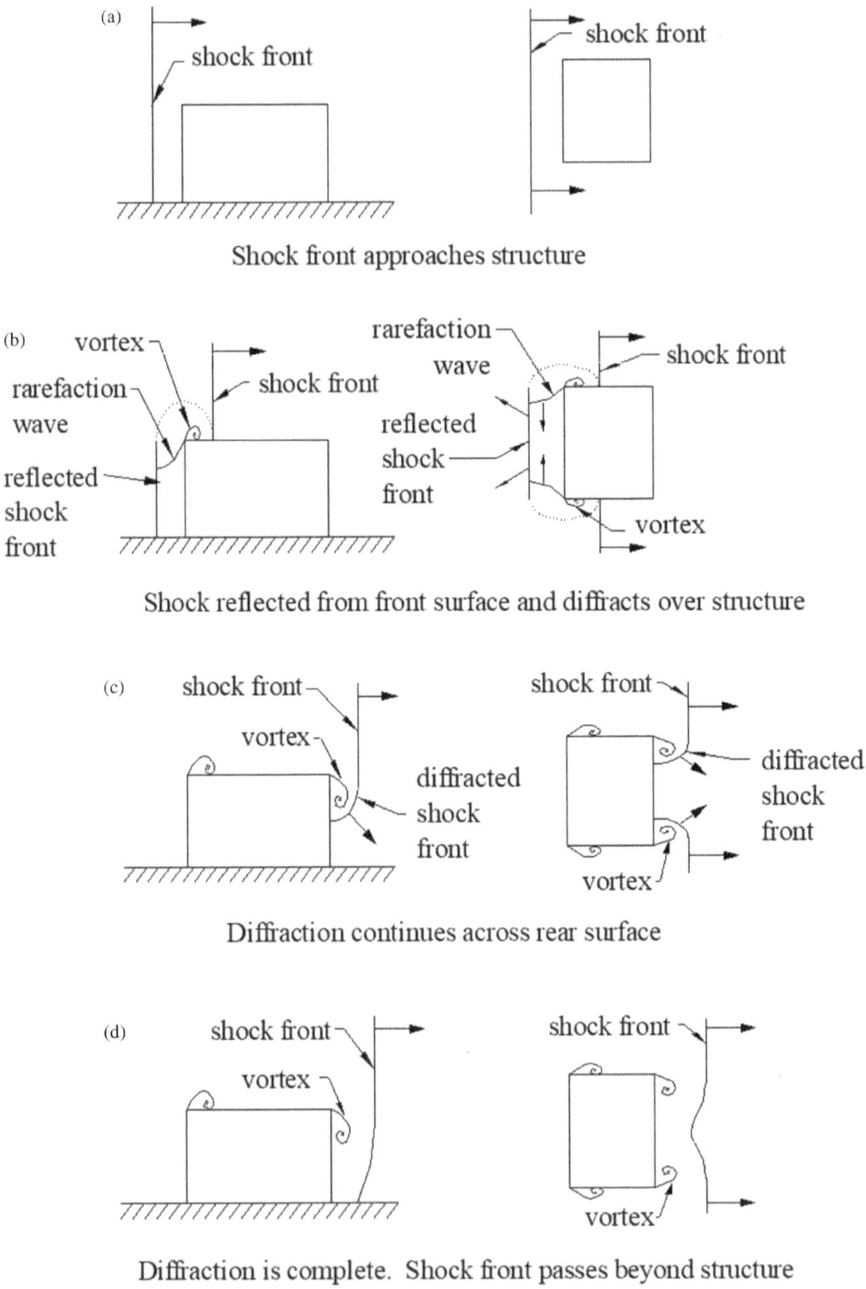

Figure 3-6. Schematic of blast wave interaction with a rectangular building.

Source: TNO Green Book.

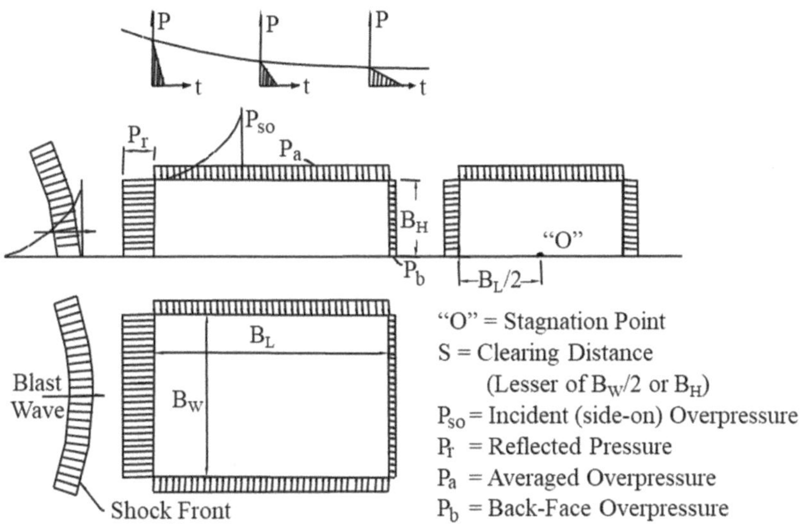

Figure 3-7. Blast loading general arrangement for a rectangular building.

Source: Forbes (1998).

building site, the blast could strike the building from other directions, and may, in the case of an elevated explosion source, slant downward toward the building.

Depending on its distance and orientation, relative to the blast source, the building and its components will experience various combinations of blast effects (reflected overpressure, side-on overpressure, dynamic pressure, and negative pressure) discussed previously. On the basis of the owner-specified side-on overpressure and duration, the design engineer can determine the blast loads for the various components of the building, as illustrated subsequently, for a closed rectangular box–shaped building.

3.5.1 Front Wall Loading

The walls facing the explosion source will experience a reflected overpressure. As discussed previously, the reflected overpressure amplification of the blast wave depends on the angle of incidence, α, and on the rise-time, t_r, of the side-on overpressure pulse. For design purposes, the normal shock reflection conditions ($\alpha = 0$, $t_r = 0$) are typically assumed unless the specified design provides a specific blast load function. In some cases, oblique reflection (about 30° to 60°) may be more critical to the overall building because the full reflected overpressure could load two adjacent sides of the building simultaneously. The reflected overpressure decays until, at the time t_c (clearing time), it reaches a value of $P_s\ (t_d - t_c)/t_d$, where P_s is the value of the stagnation pressure at time zero. P_s and the clearing time, t_c, are defined subsequently and illustrated for a shock load function in Figure 3-8.

$$P_s = P_{so} + C_d q_o \tag{3-7}$$

$$t_c = 3S / U < t_d \tag{3-8}$$

where

S = Clearing distance, for buildings sitting at grade, the smaller of B_H, or $B_W/2$;

B_H = Building height; and

B_W = Building width.

Note that for elevated buildings, the blast wave can clear from the top and the bottom of the building, and the aforementioned equation may not be accurate for this case.

As indicated in Equation (3-8) and Section 3.3.2, the duration of the reflected overpressure effect, t_c, should not exceed that of the free field positive overpressure, t_d.

To use the dynamic response charts on the basis of a triangular load, the bilinear pressure–time curve shown in Figure 3-8 can be simplified to an equivalent triangle. This equivalent load is computed by equating the impulse for each load shape and using the same peak pressure, P_r. The impulse, I_w, under the bilinear pressure–time curve is

$$I_w = 0.5(P_r - P_s)t_c + 0.5P_s t_d \tag{3-9}$$

The duration, t_e, of the equivalent triangle is determined from the following equation:

$$t_e = 2I_w / P_r = (t_d - t_c)P_s/P_r + t_c \tag{3-10}$$

It is important to note that most VCE blast load cases are deflagrations having a significant rise time to peak pressure that most likely will exceed the time to clear the reflected wall. Therefore, the procedures outlined in this section can be considered conservative in most VCE deflagration cases. Computational fluid dynamic calculations can provide the effect of clearing before reaching the maximum reflected pressures, which can remove the conservatism associated with the procedure for calculating the peak reflected pressure that assumes a shock loading. In other words, the increase in reflected pressure for a pressure

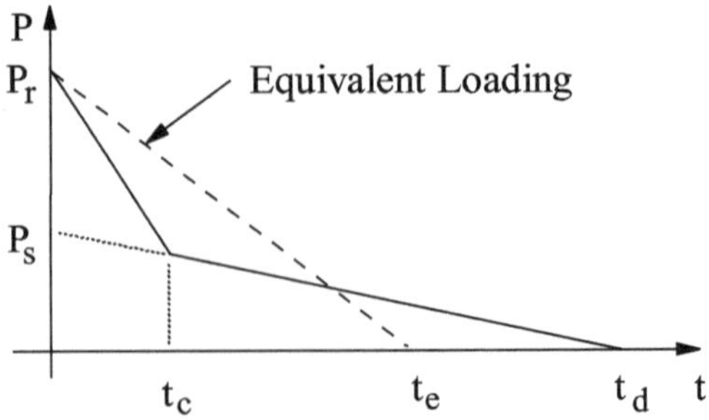

Figure 3-8. Front wall loading.

wave having a finite rise time similar to a typical deflagration pressure rise can be significantly less than that found using the reflection coefficient, C_r, provided in Equation (3-3).

3.5.2 Side Walls

The side walls are defined relative to the explosion source, as shown in Figure 3-7. These walls will experience less blast loading than the front wall because of a lack of overpressure reflection and attenuation of the blast wave with distance from the explosion source. In certain cases, the actual side wall loading is combined with other blast-induced forces (such as in-plane forces for exterior shear walls). The general form of side wall blast loading can be taken as shown in Figure 3-9.

As a blast wave travels along the length of a structural element, the peak side-on overpressure will not be applied uniformly. It varies with both time and distance. For example, if the length of the side wall equals the length of the blast wave, when the peak side-on overpressure reaches the far end of the wall, the overpressure at the near end has returned to ambient. A reduction factor, C_e, is used to account for this effect in design. Values for C_e (refer to Figure 3-10) are dependent on the length of the structural element, L_1, in the direction of the traveling blast wave. If the blast wave is traveling perpendicular to the span, then L_1 should be equal to a nominal unit width of the element. Note that the curve in Figure 3-10 differs from that presented in the previous version of this report. It was generated from the maximum effective uniform pressure acting along the side walls or roof considering a single load profile over the face (i.e., peak pressure and duration) but accounting for time delays to sweep across the face, while assuming Equation (3-6) holds.

The equation for side walls is as follows:

$$P_a = C_e P_{so} + C_d q_o \tag{3-11}$$

where P_a is the effective side-on overpressure.

Note that q_o in Equation (3-11) should be determined from the reduced side-on overpressure, $C_e\ P_{so}$, inserted into Equation (3-4).

The side wall load has a rise time equal to the time it takes for the blast wave to travel across the element being considered. The overall duration is equal to this rise time plus the duration of the free-field side-on overpressure. Other values of

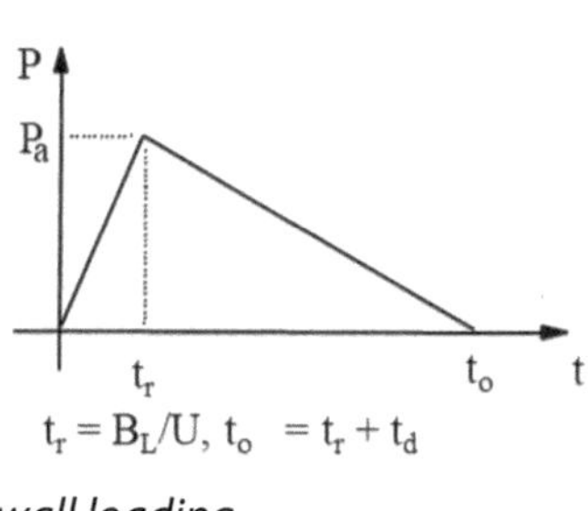

Figure 3-9. Roof and sde wall loading.

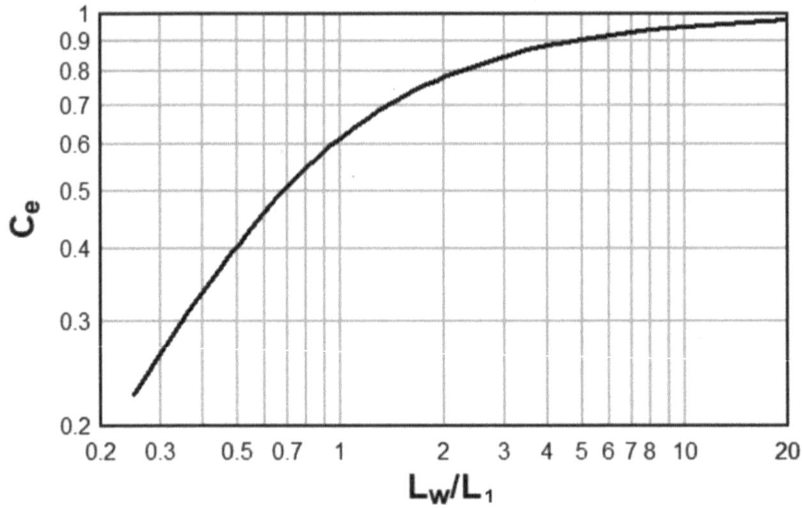

Figure 3-10. Peak equivalent uniform side wall/roof pressure reduction factors.

rise time can be calculated following the methods given in Glasstone and Dolan (1977) and UFC 3-340-01. Alternatively, the side wall load can conservatively be taken with an instantaneous rise to P_{so}, as shown in Figure 3-4.

3.5.3 Roof Loading

For a building with a relatively flat roof (pitch less than 10°), it is normally assumed that reflection does not occur when the blast wave travels horizontally. Consequently, the roof will experience the side-on overpressure combined with the dynamic wind pressure, the same as the side walls, thereby ignoring the slight reflection factors shown in Figure 3-2. The dynamic wind force on the roof acts in the opposite direction to the overpressure (upward). Also, consideration should be given to variation of the blast wave with distance and time as it travels across a roof element. The resulting roof loading, as shown in Figure 3-9, depends on the ratio of blast wave length to the span of the roof element and on its orientation relative to the direction of the blast wave. The effective peak overpressure for the roof elements can be calculated using Equation (3-11) or as otherwise discussed in Section 3.5.2 similar to the side walls.

3.5.4 Rear Wall Loading

Rear wall loading is normally used only to determine the net overall frame loading. Because the rear wall load is opposite in direction to the front wall load, its inclusion tends to reduce the net lateral blast load. For buildings where a blast load could occur from any direction, rear wall effects are many times conservatively ignored.

The shape of the rear wall loading is similar to that for side and roof loads; however, the rise time and duration are influenced by a not well-understood pattern of spillover from the roof and side walls and from ground reflection effects. The rear wall blast load is delayed as compared to the the front wall load by B_L/U, the time for the blast wave to travel the length of the building, B_L. The effective peak overpressure is similar to that for side walls and can be determined using Equation (3-11) (P_b is normally used to designate the rear wall peak overpressure instead of P_a). Available references indicate two distinct values for the rise time and positive phase duration.

TNO Green Book (1992) and ASCE MOP 42 (ASCE 1985) use criteria that appear to be based on longer-duration blast loads. The positive phase has a rise time of $4S/U$ and a total duration of t_d (Figure 3-11a). Note that for blast loads of a moderate to short duration, the rise time may approach or exceed t_d. Information is not provided on dealing with this situation.

UFC 3-340-02 (US DoD 2008) provides criteria computing the rear wall load as though it were an extension of the roof. Although graphs are provided to determine the rise time and duration, for most typical buildings, the positive phase will have a rise time of approximately S/U followed by a duration of t_d [Figure (3-11b)].

3.5.5 Frame Loading

In addition to the roof loading, the framing system for the building will experience the diffraction loading, which is the net loading on the front and rear walls taking into account the time phasing. During the time, B_L/U, that it takes the blast wave to travel from the front to the back of the building, the structural framing will be subjected to the large horizontal unbalanced pressure on the front wall. After that time, the front wall loading is partially offset by the rear wall loading. Figure 3-12 shows the general form for the lateral frame loading.

The components of a building will also experience blast load effects, opposite in direction to the primary blast load effects, because of the negative phase (suction) of the blast wave, as discussed in Sections 3.3.1 and 3.3.2, together with the rebound of the structural components from the inertial effects of the overpressure loading. However, the structural components of the building should be adequately detailed to perform satisfactorily for the rebound effects. These effects can be quantified from the time history dynamic analysis of the

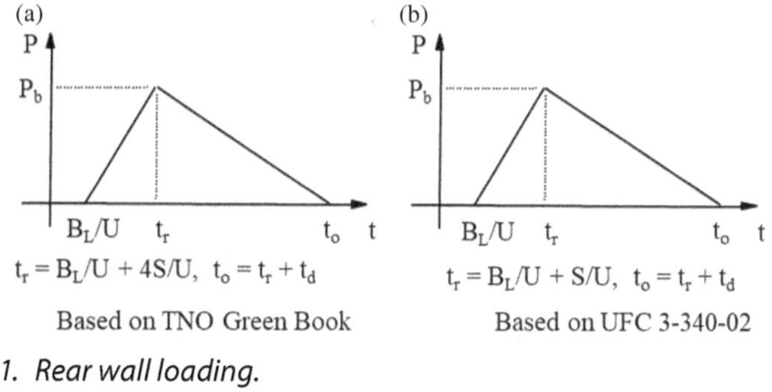

Figure 3-11. Rear wall loading.

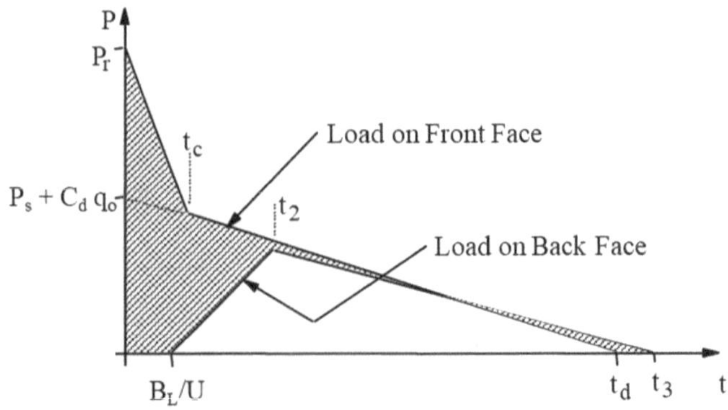

Figure 3-12. Net lateral load on a rectangular building.
Source: TNO Green Book.

structural components, as discussed in Chapter 6, "Dynamic Analysis Methods," or approximated by using design charts such as those provided in UFC 3-340-02 (US DoD 2008) or ASCE MOP 42 (ASCE 1985).

3.5.6 Negative Pressure and Rebound Loading

The components of a building will also experience blast load effects, opposite in direction to the primary blast load effects, because of the negative phase (suction) of the blast wave, as discussed in Sections 3.3.1 and 3.3.2, together with the rebound of the structural components from the inertial effects of the overpressure loading. As noted previously, the negative pressure forces are usually ignored because they can be relatively small or are unquantified for VCEs using simple blast curve methods. Computational fluid dynamic methods are one approach that can properly estimate the negative pressure histories. In certain instances, including the effects of the negative phase pressure may decrease a component's peak blast response, whereas in other instances, the negative phase may exacerbate structural response. Therefore, the structural components of the building should be adequately detailed to perform satisfactorily for rebound effects. These effects can be quantified from the time history dynamic analysis of the structural components, as discussed in Chapter 6, approximated by using design charts such as those provided in UFC 3-340-02 (US DoD 2008) or ASCE MOP 42 (ASCE 1985), or calculated using more advanced analysis methods such as computational fluid dynamics, as discussed in Section 3.6.

3.5.7 Leakage Pressures

Blast loads applied to the building exterior can potentially expand into the building through openings in the walls or roof, open doorways (roll-up doors), and early-time failed windows, personnel doors, and cladding. These are typically referred to as "leakage pressures." As a blast wave expands through an opening, the pressure level drops because of the restriction and sudden expansion into the building volume. Leakage pressures may cause direct bodily injury, injury from falling objects such as light fixtures, and equipment malfunctions. Methods are available to compute the average pressure buildup inside a structure. The reader is referred to Section 2-15.5 of UFC 3-340-02 (US DoD 2008) for detailed procedures and Section 1-11 of UFC 3-340-02 (US DoD 2008) for information on pressures causing bodily injury. Computational fluid dynamic methods can also be used to quantify leakage pressures at any discrete point within a building's envelope, such as a maintenance shop with open roll-up doors. Computational fluid dynamic methods can also be used to estimate leakage pressures underneath raised buildings with regard to applied floor uplift loading.

Interior pressures can also increase because of the flexing of the building envelope. Limited testing and CFD calculations have shown that these increases can be significant. In addition, if a hardened envelope is constructed around an existing conventionally designed building, the increase in pressure in the air gap owing to the flexural response of the hardened envelope should be estimated to assess whether the pressures in the gap will cause damage to the original building envelope.

3.5.8 Floor Loading on Elevated Buildings

If the building floor is exposed, it should be checked for uplift pressures applied from underneath the floor. An estimate of the effect can be developed by considering the two extremes; no uplift and 100% uplift.

3.6 COMPUTATIONAL FLUID DYNAMICS

Building blast loads can be determined through the use of CFD computer programs. Computational fluid dynamic usage requires a blast design situation where cost-effective results can be realized, as well as experience, time, computer resources, and validation to justify the increase in computer modeling effort. Advancements in modeling methods and computer hardware since the last publication of this document have made these applications significantly more cost-effective. The basic premise of CFD modeling is to discretize the building and surrounding area encompassing the blast source and adjacent obstacles into discrete elements (using finite volume, finite-element, or finite different methods) and then solve a form of the governing equations for conservation of mass, momentum, and energy within each element, taking into account the effects of adjacent elements.

Among other uses, CFD could be used to simulate the following:

- Propagation of shock or blast waves in an environment of obstacles, such as a building or equipment that shields adjacent buildings;
- Unusually shaped buildings that may or may not include re-entrant corners;
- Leakage through openings into buildings;

- Interior explosions; and
- Near-field explosion effects.

One of the more significant advantages of CFD is the ability to explicitly quantify both positive and negative phases of the pressure-time history on each face of a building or structure at any discrete location, while inherently including clearing and drag loading effects. Figure 3-13 illustrates the ability of CFD calculations to capture wave propagation physics such as re-entrant corners and shielding.

Traditionally, CFD has been commonly used in offshore applications where compact, complex layouts are used and platform occupants are relatively close to the blast hazards. Today, CFD is becoming more commonly used for onshore facilities to address nonstandard vapor cloud shapes, include the effects of complex geometries and terrain found throughout sites, and account for the effects of shielding and channeling caused by co-located buildings within process facilities. Where applicable, CFD could be used as a more refined solution to the more commonly used empirical methods described in UFC 3-340-02 (US DoD 2008) and Section 3.5.

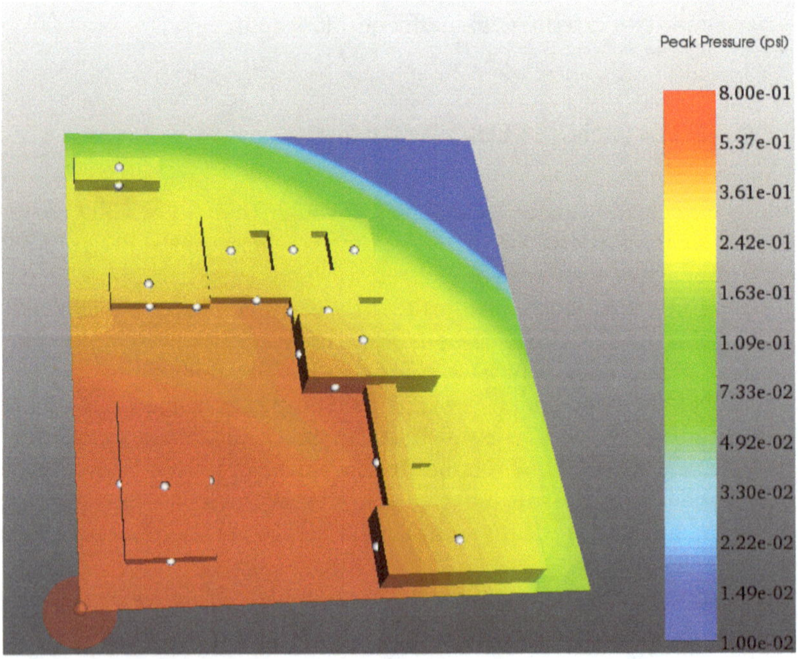

Figure 3-13. Pressure contour plot of 3D CFD calculation of a petrochemical facility capturing re-entrant corner and shielding effects.

Source: Thornton Tomasetti (2019).

In deciding to employ CFD, it should be understood that CFD results can be sensitive to modeling techniques and the software used, and thus, best practice approaches should be employed. Computational fluid dynamic programs can employ a first-principles approach that includes dispersion, turbulence modeling, and detailed combustion, or a semi-empirical approach where simplifications of the explosion source are made, on the basis of test data and guidance, to simplify and speed up the analysis. Various commercial, open-source, and computer models developed by specialty consultants are available, which can be used to leverage CFD modeling. These codes include FLACS-CFD (GEXCON 2024a), CEBAM (SciRisq 2024), and FX. The choices of computer model and approach are highly dependent on the problem, facility, and structure (target) of interest.

Furthermore, in first-principles CFD models, changes in input parameters such as ignition point in the cloud can result in significant differences (in some cases by an order of magnitude) in the results. Therefore, it is important that the user of any CFD code become familiar with its capability to accurately model blast loads in either the near field or far field depending on the building location. For further information, refer to Dhamavaram et al. (2005), *Gas Explosion Handbook* (GEXCON 2024b), Hoorelbeke et al. (2006), Hanna et al. (2004), Geng and Thomas (2007), Van Wingerden et al. (1999), and Herrmann (2005, 2006).

First-principles based CFD models can be driven by flow fields in the near field produced from turbulent combustion CFD analysis results to improve far-field blast propagation predictions. Some of these codes include ADINA (Bentley 2024), Autodyn (Ansys 2024), CTH (McGlaun et al. 1990), BWTI (BakerRisk 2024), and FacilityBlast VCE (Nikodym and Xie 2017). Computational fluid dynamic models can also be tuned to reproduce free-field pressure and impulse over a specified range from simplified methods before including obstacles in the path of the blast propagation field and can also include multiple VCE events.

A simplified two-dimensional (2D) CFD result for a 2 psi blast having a 100 ms duration is shown sweeping across a 12 ft high by 60 ft long building section in Figure 3-14. The effects of reflection, clearing, and backface loading as discussed in the aforementioned sections can be seen in the pressure contours in these CFD results. Note that the load is propagated within a 2D space, so no attenuation of the blast wave is possible. In contrast, as discussed previously, blast overpressures decay over distance. A CFD model has the ability to capture decay in pressure with distance if the blast source is allowed to expand in hemispherical three-dimensional space.

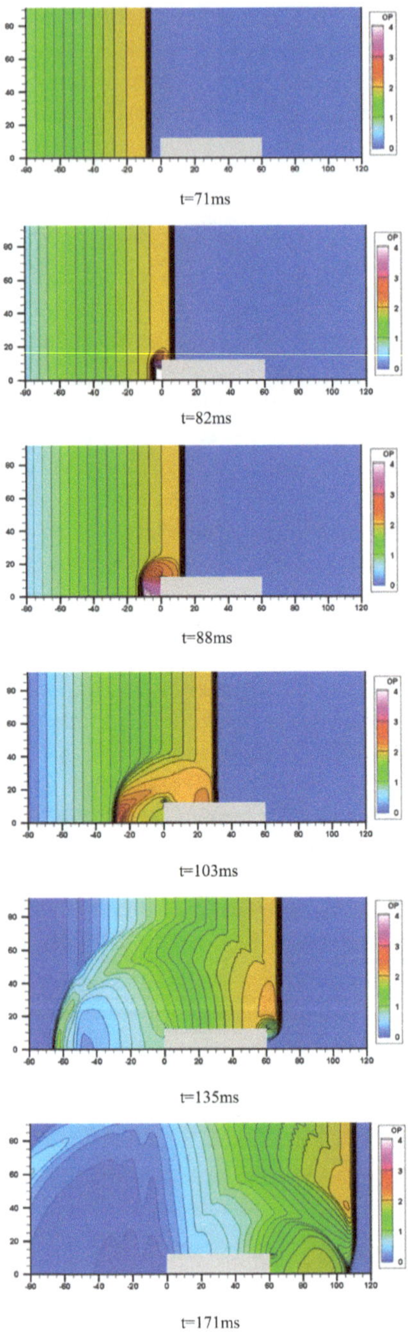

Figure 3-14. 2D CFD overpressures from a 2 psi/100 ms blast for a 12 ft-high by 60 ft-wide building.

Source: Thornton Tomasetti (2019).

APPENDIX 3A. BLAST LOAD EXAMPLE

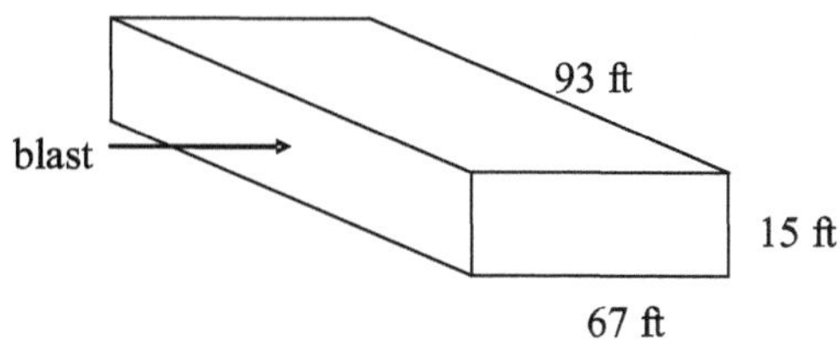

This example illustrates the calculation of blast loading on the components of a building subjected to a shock wave traveling horizontally. The building dimensions are as follows:

Width, $B_W = 93$ ft (28.3 m)

Length, $B_L = 67$ ft (20.4 m)

Height, $B_H = 15$ ft (4.6 m)

Blast Loading

A blast wave has been given and will be applied normal to the long side of the building. It is further determined that the distance to the explosion and the length of the building are such that the overpressure and duration do not change significantly over the length of the building. The blast (shock) wave parameters are as follows:

Peak side-on overpressure, $P_{so} = 6$ psi (41 kPa)

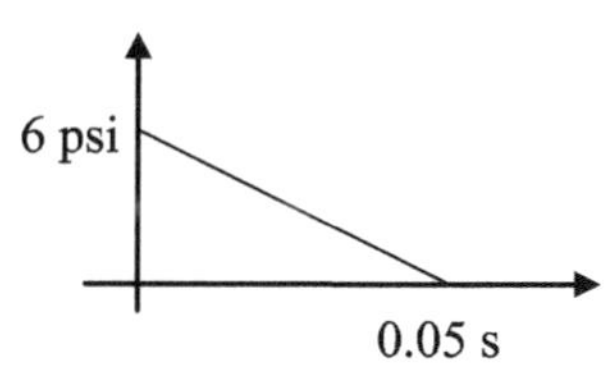

duration, $t_d = 0.05$ s
shock front velocity, [Equation (3-5)]

$$U = 1{,}130(1 + 0.058P_{so})^{0.5}$$
$$= 1{,}130[1 + 0.058(6\text{ psi})]^{0.5}$$
$$= 1{,}312\text{ ft/s}(400\text{ m/s})$$

length of pressure wave, [Equation (3-6)]

$$L_w = U(t_d) = (1{,}312\text{ ft/s})(0.05\text{ s}) = 66\text{ ft}(20\text{ m})$$

peak dynamic wind pressure, [Equation (3-4)]

$$q_o = 0.022(P_{so})^2 = 0.022(6\text{ psi})^2 = 0.8\text{ psi}(5.5\text{ kPa})$$

Front Wall Loading

The front wall is assumed to span vertically from foundation to roof. The design will be for a typical wall segment 1 ft wide.

reflected overpressure, [Equations (3-2) and (3-3)]

$$P_r = [2 + 0.05(P_{so})]P_{so} = [2 + 0.05(6\text{ psi})](6\text{ psi}) = 13.8\text{ psi}(95\text{ kPa})$$

clearing distance, (Section 3.5.1)

S = minimum of B_H or $B_W/2$ = 15 ft (4.6 m)

reflected overpressure clearing time [Equation (3-8)]

$t_c = 3\ (S/U) < t_d = 3\ (15\text{ ft})/(1{,}312\text{ ft/s}) < 0.05\text{ s} = 0.034\text{ s}$

drag coefficient, $C_d = 1.0$ (Section 3.3.3)

stagnation pressure [Equation (3-7)]

$P_s = P_{so} + C_d\,(q_o) = (6\text{ psi}) + (1.0)(0.8\text{ psi}) = 6.8\text{ psi}\ (47\text{ kPa})$

front wall impulse, [Equation (3-9)]

$$\begin{aligned} I_w &= 0.5(P_r - P_s)t_c + 0.5P_s t_d \\ &= 0.5\left[(13.8\text{ psi}) - (6.8\text{ psi})\right](0.034\text{ s}) + 0.5(6.8\text{ psi})(0.05\text{ s}) \\ &= 0.289\text{ psi}\cdot\text{s}(2\text{ kPa}\cdot\text{s}) \end{aligned}$$

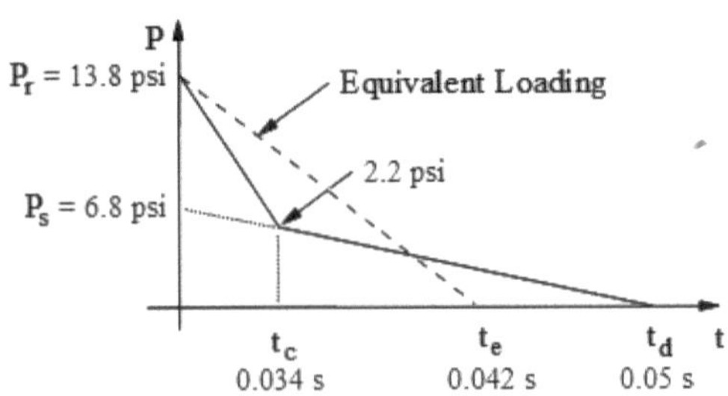

effective duration [Equation (3-10)]

$t_e = 2\ I_w/P_r = 2\ (0.289\text{ psi·s})/(13.8\text{ psi}) = 0.042\text{ s}$

Side Wall Loading

The side wall is the same as the front wall, spanning vertically from foundation to roof. Because the highest loads are on the front wall, a side wall analysis would only be necessary to check the interaction of in-plane and out-of-plane shear wall forces. This calculation will be for a wall segment, L_1, 1 ft wide (0.3 m).

drag coefficient, $C_d = -0.4$ (Section 3.3.3)

equivalent load coefficient (Figure 3-10)

$L_w/L_1 = (66\text{ ft})/(1\text{ ft}) = 66$; therefore, C_e = essentially 1.0

equivalent peak overpressure [Equation (3-11)]
$P_a = C_e P_{so} + C_d q_o = (1.0)(6 \text{ psi}) + (-0.4)(0.8 \text{ psi}) = 5.7 \text{ psi } (39 \text{ kPa})$

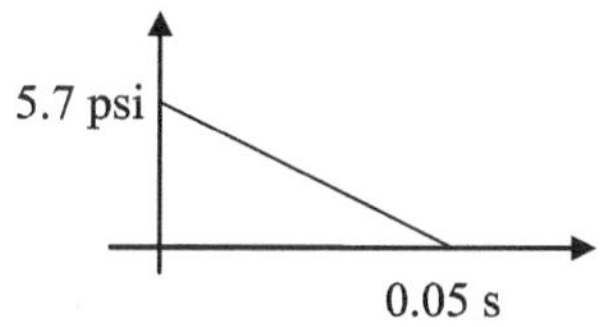

rise time (Figure 3-9)
$t_r = L_1/U = (1 \text{ ft})/(1{,}312 \text{ ft/s}) =$ essentially 0.0 s
duration, $t_d = 0.05$ s

If an average overpressure over the entire side wall is needed, the value of L_1 would then be the length of the building. The value of C_e would then be less than 1 and thus reduce the value of P_a. The rise time would become significant.

Roof Loading

The roof is a slab spanning between roof beams. For the design of the roof, a section 1 ft wide by 8 ft long will be used.

$L_1 = 8.0$ ft (2.4 m)
drag coefficient, $C_d = -0.4$ (Section 3.3.3)
equivalent load coefficient (Figure 3-10)
$L_w/L_1 = (66 \text{ ft})/(8 \text{ ft}) = 8.25$; therefore, $C_e = 0.94$
reduced dynamic wind pressure [Equation (3-4)]
$q_o = 0.022\,(C_e P_{so})^2 = 0.022\,(0.94 \times 6 \text{ psi})^2 = 0.7 \text{ psi } (5 \text{ kPa})$
equivalent peak overpressure [Equation (3-11)]
$P_a = C_e P_{so} + C_d q_o = (0.94)(6 \text{ psi}) + (-0.4)(0.7 \text{ psi}) = 5.4 \text{ psi } (37 \text{ kPa})$

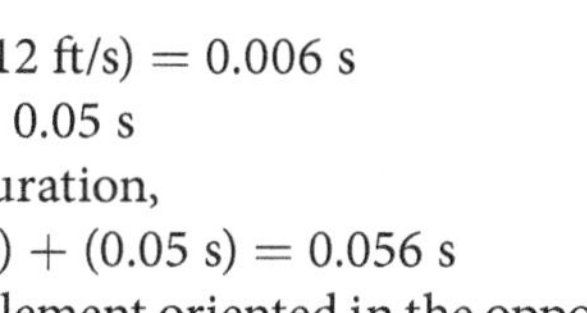

rise time (Figure 3-9)
$t_r = L_1/U = (8 \text{ ft})/(1{,}312 \text{ ft/s}) = 0.006$ s
time of duration, $t_d = 0.05$ s
total positive phase duration,
$t_o = t_r + t_d = (0.006 \text{ s}) + (0.05 \text{ s}) = 0.056$ s

For a structural roof element oriented in the opposite direction, the length of the element in the direction of the traveling wave, L_1 would be only 1 ft (0.3 m). In this case, as for the side wall panel, essentially, no averaging would be necessary.

If an average overpressure over the entire roof is needed, the value of L_1 would then be the length of the building. The value of C_e would then be reduced along with the value of P_a. The rise time would be greater.

Rear Wall Load

The rear wall is proportioned the same as the front and side walls, spanning vertically from foundation to roof. Because the highest loads are on the front wall, a rear wall analysis would only be necessary to determine a net loading on the overall building. The analysis will be for a wall segment 1 ft (0.3 m) wide.

drag coefficient, $C_d = -0.4$ (Section 3.3.3)
equivalent load coefficient (Figure 3-10)
$L_w/S = (66 \text{ ft})/(15 \text{ ft}) = 4.4$; therefore, $C_e = 0.89$
reduced dynamic wind pressure [Equation (3-4)]
$q_o = 0.022\,(C_e\,P_{so})^2 = 0.022\,(0.89 \times 6 \text{ psi})^2 = 0.63$ psi (4 kPa)
equivalent peak overpressure [Equation (3-11)]
$P_b = C_e\,P_{so} + C_d\,q_o = (0.89)\,(6 \text{ psi}) + (-0.4)\,(0.63 \text{ psi}) = 5.1$ psi (35 kPa)
time of arrival (Section 3.5.4)
$t_a = B_L/U = (67 \text{ ft})/(1{,}312 \text{ ft/s}) = 0.051$ s
rise time (UFC 3-340-02, US DoD 2008)
$t_r = S/U = (15 \text{ ft})/(1{,}312 \text{ ft/s}) = 0.011$ s

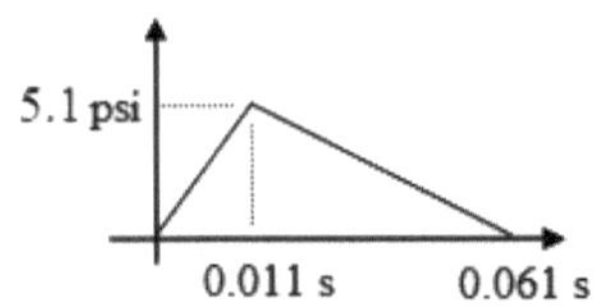

duration, $t_d = 0.05$ s
total positive phase duration,
$t_o = t_r + t_d = (0.011 \text{ s}) + (0.05 \text{ s}) = 0.061$ s

References

Anderson, T. H., B. J. Horn, J. K. Thomas, L. Magenes, et al. 2017. "Deflagration load generator blast load testing of an iso shipping container and blast resistant wood building." In *Proc., Structures Congress*. Reston, VA: ASCE.

Ansys. 2024. *Autodyn*. Canonsburg, PA: Ansys.

API (American Petroleum Institute). 2009. "API recommended practice 752 (RP 752)". 4th ed. Accessed November 20, 2024. https://www.api.org/products-and-services/standards/important-standards-announcements/rp752.

API. 2013. *Fireproofing practices in petroleum and petrochemical processing plants*, API RP 2218. Washington, DC: API.

ASCE. 1985. *Design of structures to resist nuclear weapons effects*, MOP 42. Reston, VA: ASCE.

Baker, Q. A., M. J. Tang, E. A. Scheier, and G. J. Silva. 1996. "Vapor cloud explosion analysis." *Process Saf. Prog.* 15 (2): 106–109.

Baker, W. E., ed. 1983. *Explosion hazards and evaluation*. New York: Elsevier.

BakerRisk. 2024. "*Computational fluid dynamics*." BWTI (Blast Wave Target Interaction). San Antonio, TX: BakerRisk. Accessed November 20, 2024. https://www.bakerrisk.com/services/computational-fluid-dynamics/.

Bentley. 2024. *ADINA: Finite element analysis software*. Exton, PA: Bentley.

CCPS (Center for Chemical Process Safety). 1994. *Guidelines for evaluating the characteristics of vapor cloud explosions, flash fires, and BLEVEs.* New York: CCPS.

CCPS. 1996. *Guidelines for evaluating process plant buildings for external explosions and fires.* CCPS Building Guidelines. New York: CCPS.

CCPS. 2003. *CCPS guidelines for fire protection in chemical, petrochemical, and hydrocarbon processing facilities.* New York: CCPS.

CIA (Chemical Industries Association)2020. *CIA guidance for the location and design of occupied buildings on chemical manufacturing and similar major hazard sites.* 4th ed. London: CIA.

Dhamavaram, S., S. R. Hanna, and O. R. Hansen. 2005. "Consequency analysis—Using a CFD model for industrial sites." *Process Saf. Prog.* 24 (4): 316–327.

FM Global. 2020. *Property loss prevention data sheets—Prevention and mitigation of combustible dust explosion and fire.* Johnston, RI: FM Global.

Forbes, D. J. 1998. "Protecting petroleum process plant buildings from vapor cloud explosions." In *Concrete and blast effects.* Special Publication 175, 53–86. Detroit: American Concrete Institute.

Geng, J., Lowry, W., and K. Thomas. 2018. "Drag loads from vapor cloud explosions." In *Proc., ASME Pressure Vessels and Piping Conf. PVP2018-84307.* New York: ASME.

Geng, J., T. Mander, and Q. Baker. 2015. "Blast wave clearing behavior for positive and negative phases." *J. Loss Prev. Process Ind.* 37 (2): 283—293.

Geng, J. H., and J. K. Thomas. 2007. "Simulation and application of blast wave-target interactions." In *Proc., 41st Loss Prevention Symp.* New York: American Institute of Chemical Engineers.

GEXCON. 2024a. *FLACS-CFD.* Accessed November 20, 2024. https://www.gexcon.com/software/flacs-cfd/.

GEXCON. 2024b. *Gas explosion handbook.* Gas Safety Programme 1990–1992 (GSP90-92). Bergen, Norway: Christian Michelsen Research.

Glasstone, S., and P. J. Dolan, eds. 1977. *The effects of nuclear weapons.* 3rd ed. Washington, DC: US Department of Defense/US Department of Energy.

Hanna, S. R., O. R. Hansen, and S. Dharmavaram. 2004. "FLACS-CFD air quality model performance evaluation with kit fox, MUST, prairie grass, and EMU observations." *Atmos. Environ.* 38 (28): 4675–4687.

Herrmann, D. D. 2005. "Developing a sound basis for the design of vented explosion barricades in chemical processes." *Process Saf. Prog.* 24 (1): 52–58.

Herrmann, D. D. 2006. "Explosion consequences of low velocity releases of dense flammable vapor inside a chemical manufacturing facility with a complex ventilation system." In *Proc., 40th Annual Loss Prevention Symp.* New York: American Institute of Chemical Engineers.

Hoorelbeke, P., J. R. Bakke, J. Renoult, and C. Izatt. 2006. "Vapor cloud explosion analysis of onshore petrochemical facilities." In *Proc., ASSE-MEC 7th Professional Development Conf. and Exhibition.* http://www.gexcon.com/index.php?src=welcome/publications.html

Lenoir, E. M., and J. A. Davenport. 1993. "A survey of vapor cloud explosions: second update." *Process Saf. Prog.* 12 (1): 12–33.

McGlaun, M., S. L. Thompson, and M. G. Elrick. 1990. "CTH: A three-dimensional shock wave physics." *Int. J. Impact Eng.* 10 (1–4): 351–360.

NFPA (National Fire Protection Association). 2017. *Protection of buildings from exterior fire exposures*, NFPA 80A. Quincy, MA: NFPA.

NFPA. 2018. *Standard on explosion protection by deflagration venting*, NFPA 68. Quincy, MA: NFPA.

NFPA. 2020. *Standard for the prevention of fires and dust explosions in agricultural and food processing facilities,* NFPA 61. Quincy, MA: NFPA.

Nikodym, L., and M. Xie. 2017. *FacilityBlast for vapor cloud explosions: Quick user guide.* Cupertino, CA: Thornton Tomasetti.

NIST (National Institute of Standards and Technology). 2010. *Best practice guidelines for structural fire resistance design of concrete and steel buildings.* NIST Technical Note 1681. Gaithersburg, MD: NIST.

SciRisq. 2024. *Computational explosion and blast assessment model (CEBAM).* Houston, TX: SciRisq.

Tang, M., and Q. Baker. 1999. "A new set of blast curves from vapor cloud explosion." In *Proc., 33rd Loss Prevention Symp.* Paper No. 29e. Reston, VA: ASCE.

Thornton Tomasetti. 2019. *FacilityBlast VCE.* New York: Thornton Tomasetti.

TNO. 1985. "The multi-energy method—A framework for vapor cloud blast prediction." *J. Hazard. Mater.* 12 (1): 1–10.

TNO Green Book. 1992. *Method for the determination of possible damage to people and objects resulting from releases of hazardous materials (CPR 16E).* The Hague, The Netherlands: Committee for the Prevention of Disasters due to Dangerous Substances, The Director-General of Labour.

UL (Underwriters Laboratories). 2018. *Best practice guide for passive fire protection for structural steelwork.* Northbrook, IL: UL.

US DoD (US Department of Defense). 2008. *Structures to resist the effects of accidental explosions,* UFC 3-340-02 (formerly TM 5-1300). Washington, DC: US DoD.

Van Wingerden, K., O. R. Hansen, and P. Foisselon. 1999. "Prediction of blast overpressures caused by vapor cloud explosions in the vicinity of control rooms." *Process Saf. Prog.* 18 (1): 17–24.

CHAPTER 4
Types of Construction

4.1 INTRODUCTION

The design of blast-resistant structures requires the use of established design and construction practices, as well as a knowledge of the characteristics of the blast loading and the behavior of structures and their components under these loadings. Accounting for siting considerations, the engineer participates in selecting the type of construction that is required to withstand the potential blast loading conditions. Although all types of construction provide some level of blast resistance, there are some types of construction that are more appropriate than others.

Nonstructural considerations such as safety, operation, architecture, cost, fire and toxic hazards, nonprocess hazards (i.e., hurricane, seismic, and tornado), risks, occupancy distribution, and owner preference may dictate the shape, orientation, and layout of a plant building. In establishing these, however, the engineer should also consider the requirements for blast-resistant construction.

This chapter provides blast-resisting capacity ranges for various building types for typical vapor cloud explosions. Lower, moderate, and large blast load capacities are assigned to each building type to provide a relative ranking compared to the all-building types covered in this section. These blast-resistance ranges are not intended to be used for screening purposes but to give an idea to the reader about the practical upper bounds of each building type under blast loading. The actual blast-resisting capacity of a structure is greatly influenced by, but not limited to, the story height, connection design, openings along the perimeter, geometry, and so on.

4.2 GENERAL CONSIDERATIONS

The most important feature of blast-resistant construction is the ability to absorb blast energy without causing catastrophic failure of the structure as a whole or in a portion of the building. Construction materials in blast-protective structures inherently have inertial resistance, but they should also have a degree of ductility as well as strength so that an economical building can be realized. Furthermore, in a plant explosion, a building will be exposed to a force resulting from the blast loading on exterior surfaces. For a structure to exhibit any measure of blast resistance, its frame and envelope must be capable of absorbing this large lateral load. This requirement is similar to that for earthquake-resistant design. Structural components must possess adequate deformation capacity to form a yield mechanism.

Reinforced concrete is typically considered one of the most suitable and economical construction materials for blast-resistant buildings, especially for those close to a potential blast source where they are likely to be subjected to relatively high overpressure, debris, and thermal effects in the event of an explosion and fire. Steel space–framed, blast-resistant modular (BRM) buildings, pre-engineered metal buildings, reinforced masonry buildings, and other building types, if properly designed, can be used if sited at appropriate distances from these hazards.

Brittle materials such as unreinforced concrete, brick, timber, and unreinforced masonry are not suitable for blast-resistant structures as they undergo catastrophic sudden failure under blast overload. Although brittle materials are not recommended for new construction for blast load applications, each type does have some blast capacity, which can be determined using guidelines included in this reference. The principal criterion for evaluating such construction is its mode of failure if severe overloading occurs. If brittle behavior cannot be avoided, as is sometimes the case for axially loaded reinforced concrete columns or for shear walls, their capacities should be conservatively estimated.

The profile of a blast-resistant building should be as “clean and simple” as possible. Reentrant corners and offsets, in particular, should be avoided. Such features create local high concentrations of blast loading. The orientation of the building should be suh that blast-induced loads are reduced as much as possible.

4.3 COMMON SYSTEMS FOR PETROCHEMICAL BUILDINGS

Conventional building construction may provide some level of blast resistance. However, certain features of ordinary building construction, such as large windows, unreinforced masonry walls, and weak structural connections, could make these buildings vulnerable to even low-level blast effects. Conventional construction includes pre-engineered steel framing with metal cladding and steel framing with masonry or precast concrete walls. Usually, these buildings are designed only for dead, live, wind, and seismic loads. Outlined subsequently are types of common construction in petrochemical facilities.

4.3.1 Unreinforced Masonry Wall Construction

Unreinforced masonry construction was common in petrochemical facilities several decades ago but is not recommended for blast-resistant construction in modern times. This type of building typically has load-bearing walls supporting a light metal roof with or without concrete topping. Because of the brittle response of unreinforced masonry walls, this building type will fail catastrophically once its blast capacity is exceeded. This type of construction typically has lower blast-resisting capacity. Design improvements to enhance existing blast resistance can be achieved by

- Adding vertical steel posts, spanning from the foundation to the roof structure, on the exterior walls to reduce wall span. Posts can be installed on the interior or exterior face of the walls and should be secured to the masonry,
- Grouting existing hollow masonry cells, and
- Installing new interior steel columns to support the existing roof structure.

4.3.2 Timber Construction

Timber construction is not recommended for permanent structures; however, it exists in petrochemical facilities. Timber construction can vary from wood trailers to wood-framed structures such as residential houses. Timber construction has limited ductility for blast response, so it is usually not recommended for new construction; also, it has lower blast-resisting capacities. Please refer to API RP 753 (API 2007) for capacity values of wood trailers. Timber construction is also susceptible to fire risks, which occur in petrochemical facilities.

4.3.3 Pre-Engineered Metal Building Construction

Pre-engineered metal buildings are comprised of steel frames with cold-formed steel panels supported on cold-formed steel girts and purlins, as illustrated in Figure 4-1. The steel frame is designed to resist all vertical and lateral loads in one of the two primary axes. Lateral load resistance in the other primary axis is accomplished using bracing or local moment–resisting frames. Design improvements to enhance blast resistance may be achieved by

- Specifying closer spacing of steel frames.
- Using symmetric sections (back-to-back C-shapes) for girts and purlins and reducing their spacing.
- Increasing the size of anchor bolts and strengthening wall panel connections at the foundation and at the roof.
- Increasing the number of cladding fasteners and using oversized washers to reduce tear-out of siding material to develop higher plastic strength (ultimate moment and/or shear capacities) of these structural members.
- Increasing the web thickness of the steel frame's column and girder connection to prevent panel shear failure.
- Adding additional bracings to wall and roof frames.
- Increasing the web thickness of the steel frame's connections to prevent local buckling.
- Using deeper panels with thicker gauge thicknesses.

These buildings can be designed to have moderate blast-resisting capacities.

Figure 4-1. Pre-engineered metal building.
Source: Omega Risk.

4.3.4 Reinforced Masonry Wall Construction

Reinforced masonry clad buildings are very similar to conventional commercial buildings normally constructed to resist conventional loading. A structural steel or concrete frame can be designed to support vertical loads and, in some cases, to resist lateral forces. Reinforced masonry is used for the exterior walls and is designed to span vertically and/or horizontally. The reinforced masonry walls that run parallel with a directional blast force can also be used as shear walls to transmit lateral forces to the foundation. The reinforced masonry wall is attached to the building frame to tie all components together and provide resistance to rebound forces. These buildings can be designed to have moderate blast resistances.

4.3.5 Metal Clad Construction

Metal clad buildings utilize conventional steel space frame design through the use of hot-rolled structural shapes for frame, girts, and purlins. Metal siding or insulated sandwich panels, with thicker gauge metal and more connectors, are used for exterior walls. As for pre-engineered metal buildings, the steel frame resists all vertical and lateral loads in one or both primary axes depending on whether bracing is used in one axis. The connections are enhanced to develop higher plastic strength (ultimate moment and/or shear capacities) of the structural members. These buildings typically have moderate blast-resisting capacities. With enhancements, they can be designed to have large blast-resisting capacities.

4.3.6 Precast Concrete Wall Construction

This type of construction typically uses precast concrete walls with steel or concrete frames (Figure 4-2). Load-bearing precast concrete panels carrying a precast concrete roof are also used in petrochemical facilities. Both systems typically have limited blast-resisting capacity because of connection limitations of precast concrete wall panels.

In framed systems, the frame resists all vertical loads and precast shear walls resist lateral loads. Ductile connections for precast panels are an important consideration as these connections are where failures occur under overload blast pressures. Precast panels are made with embedded steel connection devices attached to the building frame by bolting or welding. The roof is usually a concrete slab cast on top of a metal deck. The metal deck is attached to steel framing or open-web-joists by studs or puddle welds. This type of construction can be designed to withstand moderate blast loads.

4.3.7 Cast-in-Place Concrete Wall Construction

Cast-in-place concrete construction (Figures 4-3 and 4-4) is used to resist relatively high blast overpressures where precast concrete is not economical or practical. Lateral loads in cast-in-place construction are typically resisted by shear walls. The structure typically depends on a structural steel or reinforced concrete frame to support vertical loads. The thickness of the concrete walls and size and

Figure 4-2. Reinforced masonry building.

Source: BakerRisk.

Figure 4-3. Precast concrete building.
Source: BakerRisk.

Figure 4-4. Reinforced concrete cast-in-place building.
Source: BakerRisk.

placement of the reinforcing steel can be chosen to provide resistance to virtually any anticipated design blast loads.

4.3.8 Cold-Formed Steel Interlocking Panel Modular Construction

This type of construction involves the use of a structural steel building flooring with cold-formed steel interlocking panel walls, ceiling, and roof. This type of construction is extensively used for buildings that host power control equipment and instrumentation such as power control rooms (PCRs), power distribution centers (PDCs), substation buildings, remote instrument enclosures (RIEs), and local equipment rooms (LERs).

Different from site-erected construction, these buildings are fabricated at a manufacturing facility and shipped in modules to their final destination, where they are installed onto a foundation system such as a reinforced concrete mat or piers. The most typical configuration for interlocking panel modular buildings in industrial facilities is on elevated reinforced concrete pier foundations or steel frames, with power cable access from the floor of the building (Figure 4-5). Note that uplift blast pressures on the floor system design are required because of the elevated support of these modular buildings. Sway response for the elevated piers should also be evaluated for blast response.

Figure 4-5. Interlocking cold-formed steel panel building on an elevated reinforced concrete pier foundation.

Source: Powell Industries.

The floor is constructed with a steel plate that is welded onto the floor steel framing. Walls are built using interlocking cold-formed steel or aluminum panels, which comprise the framing and siding of the buildings concurrently. Ceiling and roof panels are built using the same panels and are supported by the walls. Steel thickness typically varies from 18 to 12 gauge.

Although not typically occupied during plant operation, these buildings can enclose power control equipment and instrumentation that have a critical function during an emergency or during plant operation. To protect the enclosed equipment, buildings can be designed to resist an accidental blast loading. Extra consideration may be needed for the nonstructural components that are essential for functionality but are not protected within the building. There is very little design guidance for maintaining functionality of equipment within a building that is dependent on external utilities.

This construction system can be designed to resist moderate blast loading.

4.3.9 Blast-Resistant Modular Steel–Framed Buildings

Blast-resistant modular (BRM) steel–framed buildings are commonly utilized in petrochemical facilities as alternatives to conventional in-place construction. These modular buildings are specifically designed for blast loading. These modular buildings utilize steel structural frame members with crimped or flat steel plate walls. The method of attaching the plate walls to the frame members typically does not utilize mechanical fasteners; rather, welded construction is usually used. Roof joists and floor joists (usually seismically compact steel sections) typically support flat or crimped steel plate roofs and flat steel plate floors. Thickness and corrugation of the steel walls and size and placement of the steel frame can be chosen to provide resistance to virtually any anticipated design blast loads. Nonstructural debris can be a significant hazard to occupants at higher levels of blast loading and structural response owing to the typically compact floor plans, particularly for single-wide modules. To reduce nonstructural debris hazards, these components and their connections can be designed for blast loading.

BRM buildings may be anchored or unanchored. If unanchored, they may slide, which can result in additional risks to occupants owing to acceleration and nonstructural debris.

Typical modules are 8 to 16 ft (2.2 to 4.3 m) wide by 10 to 50 ft (3.0 to 15.2 m) long. BRM buildings can range from single-module structures to multimodule and multistory, integrally connected structures, with floor areas of more than 10,000 ft^2 (929 m^2) (Figures 4-6 and 4-7).

4.4 TENTS

The term *tent* is meant to include fabric or membrane structures that do not have rigid walls or roofs. Tents may be used at petrochemical facilities as permanent or temporary shelter for personnel, to protect in situ work processes, or for material

Figure 4-6. Multiunit BRM.
Source: Hunter Buildings.

Figure 4-7. BRM installation with stacked caps and a raised foundation.
Source: RedGuard.

storage. An example of a tent at a chemical processing facility is shown in Figure 4-8. Tents can be designed to have lower to moderate blast-resisting capacities.

The types of tents common to facilities include the following:

- Pole Tents (aka circus tents): These tents do not have rigid frames and are supported by center and perimeter poles and cables. The lateral resistance of the tent is provided by guy wires.
- Light-Framed Tents: These tents are similar to circus tents but use light framing members for roof and side wall support. Light-framed tents rely on guy wires to provide lateral stability.

Figure 4-8. Tent structure.
Source: BakerRisk.

- Rigid-Frame Tents: These tents have framing similar to a pre-engineered metal building. The frames provide lateral resistance in one direction, and cross bracing provides the lateral resistance in the perpendicular direction.
- Air-Supported Structures: These tents use inflatable tubes to form the framing members or an inflated membrane. Guy wires provide the lateral stability.
- Scaffolding Ttents: These tents are comprised of scaffolding components for the framing members, which provide the lateral resistance.

API RP 756 (API 2014) covers *Management of Hazards with Location of Process Plant Tents*. Please refer to API RP 756 and Chapter 7, "Design Procedures," for further information.

4.5 OTHER SYSTEMS

Under special circumstances, the following types of construction may be considered.

4.5.1 Pre-Engineered Concrete Boxes

Pre-engineered concrete boxes can be used to provide smaller footprint buildings. These buildings are manufactured in a factory, are prewired, come with HVAC installed and are truck-delivered to the site ready to be secured to a foundation and connected to desired utilities. These buildings can be designed up to moderate overpressures.

4.5.2 Portable Buildings

Portable buildings (e.g., trailers) are often used at petrochemical facilities on a temporary basis. These structures are typically designed and manufactured for service loads (e.g., wind, seismic, snow) and transportation loads, but they seldom contain any provisions to resist blast loads. Portable buildings are typically constructed with timber framing with aluminum or light gauge steel wall and roof cladding. Wall studs are often notched to minimize section thickness, further weakening the trailer's blast resistance. The structures referred to here are quite different from steel-framed modular buildings that are specifically designed for blast loads.

A previous accidental explosion resulted in multiple fatalities from the use of portable buildings located near the explosion center. As a result, significant changes in the use of portable buildings in petrochemical plants have been made. The American Petroleum Institute (API) has published API RP 753, *Management of Hazards Associated with Location of Process Plant Portable Buildings* (API 2007), for siting of portable buildings. This document provides information regarding siting of this building type under blast conditions in explosion accidents. The reader is encouraged to consult this API document for siting of portable buildings.

References

API (American Petroleum Institute). 2007. *Management of hazards associated with location of process plant portable buildings*. API RP 753. 1st ed. Washington, DC: API.

API. 2014. *Management of hazards associated with location of process plant tents*. API RP 756. 1st ed. Washington, DC: API.

CHAPTER 5

Dynamic Material Strength and Response Criteria

5.1 INTRODUCTION

Designing structures to resist the effects of accidental explosions requires a knowledge of the dynamic material properties of the structural components as well as the allowable responses of components and systems. Materials and structural systems respond differently to dynamic loads produced by explosions than to dynamic loads produced by seismic events, and they respond much differently from that to statically applied conventional loads. Therefore, it is imperative that the engineer understand these differences. Under blast dynamic loading, materials achieve a strength increase because of high strain rate effects that can significantly enhance structural resistance. Also, blast loading is a single-cycle dynamic load event and therefore structures subjected to blast loads are typically allowed to undergo significant plastic (permanent) deformation that absorbs the explosion energy, whereas response to other types of loading with longer duration is normally required to remain in the elastic range.

Structural design for accidental explosions at industrial facilities is similar in many ways to design for high explosive detonations, nuclear weapons effects, and nuclear power accidents for which design guides are available. However, structural response criteria for petrochemical plants may allow more structural damage to be tolerated, in accordance with an owner/operator's blast protection philosophy.

This chapter provides dynamic material properties and response criteria necessary to design structures constructed of reinforced concrete, reinforced masonry, structural steel, cold-formed steel, and aluminum. Static and dynamic properties are covered for the materials used in these structures. Allowable response criteria are covered for both individual members and structural systems.

5.2 STATIC VERSUS DYNAMIC RESPONSE

Conventional loads, such as wind and live loads, are applied relatively slowly to a structure and remain constant for a relatively long period compared to the response time of the structure. Blast-loaded structures experience a very rapid application of the load and a corresponding rapid rise in member stresses. This load is transient and will normally return to ambient conditions in a short period (typically tens or hundreds of milliseconds).

In conventional design, stresses are limited to the elastic range. In blast design, yielding is acceptable and, in fact, desirable. As the structural member is stressed in the plastic region, it continues to absorb the blast energy by balancing the kinetic energy of the explosion against the strain energy of the member. The total strain energy available is a function of dynamic material properties, section properties, and the amount of plastic deformation permitted. The total amount of blast energy required to be absorbed is a function of the peak load and duration of the blast. The adequacy of a blast-loaded member is typically based on maximum deformation rather than stress level.

Material response under dynamic loads is markedly different from that under static loads. As a material is loaded rapidly, it deforms with a high strain rate that creates an increase in the stress level at which yield occurs, as well as the ultimate stress achieved prior to rupture. In general, the faster the material is deformed (strain rate), the greater the increase in strength. The resulting strength increase allows members to develop structural resistance in excess of their static capacity. This increase can be as much as 10% to 30% for some materials; thus, it is too significant to ignore these effects when computing flexural response. This effect is accounted for in blast design with the use of a dynamic increase factor, or DIF (refer to Section 5.5.4). In addition, the actual static strengths of materials are higher than the minimum specified strengths. This is included in blast-resistant design with the use of a static increase factor, or SIF, as discussed in this chapter. Connection forces and loads on supporting members will be underestimated if material strength increase associated with the SIF and DIF is ignored.

5.3 RESISTANCE–DEFLECTION FUNCTION

Structural elements resist blast loads by developing an internal resistance on the basis of material stress and section properties. To design or analyze the response of an element, it is necessary to determine the relationship between resistance and deflection. In flexural response, stress rises in direct proportion to strain in the member. Because resistance is also a function of material stress, it also rises in proportion to strain. After the material in the outer fibers reaches the yield limit, the relationship between stress and strain, and thus resistance, becomes nonlinear. As the outer fibers of the member continue to yield, stress in the interior of the section also begins to yield and a plastic hinge is formed at the locations of maximum moment in the member. If ductile response occurs (e.g., buckling is prevented), deformation continues as the member absorbs load until rupture strains occur.

Variation in internal resistance can be related to the strain because stress in a member is a function of the strain experienced at a given point. Deformation of a

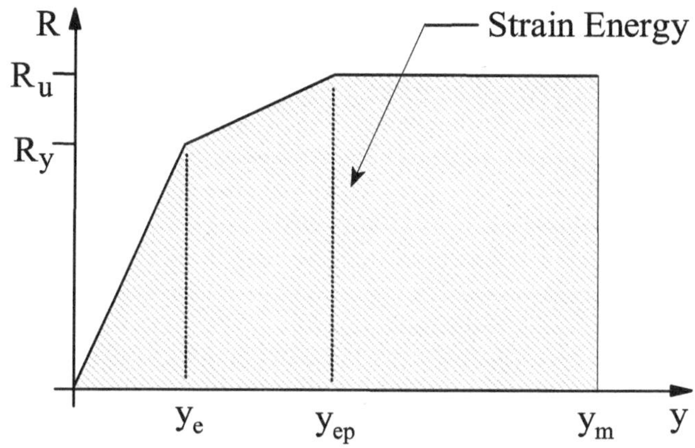

Figure 5-1. Typical resistance–deflection curve.

key point on the member can also be related to the strain producing a relationship between resistance and deflection, as shown by the curve in Figure 5-1. Elastic resistance is the level at which the material reaches yield at the location of maximum moment demand in the member. Beyond the point of first yield of a member, plastic regions are formed in the section and an elastic–plastic condition occurs. Internal resistance continues to increase as the stress demand in other locations of the member rises in response to the applied load, although at a lower slope than the elastic region. During this extended response, portions of the member are responding plastically, whereas other sections are responding elastically on the basis of cross section and location along the member. As the response continues, other critical sections reach yield and additional plastic hinges are formed. Each yield point changes the slope of the resistance–deflection curve. When the last critical moment section yields, causing the member to become a mechanism, no additional resistance is available and the resistance–deflection curve is flat, provided that ductile response with no strain hardening or membrane action is assumed. The area under this curve represents the total strain energy available to resist load at a given deflection.

5.4 MATERIAL AND STRUCTURAL ELEMENT TYPES

A brief description of the materials and structural elements used in blast design applications is covered in this section. The response of each material to blast loads is described along with typical applications.

5.4.1 Reinforced Concrete

The high resistance and mass provided by reinforced concrete structures make it particularly suited for buildings exposed to explosion sources. Concrete also provides effective resistance to fire and projectile penetration, which are also important considerations in many explosion accidents.

Reinforced concrete is a complex material to model because of the brittle nature of concrete and nonhomogeneous properties for both compression and tension. Although sophisticated methods are becoming more prevalent in finite-element analysis (FEA) where constitutive models can capture cracking and other response modes, simplified methods are still primarily used in blast design of facilities. These simplified methods are solely based on a flexural response and rely on the elimination of brittle modes of failure. To achieve a ductile response for concrete, proper proportioning and detailing of the reinforcing is necessary.

As the member is strained, the reinforcing bars yield and allow the formation of plastic hinges. Concrete in these regions is cracked on the tensile face and subsequently reaches its crushing strain on the compressive face. If rotation of the hinge increases beyond this point in a singly reinforced section, the concrete will be dislodged and will be incapable of providing a compressive component for the internal resisting couple. Additional rotation can be achieved in doubly reinforced sections, especially if flexural reinforcing is sufficiently restrained by shear reinforcing to prevent buckling. In these plastic hinge regions, the internal resistance of the section is provided by a couple formed between the longitudinal reinforcing bars. A rebound of a structural member under dynamic loads produces a reversal of the forces in the section. Additional discussion of reinforced concrete response is provided in Chapter 7, "Design Procedures."

Prevention of brittle failure modes is accomplished by limiting concrete shear stresses or by increasing material strength, section thickness, or shear reinforcing. The amount of flexural reinforcing in a member is also limited to assure that the tension reinforcing yields before concrete crushing can occur. Shear steel may be used to increase shear resistance, confine the flexural reinforcing, and prevent buckling of the bars in compression.

UFC 3-340-02 (USACE/NAVFAC/AFCESA 2014) indicates that Grade 60 reinforcing bars (No. 11 and smaller) have sufficient ductility. Bars with a higher yield strength may not have the necessary ductility for flexural resistance and shop bending; thus, straight bars should be used when possible for these materials. Welding of reinforcement is usually discouraged for blast design applications; however, it may be required for anchorage. In these cases, ASTM A706 (ASTM 2023) bars may be used.

A minimum concrete compressive strength of 3,000 psi (20.7 MPa) should be used to reduce the probability of shear failures. A value of at least 4,000 psi (27.6 MPa) is preferred.

5.4.2 Reinforced Masonry

Because of its relatively high mass and ductility, reinforced masonry buildings can be cost-competitive material for blast-resistant design. Reinforced masonry responds to dynamic loads similarly to reinforced concrete, with similar increases in dynamic strength as the strain rate increases. Limited options for placement of reinforcing and low shear strength of mortar joints are significant disadvantages compared to reinforced concrete. Although unreinforced masonry structures are common in older facilities, they typically do not have sufficient ductility to resist any significant blast load and may be deemed inadequate, especially when used as load-bearing walls.

Hollow masonry units should conform to ASTM C90 (ASTM 2024b), Grade N. Joint reinforcing should meet the requirements of ASTM A82 (ASTM 2017a) with a minimum yield stress of 70 ksi (483 MPa) and a minimum ultimate strength of 80 ksi (552 MPa). Grade 60 bars should be used for primary reinforcing.

5.4.3 Structural Steel

Low and medium carbon structural steels, such as A36 (ASTM 2019), A572 (ASTM 2006), A500 (ASTM 2017b), A1085, and A992, are sufficiently ductile for blast design applications. High-strength materials (greater than nominal 50 ksi, 350 MPa, yield) should be avoided, or used with caution, in most applications to prevent problems with decreased ductility. Cold weather conditions can cause a reduction in ductility for all steel types that should be accounted for in blast design when applicable. In certain situations, such as blast door latch bolts, high-strength steel may be required to provide the required resistance. Brittle modes of failure, such as shear, should be examined carefully in these applications.

To achieve large deformations without failure, steel members must be sufficiently laterally braced and connected at their supports to avoid buckling and instability problems. Also, compact cross sections for structural steel components should typically be used for blast design. As unstiffened elements buckle, the cross-sectional properties are reduced, and the resistance is lowered. Dynamic explicit FEA can be used to capture buckling modes of response and the reduced capacities that are attributed to members that are not adequately braced.

5.4.4 Cold-Formed Steel

For low blast pressure applications, cold-formed steel members can provide a cost-effective cladding for buildings. Cold-formed members include corrugated steel panels as well as Z and C shapes. Cold-formed steel insulated panels have not been subjected to many blast tests and are not included in this manual. However, some information from limited blast testing is available (Idriss et al. 2016). Members complying with the requirements of ASTM A653 (ASTM 2022b)Au, which is commonly used for cold-formed structural components, have yield strengths ranging from 33 ksi (228 MPa) to 65 ksi (450 MPa). Cold-formed members that have been subjected to blast testing include corrugated steel panels, girts and purlins with Z and C shapes, and metal studs.

A key consideration in the design of cold-formed members for blast is premature local buckling of the relatively thin webs and flanges. This effect is accounted for by using the same effective section properties that are used for static design to calculate the ultimate resistance of cold-formed members. However, local buckling also causes additional reduction in the resistance at deflections greater than yield deflection. A factor of 0.9 is recommended to be applied to the design resistance to model this reduction when using the SDOF methodology. Also, the compression flange must be well braced to achieve the ductility that is assumed for blast design of cold-formed members. Dynamic explicit FEA can better estimate this reduced section property.

Special precautions must be taken regarding end bearing for these members to avoid crushing of the web at peak response. If end bearing controls, the allowable response is limited to reduce the chance for nonductile failure. Connections for these members also present difficulty because of the thin web material. To develop the ultimate strength of a member, multiple fasteners may be required so that the shear strength of the material is not exceeded.

At large deflections, cold-formed metal panels, girts, and purlins respond in membrane action. In this mode, resistance to blast loads is provided by stretching of the panel rather than flexure (Figure 5-2). These components can be quite strong because this is a very efficient structural action; however, end anchorage is extremely important to achieving significant capacity. The tension membrane capacity is almost always limited by the connection capacity rather than the tension capacity of the component itself.

Insulated cold-formed panels have a more complex structural response than other types of cold-formed members that must be considered in blast-resistant design. Design guidance based on shock tube testing is available from Idriss et al. (2016).

Where fragment hazards are a concern, cold-formed panels may not be suitable because they have a very low resistance to fragment penetration.

5.4.5 Open Web Steel Joists

Conventional reinforced masonry structures as well as steel frame buildings often utilize open web steel joists to provide support for roof decks. Principal concerns for these members are buckling of the web at the ends owing to high

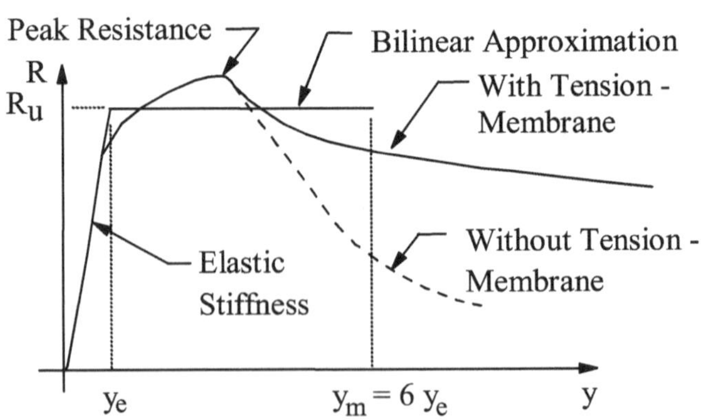

Figure 5-2. Typical membrane response.

Source: UFC 3-340-02 (USACE/NAVFAC/AFCESA 2014).

shear forces and instability in the bottom chord during rebound of the section. Older steel joists have performed surprisingly well in many explosion accidents, provided they are adequately attached at the supports. This typically requires additional welding of the chord members to the embedded plate. Bracing for the bottom chord throughout the length of the member is not normally provided for conventional designs but is crucial to achieving acceptable response to blast load.

The quality of joist welds is also critical to achieving a ductile response. Welding is performed to Steel Joist Institute standards and the lack of specific criteria may prevent the development of a predictable ultimate capacity. Special precautions must be taken to remedy this problem, such as requiring manufacture in accordance with the criteria of the American Welding Society.

5.4.6 Connections

Structural components in blast-loaded structures have reaction loads that are much higher than those from other types of loads. Therefore, much stronger connections are usually required. Welds and high-strength structural bolts are typically used for connecting blast-loaded structural steel components. Self-drilling steel screws are typically used for connecting blast-loaded cold-formed steel components. Monolithic concrete connections are ideal, but many blast-resistant buildings are constructed with precast concrete components that have connections that are similar to those used for conventional design, but with significantly higher capacity.

For the case of column supports, substantial baseplates and high-capacity anchor bolts are required to resist high compression and/or tensile reaction forces that may be combined with high shear forces. Achieving full anchorage of these bolts is of primary importance and will usually require headed bolts or plates at the embedded end of the bolts to prevent pullout. When anchor bolts are securely anchored into concrete, the failure mechanism is a ductile, tensile failure of the bolt steel. Insufficient edge distance or insufficient spacing between bolts results in a lower anchorage capacity and a brittle failure mode.

Post-installed bolts will be required at times for attachment of blast-loaded structural components and equipment that may be subjected to large accelerations during a blast. Limited blast testing and experience has provided some guidance for the best type of post-installed anchors for blast-resistant design. Epoxy anchors have shown excellent dynamic capacity and may be considered for critical applications. Wedge-type anchors are typically qualified for dynamic loads, although most of these ratings are for vibratory loads and are based on cyclic tests at low stress levels. Expansion anchors are considered least ideal for most blast design applications.

Often, anchor bolts are designed for the maximum axial and shear reactions at the base of the columns as a static load. This method requires many bolts even using dynamic material properties. The bolts will yield under tensile loads and to some degree, shear loads. That is why it is important to use ductile materials for bolts (ASTM F1554) (ASTM 2021) is commonly used] to guard against sudden failure under peak stress. It is possible to model the tensile and/or shear response of a bolt to the calculated reaction force history and take advantage of the strain energy capacity of the bolts. This allows the bolts to respond to the load-time history rather than just a peak load. However, well-validated response criteria defining the amount of dynamic response causing failure of bolts and other types of connections are not available. A dynamic analysis using conservative response criteria is warranted only for special situations such as where the reuse of existing bolts is important. Based on experience, it is sufficient to design connections using a load and resistance force design method to have a capacity equal, or greater than, the reaction force applied by the ultimate dynamic flexural load capacity of the connected component. This reaction force is typically referred to as the "equivalent static" reaction force to differentiate it from the peak dynamic reaction force calculated with a time history calculation.

5.5 DYNAMIC MATERIAL PROPERTIES

This section describes the dynamic properties of materials used in structures designed to resist blast loads at petrochemical facilities. Static properties are available from several references and are not repeated in this chapter, except to indicate minimum acceptable values. The dynamic response of these materials has been studied extensively; however, their dynamic properties are not as widely published. Procedures for obtaining these properties are covered here in enough detail to permit an accurate determination for the design and analysis of petrochemical structures.

5.5.1 Stress–Strain Relationships

The response of a material under static or dynamic load is governed by the stress–strain relationship. A typical stress–strain diagram for concrete is shown in Figure 5-3. As the fibers of a material are deformed, stress in the material is changed in accordance with its stress–strain diagram. In the elastic region, stress increases linearly with increasing strain for most steels. This relation is quantified by the modulus of elasticity of the material.

Concrete does not have well-defined elastic and plastic regions because of its brittle nature. A maximum compressive stress value is reached at relatively low strains and is maintained for small deformations until crushing occurs. The stress–strain relationship for concrete is a nonlinear curve. Thus, the elastic modulus varies continuously with strain. The secant modulus at service load is normally used to define a single value for the modulus of elasticity. This procedure is given in most concrete texts. Masonry has a stress–strain diagram similar to concrete but is typically of lower compressive strength and modulus of elasticity. For single-degree of freedom (SDOF) analysis, the modulus of elasticities of concrete and masonry during blast loading is calculated with the same equations used for static design, using the higher dynamic compressive strengths of these materials in these equations. Constitutive material models for the dynamic response of concrete can be used in finite-element analyses.

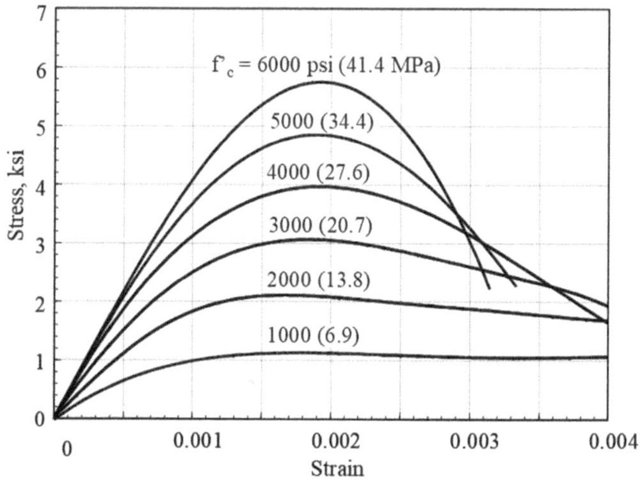

Figure 5-3. Typical stress–strain curve for concrete.
Source: ASCE MOP 42 (ASCE 1961).

For steel materials, the shape of the curve is much different from that of concrete, as can be seen in Figure 5-4. Steel is relatively ductile and can achieve large strains prior to rupture. Low-carbon structural grade steels (e.g., A36, A572) exhibit a well-defined yield point followed by a flat yield plateau. High-strength steels do not have a sharp break at the elastic limit and the yield region is highly nonlinear. Low-carbon steel materials are particularly suited to blast-resistant design because they can deform well beyond the elastic limit without rupturing. This produces a

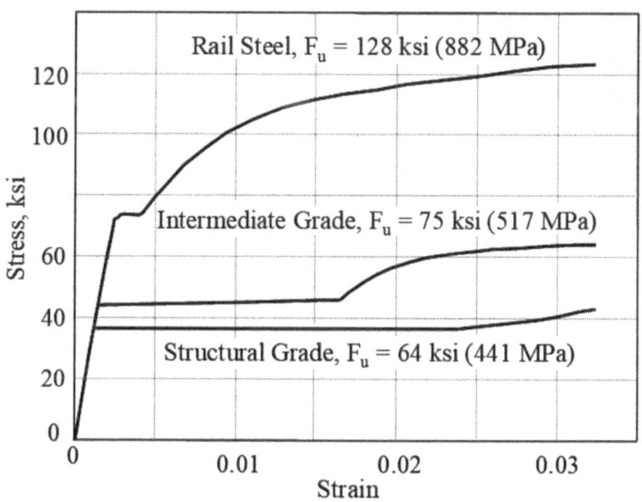

Figure 5-4. Typical stress–strain curve for steel.
Source: ASCE MOP 42 (ASCE 1961).

long resistance–deflection curve to absorb the blast energy, while avoiding brittle fracture problems. High-strength steels should be avoided for general construction because of their low ductility. Special applications, such as blast doors and shields, may require high-strength materials to achieve the desired resistance. The selection of static properties for high-strength materials should be made conservatively.

5.5.2 Strength Increase Factor

Static properties are readily available from a variety of sources and are well defined by national codes and standards organizations. Specifications referenced in the codes define minimum mechanical properties for various grades of material. In practice, the average yield strength of steel materials being installed is approximately 25% greater than the specified minimum values. A *strength increase factor* is used to account for this condition and is unrelated to the strain rate properties of the material. UFC 3-340-02 suggests using a 1.1 strength increase factor applied to the minimum yield stress for structural steel with a yield of 50 ksi (345 MPa) or less and for Grade 60 reinforcing. Several references addressing nuclear facilities suggest ignoring these strength increase factors (SIFs) to add a larger margin of safety to the design. The use of recommended SIFs is warranted for blast design where it is desired to reduce conservatism and make use of the fully available blast capacity. This also leads to more accurate calculated reaction forces, which are required to ensure that a blast-loaded component has adequate shear strength and connection strength.

Cold-formed steel also exhibits an average yield strength well in excess of the specified minimum. UFC 3-340-02 recommends a strength increase factor of 1.21 for this material.

Concrete strength is specified as the minimum compressive strength at 28 days. This value is used for design and is not typically increased to account for an increase in strength with age. For evaluating an existing structure, it may be worthwhile to determine the in situ strength of the concrete to use in the analysis. This will not make a great difference in flexural capacity, but it could be very important when examining shear resistance.

Strength increase factors are summarized in Table 5A-1. Higher increase factors may be justified for reinforcing steel based on recent research that used a large database of measured rebar yield strengths (Mander and Matamoros 2019). Higher SIF values for rebars will cause higher calculated reaction forces in blast-loaded reinforced concrete components. However, the SIF values in Appendix 5A are usually consistent with those of the conservative approach currently used in blast-resistant design to determine component shear strengths and connection strengths, at least to the extent that unexpected shear failures are not a problem noted with any frequency during blast testing on reinforced concrete components.

5.5.3 Dynamic Strength Increase

Concrete and steel experience an increase in strength under rapidly applied loads. These materials cannot respond at the same rate at which the load is applied. Thus, the yield strength increases, and less plastic deformation will occur. At a fast strain rate, a greater load is required to produce the same deformation than at a lower rate. This increase in the yield stress is quite significant for lower-strength materials and becomes less pronounced as the static yield strength increases.

For steel, the modulus of elasticity is the same in the elastic region and yield plateau for both static and dynamic responses. In the strain-hardening region, the slope of the stress–strain curve is different for static and dynamic responses, although this difference is not of significance for most structural design applications.

A strength increase is also produced at ultimate strength, F_u, for steels; however, the ratio of dynamic to static strength is less than at yield. A typical stress–strain curve describing the dynamic and static responses of steel is shown in Figure 5-5. Elongation at failure is relatively unaffected by the dynamic response of the material.

Aluminum exhibits a modest increase with strain rate, which is typically ignored. Lindholm and Bessey (1969) surveyed available test data on dynamic properties for a number of materials. This is an extremely useful resource for information on less commonly used materials.

The ultimate strength of concrete is greater under dynamic loads. Although the modulus of elasticity is also greater, this difference is small and is usually ignored. Figure 5-6 describes the relationship between dynamic and static responses for concrete.

The magnitude of dynamic increase is dependent on several factors, including static material strength and strain rate. In general, the higher the static strength of a material, the lower the increase in dynamic strength. The faster a material is strained, the higher the increase in dynamic yield and ultimate strength. Figure 5-7

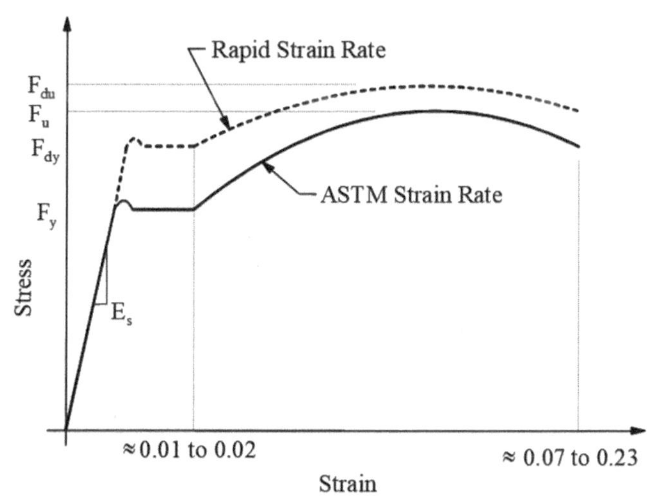

Figure 5-5. Effect of strain rate on stress–strain curve for steel.
Source: UFC 3-340-02 (USACE/NAVFAC/AFCESA 2014).

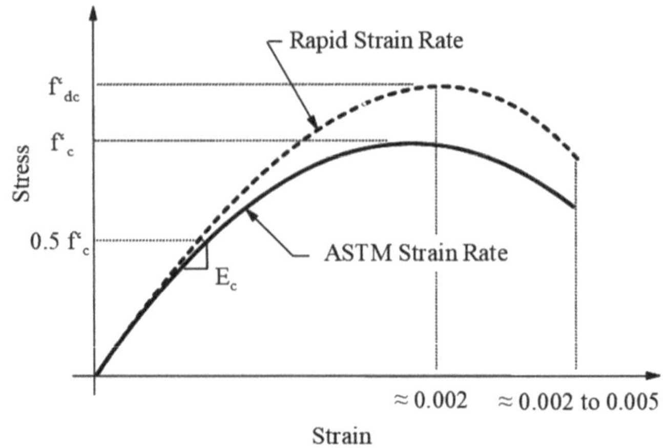

Figure 5-6. Effect of strain rate on stress–strain curve for concrete.
Source: UFC 3-340-02 (USACE/NAVFAC/AFCESA 2014).

describes the relationship between the strain rate and the ratio of dynamic to static material strength for structural steel, concrete, and reinforcing steel.

5.5.4 Dynamic Increase Factor

To incorporate the effect of material strength increase with strain rate, a dynamic increase factor is applied to static strength values using the SDOF methodology. DIFs are simply ratios of dynamic material strength to static strength and are a function of material type as well as strain rate, as described previously. DIFs are also dependent on the type of stress (i.e., flexural, direct shear) because peak values for these stresses occur at different times. Flexural stresses occur very quickly, whereas peak shears may occur relatively late in time, resulting in a lower strain rate for shear.

It is possible to determine the actual strain rate of a material when calculating the dynamic response using an iterative procedure within the SDOF method. A rate must be assumed and a DIF selected. The dynamic strength is determined by multiplying the static strength (increased by the strength increase factor) by the DIF. The time required to reach maximum response can be used to determine a revised strain rate and a revised DIF. This process is repeated until the computed strain rate matches the assumed value. There are uncertainties in many of the variables used to calculate this response, and therefore, determination of strain rates with great accuracy is usually not warranted. The effect of strain rate on material strength is embedded within given constitutive material models used in finite-element methods, and therefore, no iteration is required.

UFC 3-340-02 and other references suggest selecting DIF values for SDOF analyses based on pressure range or scaled distance to the explosion source. This method groups blast loads of less than a few hundred psi (100 psi = 690 kPa) into the low pressure category with a single DIF value for each stress type. For petrochemical facilities, most structures will fall in this low pressure category.

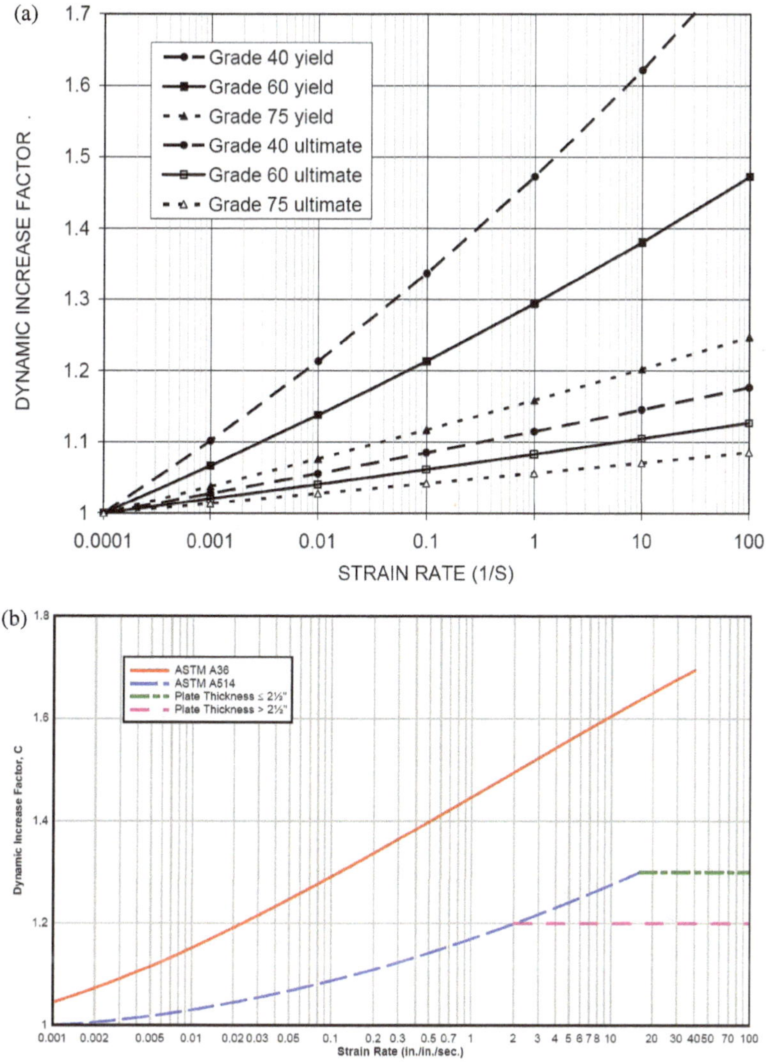

Figure 5-7. Effect of strain rate on dynamic yield strength of steel: (a) DIF for reinforcing steel (rebar), (b) DIF for ASTM A36 and A514 steels.

Source: UFC 3-340-02 (USACE/NAVFAC/AFCESA 2014).

The level of dynamic material enhancements varies for different stress types in both concrete and steel for several reasons. Flexural response is typically ductile and the dynamic enhancement values are based on actual strain rates. Shear stresses in concrete produce brittle failures and thus require a degree of conservatism to be applied to the selection of a DIF using the SDOF methodology. In addition, test data for the dynamic shear response of concrete materials are not as well established as for compressive strength. Strain rates for tension and compression in steel and concrete members are lower than those for flexure and thus dynamic material enhancement properties are necessarily lower.

Values for dynamic increase factors are presented in a variety of references, although most are based on the same data source. Additional data have been produced in various test programs but have not been assembled into a central source. Table 5A-2 provides recommended DIFs for reinforced concrete and masonry, and Table 5A-3 contains values for structural steel, cold-formed steel, and aluminum. These tables are largely based on recommended default values for DIF in UFC 3-340-02, which recommends that default DIF values for blast design can be based on an assumed strain rate of 0.1 s^{-1} for typical blast-resistant reinforced concrete and steel components. UFC 3-340-02 also provides empirical graphs of the dynamic increase factor versus strain rate for various steel materials, as shown in Figure 5-7. It also has similar plots for concrete in compression and tension. For dynamic finite-element analyses, the constitutive material properties can include the dynamic enhancement of yield strength for all strain rates. In this case, the strain rate for each finite element is calculated at each time step and then used to calculate the applicable dynamic yield strength for the element. It is very important that the correlation used between strain rate and yield strength in the constitutive material properties is well validated with test data in the applicable range of strain rates.

Dynamic increase factors for steel member connections can be conservatively ignored. UFC 3-340-02 suggests that the dynamic enhancement properties of steel material with similar yield stress to the welds or fasteners be used for the various types of connections. UFC 3-340-02 recommends a DIF value of 1.05 for ASTM A514 (ASTM 2022a) steel with a yield stress of 90 to 100 ksi (620 to 690 MPa) for tension or compression and a DIF value of 1.07 to 1.09 for bending. Therefore, if necessary, a DIF value of 1.05 can be justified for steel connections using typical welds with a yield stress of 50 to 100 ksi (345 to 690 MPa) and ASTM A325 (ASTM 2016) or ASTM A490 (ASTM 2014) high-strength bolts. Also, there are no well-validated DIF values for post-installed anchors in concrete.

5.5.5 Dynamic Design Stress

Strain-hardening effects in steel members and concrete reinforcing are included in both SDOF and FEA analyses by using a design stress that is greater than yield. During a dynamic response, the stress level at critical sections in a member varies with the strain of the section. In the elastic region, the strain across the section varies with location from the neutral axis of the member. Beyond this region, the member experiences plastic response in which the fiber stress of the entire section exceeds the elastic limit. At this point, the stress is constant over the cross section but is still changing with the total member strain. Steel members experience an increase in stress in the strain-hardening region until the ultimate dynamic material stress is reached. After this point, the fiber stress decreases with an increasing strain until rupture occurs. Concrete exhibits an increasing stress until the maximum compressive stress is reached, after which the stress level decreases with additional deformation. Because of its brittle nature, strain hardening does not occur in concrete; however, reinforcing steel will exhibit this effect.

To predict a true dynamic response, it would be necessary to continuously vary the material stress with deformation. This variation is difficult to model using SDOF analysis methods because it requires tracking a complex resistance–deflection curve at each time step. It is desirable to represent the design material stress as a bilinear stress–strain curve in which stress increases linearly with strain to yield and a constant value after yield (refer to Chapter 7, Section 7.2.5). This produces a simple, bilinear resistance–deflection curve, as shown in Figure 5-8, which can include strain-hardening effects and is relatively easy to incorporate into the SDOF analysis. To achieve this simplification, while accurately modeling the dynamic response, it is necessary to select a design stress equal to the average stress occurring in the actual response. This can be done by estimating a maximum response range and using recommendations in Tables 5A-4 and 5A-5 for reinforcing and steel members from UFC 3-340-02. Concrete constitutive material models that are validated against test data and include nonlinear and cracking in compression and tension can be used in FEA to account for these effects.

Using the SDOF methodology, at low response ranges, the maximum design stress is equal to the dynamic yield stress. At higher response ranges, the design stress is increased to account for strain hardening. In the initial portion of the response, this increased design stress will result in an overprediction of resistance. As greater deformations occur, the stress level, and thus resistance, will be underpredicted by the design stress.

Finite-element methods can incorporate complex variations in material stresses in the time-varying response. The dynamic material properties presented in this section can be used in FEM calculations; however, the simplified response limits in the next section may not be suitable. Most FEM codes contain complex failure models that are better indicators of acceptable response that should be used only if they are based on test data. Refer to Chapter 6, "Dynamic Analysis Methods," for additional information.

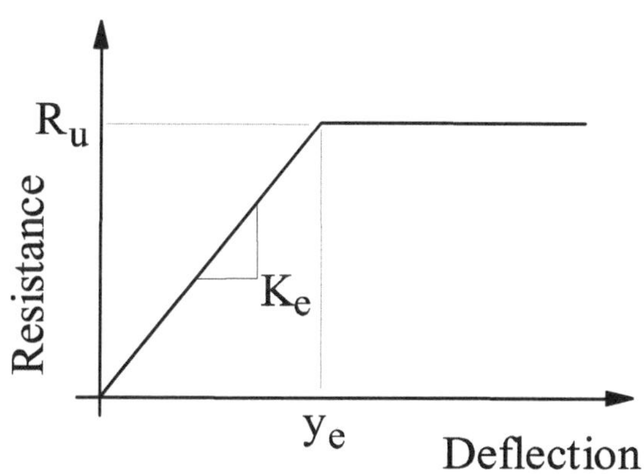

Figure 5-8. Bilinear resistance–deflection curve.

5.6 DEFORMATION LIMITS

Response deformation limits are used to ensure that adequate response to blast loads is provided. These limits are based on the type of structure or component, construction materials used, location of the structure, and desired protection level.

The primary method for determining the adequacy of a structure for conventional design is evaluation of the stress level achieved compared to the maximum stress permitted. Deflections are also checked for certain members, although this is typically done for serviceability or architectural reasons rather than structural requirements. Blast-loaded members, however, reach or exceed yield stresses to achieve an economic design. In general, the more deformation the structure or member can undergo without failure, the more blast energy that can be absorbed. As member stresses exceed the yield limit, stress level is not appropriate for judging member response as is done for static elastic analysis. In dynamic design, the adequacy of the structure is judged on maximum deformations. Limits on displacements are based on test data or other empirical evidence. A degree of conservatism is included to ensure adequate capacity because the applied loads are not "factored up" to provide a factor of safety.

The allowable response of individual frame components is less than that permitted for the same member responding as an isolated element. This is done to reduce the possibility of progressive collapse and to increase redundancy of the frame. Failure of individual girt and purlin components is not as catastrophic as failure of a frame member, and thus, a difference in criteria exists. Similarly, load-bearing walls should normally be allowed less deformation than non-load-bearing elements because of the consequences associated with failure.

The structure's performance goal becomes an important factor in the selection of maximum response values. If it is desired to provide a high degree of protection to personnel or equipment, a low response limit is chosen. This situation may be typical of a control room in which personnel are required to remain at their workstation during an emergency or for critical equipment that must be protected to implement a safe shutdown. Conversely, if a building is frequently unoccupied or contains low-value equipment, significant damage may be permitted up to the point of failure. Structures that are required to be reusable following a blast are typically designed to remain elastic under the predicted loads.

The capacity of a member to deform significantly and absorb energy is dependent on the ability of the connections to maintain strength throughout the response. If connections become unstable at large responses, catastrophic failure can occur. The resistance will drop, thereby increasing deflections. Connections often control blast capacity for structures that have been designed for conventional loads only.

Appropriate response (deformation) limits are selected on the basis of the factors discussed previously as well as company/owner safety philosophies, blast protection guidelines, and risk considerations. Risk assessments that evaluate accident probability and potential consequences can be helpful in making the appropriate selection. The deformation limits relate to a specified degree of response, which can be characterized as low, medium, or high. At the highest response limits, catastrophic failure of the structure should not occur. Points of highest stress in the members will be near incipient collapse and local failures may occur but the overall structure should remain intact. It is important to remember that predicted responses may not always account for local instabilities and the actual response can be significantly greater. The engineer must take these factors into consideration when designing or analyzing the structure to ensure that the required degree of protection is provided.

Many petrochemical companies have adopted a "neutral risk" philosophy for facilities where personnel are normally required to evacuate during an emergency. This philosophy prescribes that personnel are not to be placed in greater danger inside a building than if they were outside. Blast pressures and fragments entering the structure are not considered in the design because personnel would be exposed to these hazards outside the building. The performance goal for the structure then becomes incipient failure in which portions of the structure are damaged severely but do not tear loose and become missiles. Structural collapse is not permitted, and suspended equipment must be adequately anchored within the structure. Chapter 2, "General Considerations," contains additional discussion of protection philosophies.

5.6.1 Deformation Limit Parameters

Almost all published structural deformation limits for blast design are presented in terms of parameters that are easily compared to simplified nonlinear dynamic response calculations that involve one or several degrees of freedom models. Deformation limits are typically presented in terms of two nondimensional terms, maximum support rotation and ductility ratio, which are both based on the peak dynamic deflection of the component.

Ductility ratio is defined as the maximum displacement of the member divided by the displacement at the elastic limit and is commonly designated by the symbol μ. For indeterminate members with multiple plastic hinges, the ductility ratio is typically based on the equivalent yield deflection. The equivalent yield deflection is the ultimate resistance divided by the equivalent elastic stiffness. Equivalent elastic stiffnesses are provided in Chapter 6, Tables 6-1 to 6-3. Ductility ratio is a measure of the degree of inelastic response experienced by the member.

Support rotation is another measure of member response that relates maximum deflection to span and indirectly indicates the damage to the component in the maximum moment regions, primarily at the compressive face of the member. It is designated by the symbol θ and is defined as shown in Figure 5-9.

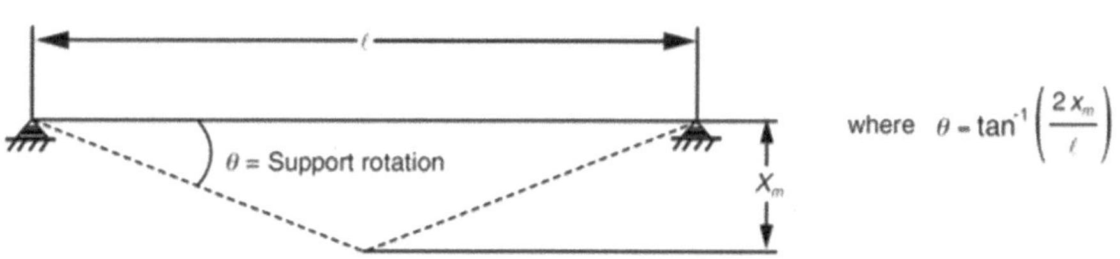

Figure 5-9. Hinge rotation.

The acceptable deformation limits for load-bearing components must consider *P*-delta effects. This is considered by reducing the deformation limits for the individual load-bearing components, if they have significant compressive forces, and with limits on acceptable sideway of the building's lateral load-resisting system. The sidesway limits provided in the tables are fairly liberal and are intended primarily for steel frames with compact cross sections that can develop significant plasticity at moment-resisting connections. In general, seismic limits for lateral drift are conservative sidesway limits for blast-resistant design because blast loading will cause much less severe cyclic loading than an earthquake load.

When given support rotations and/or ductility ratios are correlated to component response levels, or when they are defined as limits for the acceptable dynamic response for a blast design, they are referred to as "response limits," or "response criteria." These two deformation limits apply only to the global response of structural components for blast-resistant design (e.g., flexural response). They are not applicable for designing components to resist any type of local damage such as breach, spalling, fragment penetration, and others.

The response criteria discussed in this report have been developed primarily for blast design based on SDOF analysis. Finite-element analysis has become a commonly used tool for estimating structural damage under extreme dynamic loading. FEA provides significantly more response information, such as displacements and rotations, and stress and strain caused by both flexure and shear, throughout the structural continuum. The evaluation of stress or strain at a single point, or at a finite element within a member, may not indicate the ability of a component to provide the necessary level of protection from blast load. When redistribution of stresses may occur during a dynamic response, it may be more appropriate to use global parameters such as maximum support rotations and ductility ratios for the whole component as criteria for blast design. In this case, the response criteria values in Appendix 5B that have been developed for SDOF analysis can also be used for FEA. In other cases, these simplified deformation limits may not be adequate to define the limit damage state of interest. For example, design limits based on the maximum rupture strain for specific materials, or equivalent plastic strain, may be more applicable to given design cases.

For blast designs based on both SDOF and FEA analyses, response criteria based on specific applicable test data may supersede the more general response criteria published values in Appendix 5B. However, the general response criteria in Appendix 5B are appropriate for most structural components. Different limit criteria may be used if the analysis technique is sufficiently rigorous to capture potential nonlinear response modes and the response level provides the required protection.

Due care must be taken by designers to ensure that dynamic design models adequately capture the required response characteristics (local and global) of the structural component under consideration, which may include previous comparisons with applicable test data.

5.6.2 Deformation Limit Values

Maximum acceptable values for ductility and support rotation for three component response levels (i.e., low, medium, and high response) are presented in Table 5B-3. Support rotations correlate much better with response levels than ductility ratios for some component types such as reinforced concrete and masonry. The predicted response must be compared to ductility ratio and support rotation limits to ensure that neither is exceeded. The engineer must also determine whether lower limits are appropriate. The values vary with material type, section type, and required protection category. For reinforced concrete components, response limits are influenced by the reinforcement index (RI), the percent of reinforcing steel at both faces (i.e., the presence of compression face reinforcement at the maximum moment location), and shear reinforcing, as well as the type of response (i.e., flexure, shear, compression). Reinforcement at the compression face acts in compression to form a resisting moment with the tensile reinforcement after crushing of the concrete at large support rotations, allowing larger response limits for medium and high response levels. Stirrups in the maximum moment region can provide necessary lateral support to the reinforcement at the compression face after concrete crushing, but blast testing shows that reinforcement at the compression face of reinforced concrete panels with low reinforcement indexes does not require lateral support from stirrups at large support rotations.

The RI compares the reinforcing ratio of reinforced and prestressed concrete components to the balanced reinforcing ratio. Thus, components with a lower RI are more under-reinforced. Blast testing on components with low reinforcement indexes shows that these components can undergo larger support rotations for each response level.

Shear and axial compression are less ductile response modes than flexural response for all component types. In general, for elements in which the maximum dynamic load capacity of the component is controlled by shear or compression, the allowable response criteria are quite low. If better information is not available, the deformation limit for a nonductile response mode is typically assumed as a maximum ductility ratio of 1.0. A smaller ductility ratio may also be used when a low response level is required.

The deformation limit values in Appendix 5B are a combination of criteria developed for the original publication of this document, criteria in PDC-TR 06-08 Rev 1 (USACE 2008) developed by the US Army Corps of Engineers, and response limits for reinforced and prestressed concrete developed from more recent blast test data (Oswald and Bazan 2014, Mander et al. 2016, Kjosling et al. 2023). The deformation limit values for medium and high responses of cold-formed steel girts and purlins and reinforced masonry components are based on similar values for moderate and heavy component damage levels, respectively, in PDC-TR 06-08 Rev 1. The response criteria for aluminum secondary components are based on similar values for moderate and heavy component damage levels, respectively, for aluminum window mullions in PDC-TR 10-02 (USACE 2012). Values for low response of these component types are approximately one-half the values for medium response because the lowest PDC deformation limit values apply to superficial component damage rather than low damage. There has been a significant amount of blast and shock tube testing of reinforced and prestressed concrete panels since the previous publication of this document and the PDC-TR 06-08 Rev.1 report that have been used to develop the response criteria for these component types. The deformation limit values for all other component types in Appendix 5B are the same, or very similar, to deformation limit values developed for the original publication of this report.

The response limits for reinforced concrete and masonry shear walls with an ultimate resistance controlled by flexural response (i.e., flexural control) were changed in this edition so that they are based on the maximum support rotation of the shear wall rather than the maximum ductility ratio. This causes these response limits to be comparable to the response limits for deep beams in blast design guidance from the US Department of Defense.

The response limits for reinforced concrete diaphragms with an ultimate resistance controlled by flexural response were not modified because of a concern arising from the total lateral deflection from a long-span diaphragm. In addition, the supporting shear walls could cause significant *P*-delta effects for load-bearing columns and walls supported by the diaphragm that would not be considered in the dynamic response of these components to blast loads.

Deformation limits for different response levels of other component types, including wood and metal stud walls, are provided in PDC-TR 06-08 Rev 1. Limits on the actual deformation values should be used when there is a risk of a structural component (i.e., wall panel) impacting critical equipment. These limits must be imposed by the design engineer on a case-by-case basis, as applicable, in addition to the response criteria in Appendix 5B. PDC-TR 06-08 Rev. 1 also contains a methodology to determine an overall building level of protection based on the highest component damage levels of primary and secondary type components in the building, as defined in the document. This approach may be helpful for evaluating overall building blast damage.

Deformation limit values for blast-resistant design are also provided in several other publications by the US government and industry committees. These limit values tend to be in the same range as the values in Appendix 5B but are not identical. This reflects different definitions for damage or design response levels, different amounts of conservatism in the limits, considerations of different available component damage databases, and the approximate nature of all deformation limit values. Summaries and correlations between component support rotations and ductility ratios and component damage levels observed in specific blast tests, other dynamic tests, and static tests are available in numerous references. A summary of available blast test information is provided in Oswald (2005).

The deformation limit values in Appendix 5B apply for the typical case where the component is designed using a procedure that explicitly considers the dynamic component response, such as the SDOF methodology described in this document or a method based on dynamic FEA. Some components that are not directly loaded by blast and have a short response time compared to the expected rise time of their dynamic load, such as connections, shear walls and diaphragms, and primary framing components in pure axial loading, are usually designed using an equivalent static load approach. The equivalent static load is equal to the ultimate resistance of the supported components multiplied by their tributary areas and the connection or lateral load-resisting or axial component must have an ultimate dynamic load capacity that is equal or greater than this load. The dynamic load capacity is calculated in the same manner as the static load capacity except that the yield strength is increased by applicable static and dynamic increase factors for blast design. The load capacity may include applicable strength reduction factors (i.e., ϕ factors) used in conventional static design at the discretion of the design engineer. If the dynamic analysis shows that the supported component does not yield, the equivalent static load may be based on the maximum resistance of this component. However, this approach should be used with caution because it will be unconservative if the actual blast loads exceed the design blast load.

Appendix 5B also shows two sets of building damage levels (BDLs). BDLs can be applied to any building type. Each damage level is described qualitatively in terms of damage to building components. Table 5B-1 contains general building damage descriptions that include an estimate of corresponding repair costs. Table 5B-2 contains BDLs that are sometimes used by industry for occupant vulnerability assessments, where these BDLs can be correlated to the likelihood of occupant injuries. The information in this table is closely related to BDL definition from the Center for Chemical Process Safety (CCPS 2012). The CCPS document provides photographs of buildings with each BDL. Other approaches for defining BDLs may also be used.

The BDL for the building can be used to describe an assessment of blast damage to an existing building, the maximum expected blast damage to a building after it has been structurally upgraded, or the maximum expected blast damage to a new blast-resistant building. Specifically in the latter two cases, there is usually a direct correlation between maximum allowable component response level to structural components in each part of the building and the BDL for the building. These correlations may be based on engineering judgment or criteria established by the building owner. Table 5B-2 contains some specific reference to component response levels for each BDL. As given in this table, all structural components in a building with a given BDL do not necessarily have the same response/damage level. Alternatively, a maximum allowable component response level can be defined for all structural components in a new or upgraded building without any reference to an overall BDL.

APPENDIX 5A. SUMMARY TABLES FOR DYNAMIC MATERIAL STRENGTH

Table 5A-1. Strength Increase Factors.

Material	*Strength increase factors*
Structural steel ($F_y \leq$ 50 ksi or 345 MPa)	1.1
Reinforcing steel ($F_y \leq$ 60 ksi or 414 MPa)	1.1
Cold-formed steel	1.21
Concrete*	1.0
6061 T6 aluminum	1.14
6063 T5 aluminum	1.31
6063 T6 aluminum	1.24

*The results of compression tests are usually well above the specified concrete strengths and may be used in lieu of the aforementioned factor. Some conservatism may be warranted because concrete strengths have more influence on shear design than bending capacity.

Table 5A-2. Dynamic Increase Factors for Reinforcing Bars, Concrete, and Masonry.

	Dynamic increase factors			
	Reinforcing bars		*Concrete*	*Masonry*
Stress type	F_{dy}/F_y	F_{du}/F_u	f'_{dc}/f'_c	f'_{dm}/f'_m
Flexure	1.22*	1.05	1.19	1.19
Compression	1.13	1.00	1.12	1.12
Diagonal tension	1.00	1.00	1.00	1.00
Bond	1.22	1.05	1.00	1.00

*Modified consistently with Fig. 5.7a with a loading strain rate in the range of 0.1–0.3 s^{-1}.

Table 5A-3. Dynamic Increase Factors for Structural Steel, Cold-Formed Steel, and Aluminum.

	Dynamic increase factors		
	Yield stress		
	Bending/ shear	Tension/ compression	Ultimate stress
Material	F_{dy}/F_y	F_{dy}/F_y	F_{du}/F_u
ASTM A36	1.29	1.19	1.10
ASTM A588, ASTM A992, ASTM A572*	1.19	1.12	1.05
ASTM A514	1.09	1.05	1.00
ASTM A653	1.10	1.10	1.00
SAE AMS5501 (stainless steel)	1.18	1.15	1.00
Aluminum (all grades)	1.02	1.00	1.00

*Includes all ductile steels with a yield strength of 50 ksi or similar.

Table 5A-4. Dynamic Design Stress for Concrete Reinforcing Steel.

Type of stress	Type of reinforcement	Maximum support rotation	Dynamic design stress (F_{ds})
Bending	Tension and compression	$0 < \theta_1 \leq 2$	F_{dy}
		$2 < \theta_1 \leq 5$	$F_{dy} + (F_{du} - F_{dy})/4$
		$5 < \theta_1 \leq 12$	$(F_{dy} + F_{du})/2$
Diagonal tension	Stirrups		F_{dy}
Direct shear	Diagonal bars	$0 < \theta_1 \leq 2$	F_{dy}
		$2 < \theta_1 \leq 5$	$F_{dy} + (F_{du} - F_{dy})/4$
		$5 < \theta_1 \leq 12$	$(F_{dy} + F_{du})/2$
Compression	Column	All	F_{dy}

Table 5A-5. Dynamic Design Stress for Structural Steel

Type of stress	Maximum ductility ratio	Dynamic design stress (F_{ds})
All	$\mu \leq 10$	F_{dy}
All	$\mu > 10$	$F_{dy} + (F_{du} - F_{dy})/4$

APPENDIX 5B. SUMMARY TABLES FOR BUILDING DAMAGE LEVELS AND COMPONENT RESPONSE CRITERIA

Table 5B-1. General Building Damage Levels

Damage level	*Description*
Low	Localized component damage. The building can be used; however, repairs are required to restore the integrity of the structural envelope. The total cost of repairs is moderate.
Medium	Widespread component damage. The building should not be occupied until repaired. The total cost of repairs is significant.
High	Key components may have lost structural integrity and building collapse caused by environmental conditions (i.e., wind, snow, rain) may occur. The building should not be occupied. The total cost of repairs approaches the replacement cost of the building.

Table 5B-2. Building Damage Levels Used for Vulnerability Assessments.

Damage level	*Description*
1	Minor damage to roof and walls
2A	Walls facing explosion may have high response requiring replacement. Other walls, roof, and load-bearing components have medium to low response that is usually repairable.
2B	Walls facing explosion may experience severe damage or collapse. Other walls, roof, and load-bearing components have medium to high response.
3	Walls facing explosion have collapsed. Other walls, roof, and load-bearing components have high response that is near incipient collapse.
4	Failure of load-bearing elements and roof. Failure of all or most walls.

Table 5B-3. Component Response Levels

Response level	*Description*
Low	The component has none to slight visible permanent damage.
Medium	The component has some permanent deflection. It is usually repairable, if necessary, although replacement may be more economical and aesthetic.
High	The component has not failed, but it has significant permanent deflections, causing it to be unrepairable.

Note: These response-level descriptions apply to the response limit ranges mentioned in Table 5B-4 through Table 5B-6.

Table 5B-4. Response Limits for Steel and Aluminum Components

	Low response		Medium response		High response	
Component	μ_a	θ_a	μ_a	θ_a	μ_a	θ_a
Hot-rolled steel compact secondary components (beams, girts, purlins)[a]	3	2	10	6	20	12
Steel primary frame components (with significant compression)[a,b,c]	1.5	1	2	1.5	3	2
Steel primary frame components (without significant compression)[a,b,c]	1.5	1	3	2	6	4
Steel plates[d]	5	3	10	6	20	12
Open-web steel joists	1	1	2	3	4	6
Cold-formed light gauge steel panels (with secured ends)[e,f]	1.75	1.25	3	2	6	4
Cold-formed light gauge steel panels (with unsecured ends)[f,g]	1.0	—	1.8	1.3	3	2
Cold-formed light gauge steel beams, girts, purlins, and noncompact secondary hot rolled components[f]	2	1.5	3	3	12	10
Aluminum secondary components (window mullions, girts, purlins)	2	1.5	5	3	7	6

[a]Primary components are components whose loss would affect several other components supported by that component and whose loss could potentially affect the overall structural stability of the building in the area of loss. Secondary components are those supported by primary framing components.

[b]Significant compression is when the axial compressive load is more than 20% of the dynamic axial capacity of the component. Axial compression should be based on the ultimate resistance of the supported components exposed to the blast pressure. Refer to PDC-TR 06-08 for detailed examples of calculation for axial compression and dynamic axial column capacity. Refer to Section 5.6.2 for an analysis of components in pure axial compression.

[c]Sidesway limits for moment-resisting structural steel frames: low = (height)/50, medium = (height)/35, high = (height)/25.

[d]Steel plate criteria can also be applied to cold-formed components, such as corrugated (crimped) plates, if local buckling and other response modes, as discussed in Section 5.4.4, are accounted for in the analysis. The analysis should show that the cold-formed component can absorb at least as much strain energy at the given response limits as an "ideal" ductile component, such as steel plate, with an elastic-perfectly plastic response.

[e]Panels must be attached on both ends with screws or spot welds.

[f]Light gauge refers to material that is less than 0.125 in. (3 mm) thick.

[g]Panels are not attached on both ends (e.g., standing seam roof panels).

Notes: Response limits define upper limits for each response level and are for components responding primarily in flexure, unless otherwise noted. If the shear resistance, or connection strength, is less than the reaction loads caused by the ultimate dynamic flexural load capacity of the component, then the component should be analyzed with an ultimate resistance based on its shear or connection capacity and a maximum ductility ratio of 1.0. A higher ductility ratio can be used if supported by test data or analysis of the connection.

Table 5B-5. Response Limits for Reinforced Concrete and Masonry Components in Flexural Response.

				Low response		Medium response		High response	
Material	*Cross section and reinforcement type*	*Reinforcement details*[a]	*Reinforcement index (RI)*[b]	μ_a	θ_a	μ_a	θ_a	μ_a	θ_a
Reinforced concrete	Solid with conventional rebar	Rebar at each face	RI $<$ 0.05		2		4		8
		Rebar at each face	RI $\geq$ 0.05		1		2		5
		Rebar at each face with stirrups[c]			1		4		6
		Single layer of rebar	All		1		2		5
	Solid with prestressed reinforcement	Single layer or layer at each face	RI $<$ 0.12		1		2.5		5
			0.12 $\leq$ RI $<$ 0.25		1		2		3
			RI $\geq$ 0.25	0.8		0.9		1.0	
	Sandwich cross section[d]	Prestressed or conventional rebar at each wythe	RI $<$ 0.06		1		2.5		5
			RI $\geq$ 0.06		1		1.5		2.5
	Any cross section	Welded wire fabric	All		1		2		2.5
Reinforced masonry	All reinforced cells are fully grouted	Single bar per cell max. 0.75 in. dia. rebar[e]	All		1		2		4

[a]No stirrups required unless noted. All splices are full-tension splices placed in low-stress areas.
[b]See Equation (5-1) for the definition of RI.

$$RI = \frac{A_s}{bd}\frac{f_{ds}}{f'_{dc}} \tag{5-1}$$

where

RI = Reinforcement index,
A_s = Tensile reinforcing steel area,
= Area of conventional reinforcing steel for nonprestressed panels,
= Area of prestresssed reinforcing steel and any conventional reinforcing steel,
b = Spacing of tensile reinforcing steel,
d = Depth of tensile steel reinforcement,
f_{ds} = Dynamic yield strength of reinforcing steel,
= 81 ksi for conventional A615 (ASTM 2024a) Grade 60 rebar and welded wire fabric,
= 265 ksi for prestressed steel,
f'_{dc} = Dynamic compression strength of concrete = 1.19 f'_c .

[c]Stirrups or ties must enclose the top and bottom flexural reinforcement and meet minimum ACI requirements for tie reinforcement in columns.
[d]Panel with two exterior reinforced concrete wythes and interior insulation, where shear connectors between wythes are designed to cause the panel to act as a fully composite section for flexural response.
[e]Recommended details for walls 8 in. thick or less with spliced rebars. Do not splice rebars in the maximum moment region. Also, specify IBC Level 2 inspection, or equivalent.
Notes: Response limits define upper limits for each response level and are for components responding primarily in flexure, unless otherwise noted. If the shear resistance, or connection strength, is less than the reaction loads caused by the ultimate dynamic flexural load capacity of the component, then the component should be analyzed with an ultimate resistance based on its shear or connection capacity and a maximum ductility ratio of 1.0. A higher ductility ratio can be used if supported by test data or analysis of the connection.

Table 5B-6. Response Limits for Reinforced Concrete (R/C) and Masonry (RM) Shear Walls and Load-Bearing Components

Component	Low response		Medium response		High response	
	μ_a	θ_a	μ_a	θ_a	μ_a	θ_a
R/C and R/M walls, slabs, and columns (in flexure and axial compression load)*		1		2[a]		2[a]
R/C and R/M diaphragms (flexural control)	3		3		3	
R/C and R/M shear walls (flexural control)		1		1.5		2
R/C and R/M components (shear control, without shear reinforcement)	1.3		1.3		1.3	
R/C and R/M components (shear control, with shear reinforcement)	1.6		1.6		1.6	

*A support rotation of 4 degrees is allowed for R/C components that have compression face steel reinforcement and shear reinforcement in maximum moment areas. Shear reinforcement must meet minimum ACI requirements for tie reinforcement in columns.

Notes: Applicable when the axial compressive load is more than 10% of the dynamic axial capacity of the component. Axial compression should be based on the ultimate resistance of the supported components exposed to the blast pressure. Refer to PDC-TR 06-08 Rev.1 for detailed examples of calculation for axial compression and dynamic axial column capacity. Refer to Section 5.6.2 for an analysis of components in pure axial compression. Response limits define upper limits for each response level.

References

ASCE. 1961. *Design of structures to resist nuclear weapons effects.* MOP 42. Reston, VA: ASCE.

ASTM International. 2005. *Standard specification for high-strength low-allow structural steel with 50 ksi [345 MPa] minimum yield point to 4-in [100 mn] thick.* ASTM A588. West Conshohocken, PA: ASTM.

ASTM. 2006. *Standard specification for high-strength low-alloy columbium-vanadium structural steel.* ASTM A572. West Conshohocken, PA: ASTM.

ASTM. 2014. *Standard specification for structural bolts, alloy steel, heat treated, 150 ksi minimum tensile strength.* ASTM A490-12. West Conshohocken, PA: ASTM.

ASTM. 2016. *Standard specification for structural bolts, steel, heat treated, 120/105 ksi minimum tensile strength (withdrawn 2016).* ASTM A325-14. West Conshohocken, PA: ASTM.

ASTM. 2017a. *Standard specification for steel wire, plain, for concrete reinforcement.* ASTM A82-01. West Conshohocken, PA: ASTM.

ASTM. 2017b. *Standard specification for cold-formed welded and seamless carbon steel structural tubing in rounds and shapes.* ASTM A500-03. West Conshohocken, PA: ASTM.

ASTM. 2019. *Standard specification for carbon structural steel.* ASTM A36/A36M-19. West Conshohocken, PA: ASTM.

ASTM. 2021. *Standard specification for anchor bolts, steel, 36, 55, and 105-ksi yield strength.* ASTM F1554-20. West Conshohocken, PA: ASTM.

ASTM. 2022a. *Standard specification for high-yield-strength, quenched and tempered alloy steel plate, suitable for welding.* ASTM A514/A514M-22. West Conshohocken, PA: ASTM.

ASTM. 2022b. *Standard specification for steel sheet, zinc-coated (galvanized) or zinc-iron alloy-coated (galvannealed) by the hot-dip process.* ASTM A653/A653M-20. West Conshohocken, PA: ASTM.

ASTM. 2023. *Standard specification for deformed and plain low-alloy steel bars for concrete reinforcement.* ASTM A706/A706M-22. West Conshohocken, PA: ASTM.

ASTM. 2024a. *Standard specification for deformed and plan carbon-steel bars for concrete reinforcement.* ASTM A615/A615M-22. West Conshohocken, PA: ASTM.

ASTM. 2024b. *Standard specification for loadbearing concrete masonry units.* ASTM C90-24. West Conshohocken, PA: ASTM.

CCPS (Center for Chemical Process Safety). 2012. *Guidelines for evaluating process plant buildings for external explosions, fires, and toxic releases.* 2nd ed. New York: CCPS.

Idriss, J. S., J. Fritz, and J. A. Schmidt. 2016. "Dynamic analysis of insulated metal panels for blast effects." Accessed November 21, 2024. https://www.structuremag.org/article/dynamic-analysis-of-insulated-metal-panels-for-blast-effects/.

Kjosling, E., M. Weaver, C. Oswald, and J. Geringer. 2023. "Updating and documenting single-degree-of-freedom response limits for anti-terrorism design." In *Proc., 6th Int. Conf. on Protective Structures (ICPS6).* Accessed November 21, 2024. *https://www.kcse.com/updating-and-documenting-single-degree-of-freedom-response-limits-for-antiterrorism-design/.*

Lindholm, U. S., and R. L. Bessey. 1969. *A survey of rate dependent strength properties of metals.* San Antonio, TX: Southwest Research Institute.

Mander, T. J., B. L. Bingham, M. J. Lowak, M. A. Polcyn. 2016. *Development of a simplified blast design procedure and response limits for load-bearing precast wall panels subject to blast loads.* Phase II Final Report. Chicago: Precast/Prestressed Concrete Institute.

Mander, T., and B. Matamoros. 2019. Constitutive modeling and overstrength factors for reinforcing steel. *Struct. J.* 116 (3): 219–232.

Oswald, C. J. 2005. *Component Explosive Damage Assessment Workbook (CEDAW).* Contract No. DACA45-01-D-0007-0013. Washington, DC: Baker Engineering and Risk Consultants for the US Army Corps of Engineers, Protective Design Center.

Oswald, C. J., and M. Bazan. 2014. "Performance and blast design for non-load bearing precast concrete panels blast." In *Proc., ASCE/SEI Structures Congress.* https://ascelibrary.org/doi/10.1061/9780784413357.014.

SAE International. 2002. *Steel, corrosion resistant, sheet, strip, and foil 19Cf 9.5Ni (304) cold rolled, 125 ksi (862 MPa) tensile strength.* SAE AMS5501. Warrendale, PA: SAE.

SAE. 2003. *Aluminum alloy extruded profiles 1.0Mg 0.6Si 0.28Cu 0.20Cr (6061-T6) solution and precipitation heat treated.* SAE AMS4113. Warrendale, PA: SAE.

USACE (US Army Corps of Engineers). 2008. *Single degree of freedom structural response limits for antiterrorism design.* PDC-TR 06-08 Rev 1. Washington, DC: USACE.

USACE. 2012. *Blast resistant design methodology for window systems designed statically and dynamically.* PDC-TR 10-02. Washington, DC: USACE.

USACE/NAVFAC/AFCESA (US Army Corps of Engineers/Naval Facilities Engineering Command/Air Force Civil Engineer Support Agency). 2014. *Structures to resist the effects of accidental explosions.* UFC 3-340-02. Washington, DC: National Institute of Building Sciences.

CHAPTER 6

Dynamic Analysis Methods

6.1 INTRODUCTION

This chapter discusses various analysis methods for determining the dynamic response or damage of structural members subjected to blast loading. To perform dynamic analyses, it is necessary to have previously defined the load-time history as well as member properties such as structural capacity, stiffness, and mass. The design of new structures can involve several iterations of the analysis, where trial member sizes are used and the resulting response quantities are compared against the acceptance criteria defined in Chapter 5, "Dynamic Material Strength and Response Criteria."

Several dynamic analysis methods are used for blast-resistant design, ranging from hand calculations and graphical solutions to computer-based applications. One of the purposes of this chapter is to convey analysis methods that provide the necessary balance between sufficient accuracy and calculation approach.

6.2 KEY CONCEPTS

Several key concepts relating to the dynamic analysis of structures for blast loading are discussed subsequently. The main objectives of the analysis are discussed, followed by a general discussion on the level of accuracy used in typical blast design applications. The approach for separating integrally connected structural members into manageable parts for analysis purposes is described. A brief discussion on the treatment of live loads is also given.

6.2.1 Objectives

The overall objective of a dynamic blast analysis is to evaluate the capability of a structure to resist a specified blast load-time history. To accomplish this goal, the analysis should be able to predict, with a fair degree of accuracy, the linear and nonlinear dynamic responses of the members that comprise a structure. The analysis of a typical member begins with a given structural framing configuration, which includes the type of material, span length, support conditions, and applied loading history, which can be incorporated into simplified methods such as single-degree-of-freedom (SDOF) or beam element–based finite-element analysis (FEA). Member properties are then used to estimate member stiffness, mass, and section capacities. A more complex FEA modeling of member sections using shell and brick elements is another option to estimate member stiffness, mass, and capacity in pursuit of the dynamic response. Determinations of member stiffness and section capacities to be used in simplified analytical methods are described in Chapter 7, "Design Procedures." These simplified methods develop a resistance function, or applied force versus displacement relationship, on the basis of assumed failure mechanisms, the member configuration, and estimated section capacities. Both simplified analysis and discrete FEA can determine the response to a given blast load. Specifically, the analysis should provide

1. Maximum relative deflections of each structural element,
2. Relative rotation angles at plastic hinge locations,
3. Dynamic reactions transmitted to the supporting elements,
4. Deflections and reactions caused by rebound, and
5. Maximum plastic strains when using discrete FEA models.

Once the analysis is complete, the design can proceed to determine the adequacy of the member through the application of the acceptance criteria (e.g., support rotation, ductility, and plastic strain limits).

6.2.2 Accuracy

A blast analysis may contain several approximations that affect the accuracy of the results. Some of the approximations involve the following:

1. As previously stated in Chapter 3, "Determination of Loads," overpressure calculations can have a certain degree of uncertainty and be subject to large assumptions. Some types of explosions such as VCEs are more subject to uncertainty in the load calculations, whereas for other types of facilities, such as munitions plants, the blast overpressures may be accurately predicted on the basis of a known quantity and type of explosive. Although load calculation uncertainty is not directly associated with the structural analysis, it is important to note that there can be variability in the input.
2. Structural modeling of uncoupled SDOF system analyses for interconnected structural members ignores the deformation compatibility, shared inertial resistance, and equilibrium of forces at contact or connection points between members. In other words, dynamic interaction effects that may increase or decrease the calculated responses may not be considered.
3. Approximate dynamic properties of the structural materials combined with simplified bilinear resistance–deflection curves are commonly used with an equivalent SDOF system methodology. The level of solution accuracy decreases for more complex materials and member configurations for both brittle and ductile responding members.

4. Multiple-degree-of-freedom (MDOF) methods represent a series of SDOF components supported from one another that can improve the interaction of structural framing systems under dynamic loading.
5. Finite-element analysis models consisting of interconnected one-dimensional (1D) beam elements can account for MDOF behavior. Because of the number of nodes that make up each finite beam element representing columns and beam components, the blast load can be more accurately applied to the building envelope with respect to time of arrival and magnitude.
6. Discrete FEA models consist of 1D, two-dimensional (2D), and three-dimensional (3D) elements that can describe the geometrical section properties of steel sections (e.g., flanges, webs, corrugated panels) or reinforced masonry and concrete sections (e.g., slabs, beams, columns). Two-dimensional and 3D models include a finer nodal mesh, which subsequently allow for more accuracy in sweeping the blast load compared to 1D FEA modeling approaches. This approach also requires more sophisticated constitutive material models that capture the high strain-rate material behavior in tension and compression under various confined and unconfined stress states.

The degree of complexity of the structural representation and analyses can vary considerably, depending on the effort to which the engineer determines is necessary to achieve a safe and economical design. Except for those related to and/or limited by blast load definition, each of the aforementioned approximations could be improved by using more complex procedures. However, the amount of improvement ultimately may or may not be significant when compared to results achieved using relatively simplified calculations. The increased complexity can possibly lead to significant modeling errors, depending on the analyst's capabilities.

6.2.3 Interaction of Structural Elements

For fully enclosed building envelopes, the blast loads are typically applied to the exterior walls and roof and are transmitted through various structural members to the foundation. The energy of the blast is absorbed through inertial resistance and elastic and, more important, plastic deformation of the responding structural members. The remaining portion of blast energy not absorbed by the responding structure is transmitted into the ground. It is therefore necessary to establish a continuous load path with consistent tracking of the dynamic loads through the structure to the foundation to ensure a safe design.

A typical approach when using the SDOF method is to analyze a structure using a member-by-member approach. The envisioned load path, established using engineering judgment and experience, forms the basis for determining the member-by-member analysis sequence. Tracking of the member dynamic reactions and loads throughout the structure is performed manually. This basic approach is similar to the practice used in conventional static analyses. The major difference is the consideration of inertia forces, which may act in any direction compared to only a downward gravity load. This approach, however, requires the reaction forces to be in the form of a harmonic time history. An alternative approach by many practitioners is to use the tributary area method. In this method, the forcing function is the blast time history of the tributary area supported by that element.

This uncoupled member-by-member SDOF approach approximates the actual dynamic response because dynamic interaction effects between structural elements are not considered. Resulting calculated dynamic responses, which include deflections and support reactions, may be underestimated or overestimated, depending on the dynamic characteristics of the loading and the structural framing. This approximation occurs when performing an uncoupled dynamic analysis.

Separate SDOF dynamic analyses are also typically performed for each of the structural components. For example, a typical roof system consists of a roof slab supported on structural steel roof beams, which are, in turn, supported by roof girders. Separate SDOF dynamic analyses are performed for the slab, beams, and girders using either the full blast load based on the tributary area or the reaction time history of the supported member as loading input to the supporting member.

The same member-by-member approach can be used for lateral analyses of buildings, as illustrated in Figure 6-1. Front walls facing the blast are typically designed as a unit width, one-way member spanning vertically. Either the full blast load based on the tributary area or the reaction time history of a representative wall strip is used as the loading input to the horizontal roof diaphragm, which is supported by side walls oriented parallel to the direction of the blast. These walls are typically reinforced concrete shear walls or braced steel frames. The analysis proceeds from the front wall to the roof diaphragm to the side walls and finally to the foundation. A consistent, continuous load path is thus established.

Dynamic interaction effects are commonly ignored. Under certain circumstances, unconservative answers could result from ignoring the effects of coupling. Although some simple parametric studies can be made to evaluate these effects, coupling is normally expected to be negligible if the natural frequencies (or natural periods) of connected elements differ by a factor of 2 or more according to the technical guidance given by Biggs (1964). Frequencies of interconnected members are sometimes separated by changing their stiffness or weight. If ignoring dynamic interaction effects cannot be justified, the connected members can be analyzed as an MDOF system in which these effects are considered inherently.

Some studies on dynamic interaction effects for two degree-of-freedom systems have been done by Baker (1983). Although these studies were done using a limited range of variables, the results indicate that conservative responses can be obtained using uncoupled SDOF system approximations versus a coupled approach even when the periods do not differ by a factor of 2.

A more comprehensive analysis approach can be used to analyze the structure as a whole. For example, an FEA of an entire building may be performed using

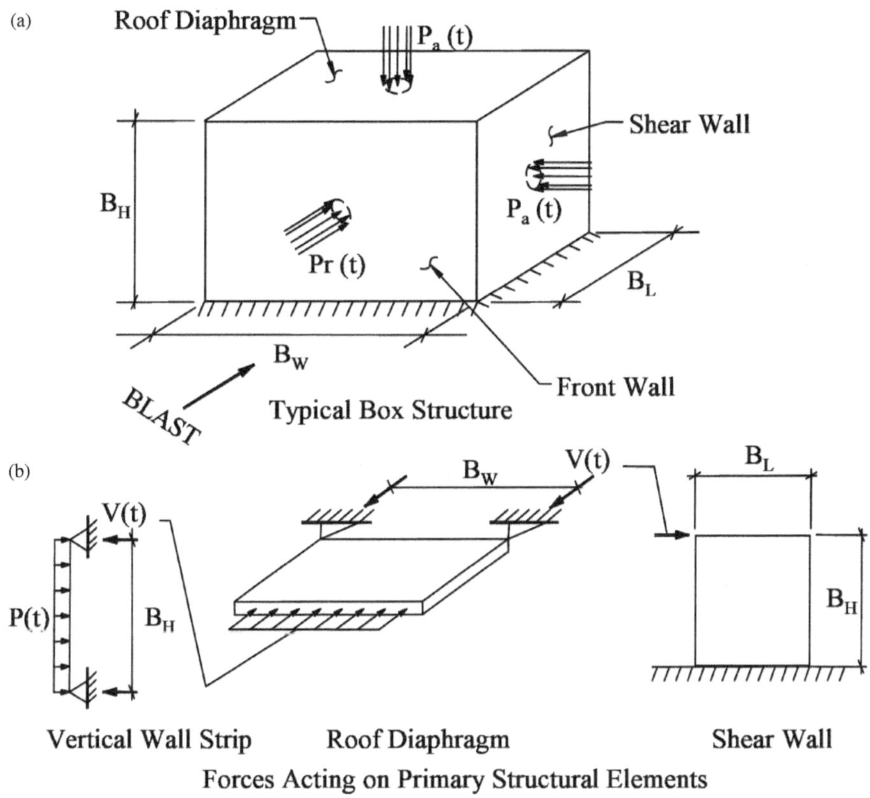

Figure 6-1. Forces acting on primary structural elements: (a) typical box structure, (b) forces acting on primary structural elements.

1D beam elements. Obviously, the load path need not be predetermined when such global analysis methods that include full or partial multi-framing elements are used. However, the load path is influenced by the type and level of detail of the modeling so that engineering judgment and experience are also necessary to achieve a safe and economical design.

6.2.4 Live Loads

Live loads that may be swept away by a blast wave or that would not contribute to the inertia of a supporting member should not be included in the mass calculation of any framing members. In addition, some judgment is needed to estimate the portion of a design live load that is normally present. For example, snow loads and partition loads may be present for relatively long durations and a portion of this live load should be included in the mass calculation. In contrast, a floor live load representing personnel and furnishings should not be included in the mass calculation.

6.2.5 Confirmation of Assumed Failure Mechanisms

In establishing the model used to represent a structure, the more simplified approach may first need to assume the locations of plastic hinges before the analysis is carried out. This approach is essentially an upper-bound analysis, which provides a predicted collapse load that is either correct or higher than the actual collapse load. In a majority of cases, simple structural models are developed where it is obvious that the assumed mechanism is correct or approximately so. For those cases involving irregular structural configurations, mass distributions, and loading, a separate check that includes advanced analysis such as FEA should be conducted to confirm that no other possible failure mechanisms exist that may result in lower predicted collapse loads.

6.3 SINGLE-DEGREE-OF-FREEDOM SYSTEMS

The basic analytical model used in a majority of blast design applications is the single-degree-of-freedom (SDOF) system. A discussion on the fundamentals of dynamic analysis methods for SDOF systems is given subsequently, which is followed by descriptions on how to apply these methods to structural members.

6.3.1 Basics

All structures, regardless of how simple the construction is, possess more than one degree of freedom. However, many structures can be adequately represented as a series of SDOF systems for analysis purposes. The accuracy obtainable from an SDOF approximation depends on how well the deformed shape of the structural members and their structural and inertial resistances can be represented with respect to time. In many cases, sufficiently accurate results can be obtained for primary load–carrying components of structures such as beams, girders, columns, wall panels, diaphragm slabs, and shear walls.

A majority of dynamic analyses performed in blast-resistant design of petrochemical facilities use the SDOF methodology. Common types of construction, such as steel beams, concrete slabs, single-story plane frames, cantilever barrier walls, and compact boxlike buildings, can be approximated as SDOF systems. Several examples of such structures are illustrated in Figure 6-2.

The dynamic equilibrium of the damped SDOF system illustrated in Figure 6-3 is expressed mathematically as follows:

$$Ma + Cv + Ky = F(t) \tag{6-1}$$

where

M = Mass,
a = Acceleration,
C = Viscous damping constant,
v = Velocity,

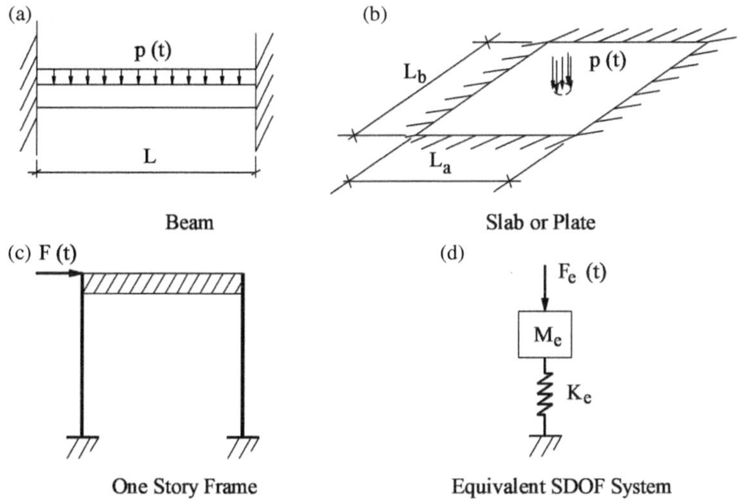

Figure 6-2. Typical structures represented as equivalent SDOF systems: (a) beam, (b) slab or plate, (c) one-story frame, (d) equivalent SDOF system.

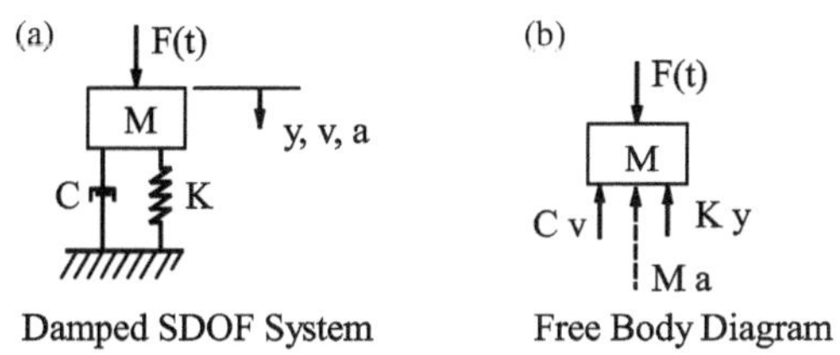

Figure 6-3. SDOF model for dynamic analysis: (a) damped SDOF system, (b) free body diagram.

K = Stiffness,
y = Displacement,
F = Blast force, and
t = Time.

Damping is usually conservatively ignored in blast-resistant design. Because of the short time in which the structure reaches its maximum response, damping typically has little effect on peak displacements.

When damping is ignored, the three forces then acting on the mass are the resistance, Ky, the inertia force, Ma, and the external applied force, F. The dynamic equilibrium equation for the undamped, elastic system then becomes

$$Ma + Ky = F(t) \tag{6-2}$$

In blast analyses, the resistance is usually specified as a nonlinear function to simulate elastic, perfectly plastic and, if significant, softening behavior of the structure. The ultimate resistance, R_u, is reached upon the formation of a collapse mechanism in the member. When the resistance is nonlinear, the dynamic equilibrium equation becomes

$$Ma + R = F(t) \tag{6-3}$$

where R is = lesser of Ky or R_u.

Solutions for Equation (6-3) can be obtained by various methods, depending on the complexity of the loading function, $F(t)$.

More rigorous analyses of SDOF systems are usually not required or warranted in typical blast design applications. However, special cases may arise where a more-sophisticated and less-conservative solution is justified, perhaps to analytically qualify an existing structure for new increased loading conditions. Refinements can be made in the analyses in areas such as strain hardening, progressive hinge formation, equivalent replacement of arbitrary pulse loading, MDOF effects, and large deformations. Discussion of these methods can be found in Stronge and Yu (1993), ASCE MOP 42 (ASCE 1961), and Krauthammer et al. (1986, 1990).

6.3.2 Transformation Factors

Examples of some typical SDOF approximations were briefly introduced in Section 6.3.1 and illustrated in Figure 6-2. These SDOF models greatly simplify the dynamic analysis effort compared to that of structures having distributed mass and stiffness. For structures having a single concentrated mass, the SDOF system can be defined without an approximation.

The procedure for obtaining an equivalent SDOF approximation for a structural component is based on its deformed shape under the applied loading and the strain energy equivalence between the actual structure and the SDOF approximation. The deformed shape of the member is usually dominated by blast loading rather than by normal design loads. In addition to strain energy equivalence, the motion of the SDOF system (displacement, velocity, and acceleration) is equivalent to a selected control point on the actual structure. The control point is usually selected at a point of maximum response such as a plastic hinge location within the span. However, the spring force is not equal to the support reactions of the actual member.

Equivalent mass, stiffness, and loading are obtained by using transformation factors. Several widely used texts on blast design such as Biggs (1964) (Chapter 5) and UFC 3-340-02 (US DoD 2008) (Chapter 3) contain tabulated transformation factors for typical structural elements such as beams and slabs. The derivations of the equations for these transformation factors are also given by these references. The transformation factors used to obtain appropriate properties for the equivalent SDOF system are as follows:

$$\text{Equivalent stiffness,} \quad K_e = K_L K \tag{6-4a}$$

$$\text{Equivalent mass,} \quad M_e = K_M M \tag{6-4b}$$

$$\text{Equivalent force,} \quad F_e = K_L F \tag{6-4c}$$

$$\text{Equivalent resistance,} \quad R_e = K_L R \tag{6-4d}$$

where K_L is the load or stiffness transformation factor, and K_M is the mass transformation factor.

The dynamic analysis can be performed using these equivalent parameters in place of the corresponding actual values. The alternate form of the bilinear dynamic equilibrium equation [Equation (6-3)] then becomes

$$M_e a + R_e = F_e \tag{6-5}$$

For convenience, Equation (6-5) is sometimes simplified by using a single load-mass transformation factor, K_{LM}, as follows:

$$K_{LM} M a + R = F(t) \tag{6-6}$$

where

$$K_{LM} = K_M / K_L$$

Shape functions, $\Phi(x)$, used in deriving the aforementioned transformation factor equations are changed according to the stress range of the member. Figure 6-4 illustrates these changes for a simply supported beam with a uniform mass and uniform pressure loading and provides the resulting transformation factors. Note that the resistance in the aforementioned equations also changes depending on whether the response is in the elastic or plastic range.

Engineering judgment is used to select the appropriate factors, depending on the anticipated predominant response mode. A trial-and-error approach may be used to evaluate the response mode behavior and its effect on the response

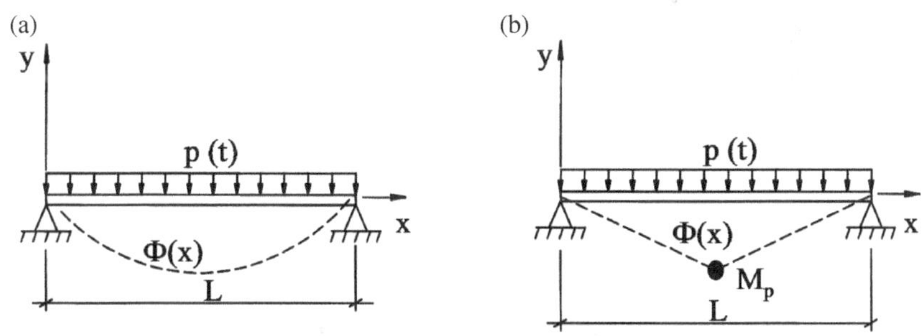

$$\Phi(x) = \frac{16}{5L^4}(L^3x - 2Lx^3 + x^4)$$

$K_L = 0.64$, $K_M = 0.50$, $K_{LM} = 0.78$

Elastic Response

$$\Phi(x) = \frac{2x}{L}, \quad x \le L/2$$

$K_L = 0.50$, $K_M = 0.33$, $K_{LM} = 0.66$

Plastic Response

Figure 6-4. Shape function and transformation factors for a simply supported beam: (a) elastic response, (b) plastic response.

results. An average of the elastic and plastic transformation factors is sometimes used for simplicity.

Transformation factors for common one-way and two-way structural members are readily available from several sources such as Biggs (1964) and UFC 3-340-02 (US DoD 2008). Refer to Tables 6-1 through 6-3 for a summary of such factors for one-way members.

The mass of the structure includes its self-weight and the weight of permanently attached components or equipment. Mass is calculated simply by dividing weight by gravity. Approximations are sometimes used in determining mass distributions of members analyzed as SDOF systems to be able to use readily available tabulated transformation factors.

When performing dynamic analyses of a series of SDOF systems representing a structure, an estimate of the amount mass "riding along" with a supporting member often must be made. For example, a roof girder supports a portion of the mass of the roof beams it supports, which needs to be added to the girder's mass as illustrated in Table 6-1.

Engineering judgments are often used in lieu of more rigorous mathematical procedures. One recommendation for a continuous reinforced concrete slab and beam-type construction given by UFC 3-340-02 (US DoD 2008) is to include 20% of the supported member's mass with the mass of the supporting member. This would correspond to a supported member that is relatively flexible in comparison with the supporting member. For the structure illustrated in Figure 6-5, the full mass of the portion of the beam supported by the girder (i.e., 50% of the beam's total mass in this case) is lumped with the girder mass at the midspan of each girder. In this example, the beam is considered to be rigid in comparison with each girder. Each case is judged individually.

General transformation factor equations for distributed mass systems and MDOF systems are given by Biggs (1964) (Chapter 5) and Clough and Penzien (1993) (Chapter 2). These general methods can be used in determining transformation factors for nonprismatic members or members that have non-uniform mass distributions.

6.3.3 Graphical Solution Methods

Blast loadings act on a structure for relatively short durations of time and are therefore considered as transient dynamic loads. Solutions for Equation (6-3) are available in the form of nondimensional charts and graphs in UFC 3-340-02 (US DoD 2008, Biggs 1964).

A typical graphical solution for a triangular pulse load (i.e., simplified shock blast load) with an elastic-plastic resistance function is shown in Figure 6-6. Additional charts covering other loading conditions and elastic rebound are available in Biggs (1964), ASCE MOP 42 (ASCE 1961), and UFC 3-340-02. Such charts can be used to determine the maximum ductility demand, μ_d, and the time of maximum response, t_m. The parameters needed to enter Figure 6-6 include the maximum applied force, F_o, loading duration, t_d, ultimate resistance, R_u, and natural period, t_n, of the equivalent SDOF system.

Table 6-1. Transformation Factors for One-Way Members and Simply Supported Boundary Conditions.

Loading diagram	*Strain range*	*Load factor K_L*	*Lumped mass factor, K_M*[a]	*Uniform mass factor, K_M*	*Bending resistance, R*[b]	*Spring constant, K*	*Dynamic reaction, V*
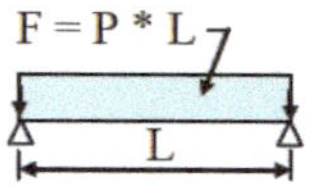	Elastic	0.64	—	0.50	$8M_{pc}/L$	384 *EI*/5*L*3	$0.39R + 0.11F$
	Plastic	0.50	—	0.33	$8M_{pc}/L$	0	$0.38R_u + 0.12F$
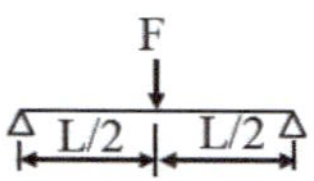	Elastic	1.00	1.00	0.49	$4M_{pc}/L$	48 *EI*/*L*3	$0.78R - 0.28F$
	Plastic	1.00	1.00	0.33	$4M_{pc}/L$	0	$0.75R_u - 0.25F$
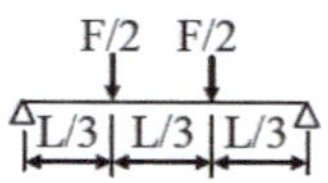	Elastic	0.87	0.76	0.52	$6M_{pc}/L$	56.4 *EI*/*L*3	$0.525R - 0.025F$
	Plastic	1.00	1.00	0.56	$6M_{pc}/L$	0	$0.52R_u - 0.02F$

[a]Equal portions of the concentrated mass are lumped at each concentrated load.

[b]M_{pc} is the ultimate moment capacity at midspan.

Source: Used with permission of McGraw Hill LLC, from *Introduction to Structural Dynamics*, Biggs (1964); permission conveyed through the Copyright Clearance Center, Inc.

Table 6-2. Transformation Factors for One-Way Members and Fixed Boundary Conditions.

Loading diagram	*Strain range*	*Load factor K_L*	*Lumped mass factor, K_M*[a]	*Uniform mass factor, K_M*	*Bending resistance, R*[b]	*Spring constant, K*	*Effective spring constant, K_e*	*Dynamic reaction, V*
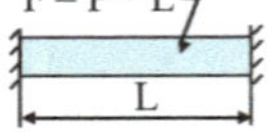	Elastic	0.53	—	0.41	$12M_{ps}/L$	$384\ EI/L^3$	$307\ E_s I_x/L^3$	$0.36R + 0.14F$
	E-P[c]	0.64	—	0.50	$8(M_{ps} + M_{pc})/L$	$384\ EI/5L^3$		$0.39R + 0.11F$
	Plastic	0.50	—	0.33	$8(M_{ps} + M_{pc})/L$	0		$0.38R_u + 0.12F$
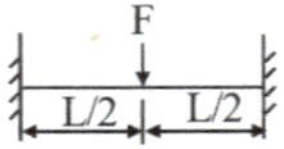	Elastic	1.00	1.00	0.37	$4(M_{ps} + M_{pc})/L$	$192\ EI/L^3$	—	$0.71R - 0.21F$
	Plastic	1.00	1.00	0.33	$4(M_{ps} + M_{pc})/L$	0		$0.75R_u - 0.25F$

[a]Equal portions of the concentrated mass are lumped at each concentrated load.
[b]M_{pc} is the ultimate moment capacity at midspan; M_{ps} is the ultimate moment capacity at support.
[c]*E-P* is elastic-plastic.
Source: Used with permission of McGraw Hill LLC, from *Introduction to Structural Dynamics*, Biggs (1964); permission conveyed through the Copyright Clearance Center, Inc.

Table 6-3. Transformation Factors for One-Way Members and Fixed-Simple Boundary Conditions.

Loading diagram	*Strain range*	*Load factor K_L*	*Lumped mass factor, K_M*[a]	*Uniform mass factor, K_M*	*Bending resistance, R*[b]	*Spring constant, K*	*Effective spring constant, K_e*	*Dynamic reaction, V*
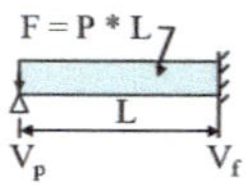	Elastic	0.58	—	0.45	$8M_{pc}/L$	185 EI/L^3	160 EI/L^3	$Vp = 0.26R + 0.12F$
	E-P[c]	0.64	—	0.50	$4(M_{ps} + 2M_{pc})/L$	384 $EI/5L^3$		$Vf = 0.43R + 0.19F$ $0.39R + 0.11F \pm M_{ps}/L$
	Plastic	0.50	—	0.33	$4(M_{ps} + 2M_{pc})/L$	0		$0.38R_u + 0.12F \pm M_{ps}/L$
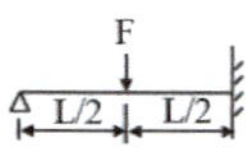	Elastic	1.00	1.00	0.43	$16M_{ps}/3L$	107 EI/L^3	106 EI/L^3	$V_p = 0.25R + 0.07F$
	E-P	1.00	1.00	0.49	$2(M_{ps} + 2M_{pc})/L$	48 EI/L^3		$V_f = 0.54R + 0.14F$ $0.78R - 0.28F \pm M_{ps}/L$
	Plastic	1.00	1.00	0.33	$2(M_{ps} + 2M_{pc})/L$	0		$0.75R_u - 0.25F \pm M_{ps}/L$
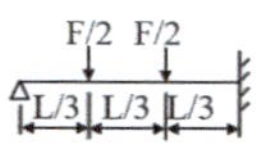	Elastic	0.81	0.67	0.45	$6M_{pc}/L$	132 EI/L^3	122 EI/L^3	$V_1 = 0.17R + 0.17F$
	E-P	0.87	0.76	0.52	$2(M_{ps} + 3M_{pc})/L$	56 EI/L^3		$V_2 = 0.33R + 0.33F$ $0.525R - 0.025F \pm M_{ps}/L$
	Plastic	1.00	1.00	0.56	$2(M_{ps} + 3M_{pc})/L$	0		$0.52Ru - 0.02F \pm M_{ps}/L$

[a]Equal portions of the concentrated mass are lumped at each concentrated load.
[b]M_{pc} is the ultimate moment capacity at midspan; M_{ps} is the ultimate moment capacity at support.
[c]*E-P* is elastic-plastic.
Source: Used with permission of McGraw Hill LLC, from *Introduction to Structural Dynamics*, Biggs (1964); permission conveyed through the Copyright Clearance Center, Inc.

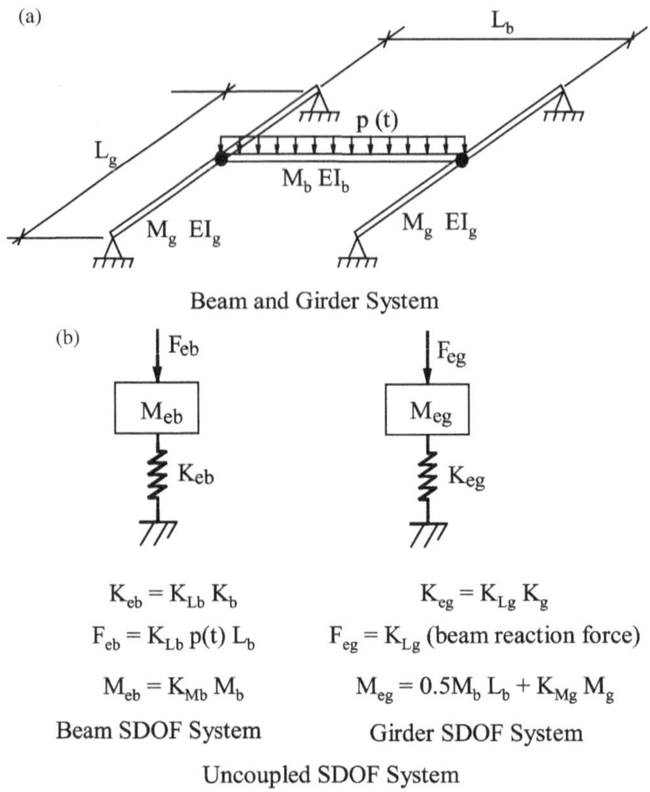

Figure 6-5. Mass distribution of a typical multimember system: (a) beam and girder system, (b) uncoupled SDOF system.

This natural period is based on the deformed shape of the member and is independent of the loading. The natural frequency of the SDOF system is expressed in cycles per second:

$$f = \frac{1}{2\pi}\sqrt{K_e/M_e} \tag{6-7}$$

and the natural period is expressed in seconds as follows:

$$t_n = \frac{1}{f} = 2\pi\sqrt{M_e/K_e} \tag{6-8}$$

This method is suitable for obtaining maximum responses of elastic-plastic SDOF systems subjected to simple loading functions. It is usually not practical to develop solution charts when loads become more complex (e.g., including both positive and negative phases). Another shortcoming of this method is that the time history of the response is not available to evaluate support reactions and rebound effects.

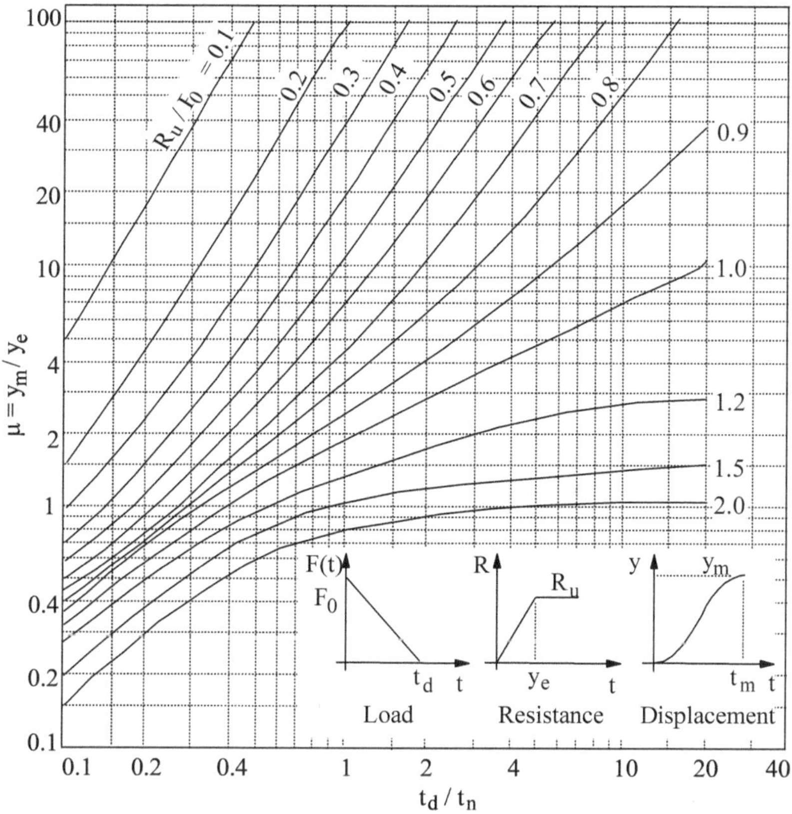

Figure 6-6. Typical graphical solution chart for an elastic-plastic SDOF system.
Source: Data from UFC 3-340-02 (US DoD 2008).

Another graphical method that is often used in the evaluation of SDOF structural elements for blast loading is the pressure–impulse (*P–I*) diagram method. The *P–I* diagram method combines both dynamic analysis and design evaluation into a single procedure that can be used to rapidly evaluate potential damage levels, as described in Chapter 5, for certain types of structural members such as reinforced concrete panels, steel beams, masonry walls, and other common building elements.

The basic concept of the *P–I* diagram method is to mathematically relate a specific damage level to a range of blast pressures and corresponding impulses for a given structural element. The *P–I* relationships, which may be theoretical or empirical, are plotted in graphical format, as illustrated in Figure 6-7. *P–I* curves can be developed based on Baker (1983) or Mays and Smith (1995). Knowing the blast pressure and impulse at a specific structure's location relative to the blast source enables the user to read the damage level directly from the *P–I* damage curves.

Two basic types of *P–I* diagrams are commonly used. Traditionally, nondimensionalized P_{bar} and I_{bar} terms were used to define the ordinate and

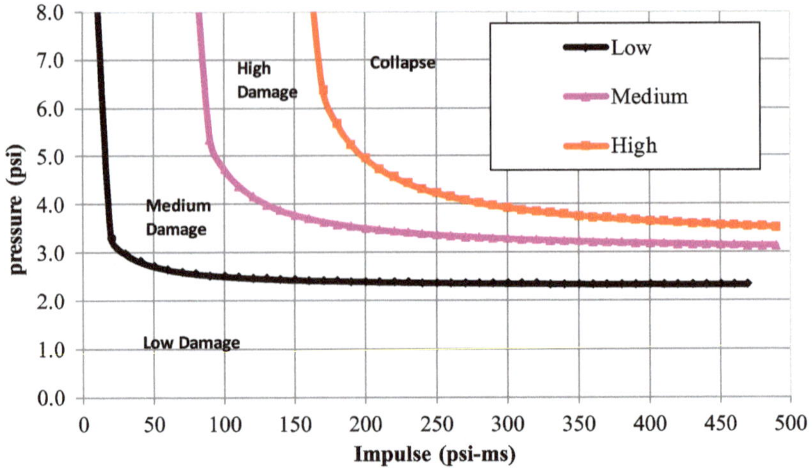

Figure 6-7. P–I curves with structural damage regions.

abscissa values of the diagram. These terms contained parameters defining the stiffness, resistance, and mass for a given type of member. Refer to Baker (1983) and the CEDAW workbook (2008), which define P_{bar} and I_{bar} terms for common structural member types. More recently, *P–I* diagrams like the one shown in Figure 6-7, in which the ordinate and abscissa values are given directly in terms of pressure and impulse, have come into use for the evaluation of building components and, in some cases, an entire structure. The curves shown in Figure 6-7 define combinations of pressure and impulse that produce a constant level of damage. Four regions defined by the constant damage curves in this particular figure are designated as low damage, medium damage, high damage, and collapse. More or less refinement may be used in defining damage levels.

Theoretical solutions usually tend to underestimate blast resistance capacities of actual structures. Blast testing is therefore sometimes used to establish a series of data points for the purpose of developing realistic damage curves. However, when using test data to establish the damage curves, test scatter inevitably requires the introduction of some conservatism to produce smooth boundaries between the damage regions of the *P–I* diagram. Also, qualitative interpretations of the test specimen responses introduce some uncertainties in the definition of the damage levels. For these reasons, semi-empirically based *P–I* diagram methods have been used primarily as a conservative screening tool for assessing existing buildings potentially subjected to blast loading.

6.3.4 Closed-Form Solutions

Closed-form solutions (i.e., equations) are available only for some simple loading cases for SDOF systems (Biggs 1964, Clough and Penzien 1993, Paz 1991). Published solutions exist for both elastic and elastic–plastic responses and for triangular and rectangular load pulses. The analysis can also be greatly simplified when the duration of the loading, t_d, is either very short or extremely long compared to the natural period, t_n.

When the loading duration is short compared to the member's natural period, $t_d/t_n < 0.1$, the shape of the load-time function becomes insignificant. The maximum response can be calculated using the impulse-momentum principle. The ductility demand, μ_d, can be determined in terms of the impulse, I_o, and the maximum resistance of the member:

$$\mu_d = 0.5[(I_o 2\pi f / R_m)^2 + 1] \tag{6-9}$$

In the other extreme case, when the loading duration is long compared to the natural period, $t_d/t_n > 10$, the system responds as though the load were suddenly applied and constant. Again, the maximum ductility demand can be expressed in a convenient form:

$$\mu_d = 1/[2(1 - F_o / R_m)] \tag{6-10}$$

Empirical formulas have been developed to transition between these two extreme dynamic response cases. ASCE MOP 42 provides the following relationship over the full response range of $\tau = t_d/t_n$:

$$F_o / R_m = \frac{\sqrt{(2\mu_d - 1)}}{\pi(t)} + \frac{(2\mu_d - 1)(\tau)}{2\mu_d(\tau + 0.7)} \tag{6-11}$$

Comparisons with more exact solutions show that this relationship yields results to within 5%, which is usually accurate enough for most applications. This formula does not lend itself to a direct calculation of ductility demand in terms of the other parameters. However, μ_d can be solved for by trial iterations.

6.3.5 Numerical Integration

A versatile alternative to simple graphical, closed-form, and empirical solutions is to perform a numerical time integration. This method is also known as the time history method. Most texts on structural dynamics (Biggs 1964, Clough and Penzien 1993, Paz 1991) provide extensive coverage on numerical solution methods for nonlinear SDOF systems.

A brief summary of the Newmark numerical integration procedure, which is commonly used to obtain the time history response for nonlinear SDOF systems, is given in Appendix 6A. It is most commonly used with either constant-average or linear-acceleration approximations within the time step. An incremental solution is obtained by solving the dynamic equilibrium equation for the displacement at each time step. The results of previous time steps and the current time step are used with recurrence formulas to predict the acceleration and velocity at the current time step. In some cases, a total equilibrium approach (Paz 1991) is used to solve for the acceleration at the current time step.

To ensure an accurate and numerically stable solution, a small time increment must be selected. A rule of thumb is to use a value less than or equal to 1/10th of either the natural period of the structure or the load duration, whichever is smaller. Computer programs using numerical time integration methods for nonlinear analyses of SDOF systems are available. Refer to Appendix 6A, Chapter 10, and Chapter 11 for example implementations of numerical integration in a blast design.

6.3.6 Support Reactions

Perhaps the most commonly overlooked aspect of using SDOF approximations is the determination of the dynamic reactions for the actual member. The spring force in the SDOF system is not equal to the support reaction. To determine the dynamic reactions, the distribution of the inertia force within the member must be considered (Biggs 1964) (Chapter 5). The basic approach, as illustrated in Figure 6-8, is to express the dynamic forces acting on the member, or a segment of the member, in terms of the displacement and acceleration at the control point. This displacement, *y*(*t*), is determined in the solution of the time history analysis of the equivalent SDOF system.

Equations for the dynamic reactions of typical structural members are available from Biggs (1964) (Chapter 5) and UFC 3-340-02 (Chapter 3) (refer to Tables 6-1 through 6-3). These equations express the dynamic reaction in terms of the resistance and applied load, both of which vary with time.

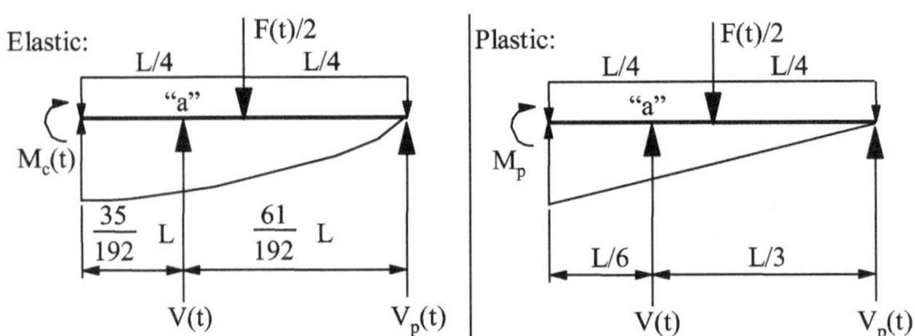

Elastic:

Sum moments at "a" (clockwise +)

$$0 = M_c(t) + \frac{F(t)}{2}\left(\frac{L}{4} - \frac{35L}{192}\right) - V_p(t)\left(\frac{61L}{192}\right)$$

Substitute $M_c(t) = R(t)\,L/8$

$$\frac{R(t)}{8}\,\not{L} + \frac{F(t)}{2}\left(\frac{1}{4} - \frac{35}{192}\right)\not{L} - \frac{61}{192}\,\not{L}\,V_p(t) = 0$$

$$\frac{1}{8}R(t) + \frac{13}{384}F(t) - \frac{61}{192}V_p(t) = 0$$

$$V_p(t) = \frac{192}{61}\left(\frac{1}{8}R(t) + \frac{13}{384}F(t)\right)$$

$$= 0.39\,R(t) + 0.11\,F(t)$$

Plastic:

Sum moments at "a" (clockwise +)

$$0 = M_p + \frac{F(t)}{2}\left(\frac{L}{4} - \frac{L}{6}\right) - V_p(t)\left(\frac{L}{3}\right)$$

Substitute $M_p = R_u\,L/8$

$$\frac{R_u}{8}\,\not{L} + \frac{F(t)}{2}\left(\frac{1}{4} - \frac{1}{6}\right)\not{L} - \frac{\not{L}\,V_p(t)}{3} = 0$$

$$\frac{R_u}{8} + \frac{F(t)}{24} - \frac{V_p(t)}{3} = 0$$

$$V_p(t) = 3\left(\frac{R_u}{8} + \frac{F(t)}{24}\right)$$

$$= 0.38\,R_u + 0.12\,F(t)$$

Figure 6-8. Reactions for a flexural member with distributed mass and load.

6.4 HIGHER-ORDER METHODS

This section provides general technical guidance for simplified analysis methods for multistory building frames, slab/beam/girder framing systems, and structures having multiple concentrations of significant lumped masses compared to FEA methods. An example of a typical MDOF structure is shown in Figure 6-9. This two-story building is subjected to lateral impulse forces at the top of the first and second stories, as would be the case during blast loading. The methods and procedures for the nonlinear FEA approach are described in Section 6.4.2. The extension of the dynamic analysis methods described in Section 6.3 for SDOF systems to MDOF systems may be used if previous comparisons have shown acceptable correlations to test data or FEA results. It should be pointed out that MDOF systems do not inherently consider the effects of local, as well as overall global, instability; these effects should be checked during design.

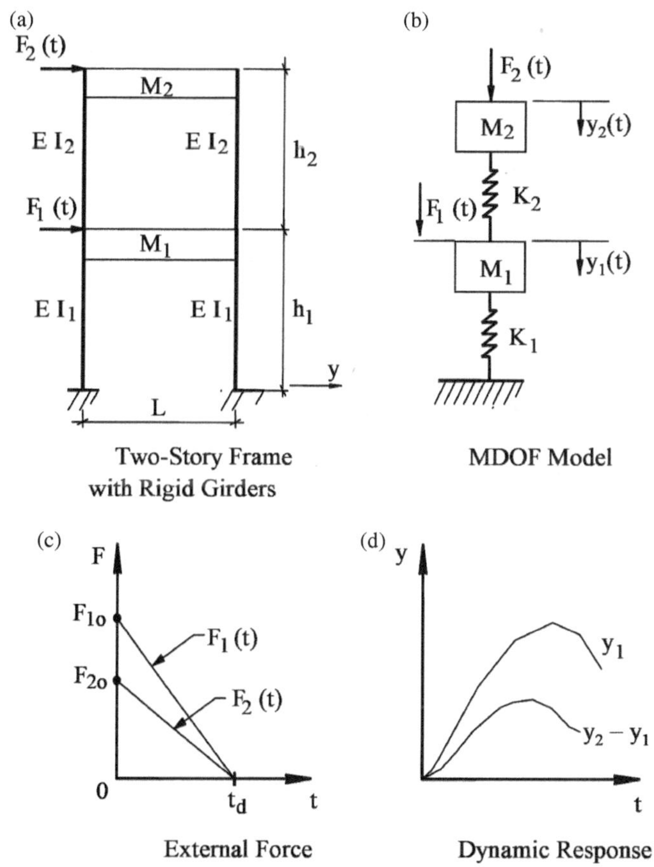

Figure 6-9. Typical MDOF structure: (a) two-story frame, (b) MDOF model, (c) external force, (d) dynamic response.

6.4.1 Multiple-Degree-of-Freedom Methods

When the structural configuration is complex or significant dynamic interaction between interconnected members cannot be avoided, a coupled analysis approach can be used. The coupled analysis approach can include as few as two degrees of freedom to represent a structural system or it can involve the use of many degrees of freedom in a single, comprehensive dynamic analysis of the entire superstructure using either simplified MDOF or FEA methods.

The MDOF approach will usually require the use of a computer program to perform the structural dynamic analyses because of the extensive computations. Frame analysis–type programs using beam elements may be used if the structural configuration lends itself to this type of modeling. The use of a general-purpose FEA program may be necessary to accurately represent the structure with the appropriate type of element, such as plate and shell elements for continuum-type structures.

A coupled analysis need not be all-encompassing. For example, a 2D plane frame analysis of a building employing two or more degrees of freedom can be considered a coupled analysis approach. Separate plane frames for each orthogonal horizontal direction can also be used in lieu of a single comprehensive 3D model. Refer to Section 6.5.2 for a discussion on modeling considerations for this type of structure.

Responses of MDOF systems are determined from the solution of the dynamic equilibrium equation presented as Equation (6-12). This equation is the matrix form of the equilibrium equation for an SDOF system [Equation (6-1)].

$$[M]\{a\}+[C]\{v\}+[K]\{y\}=\{F(t)\} \tag{6-12}$$

For practical purposes, manual solutions of this equation can be obtained for only two or possibly three degrees of freedom. An example of an elastoplastic, two-degree-of-freedom system analysis is given by Biggs in Chapter 5 (Biggs 1964). Even this simple problem involves significant effort.

Solutions for MDOF systems are usually obtained by using finite-element procedures. Because of nonlinearities associated with plasticity and possibly large displacements (Lagrangian), the direct time integration method should be used. Various direct integration methods for time integration can be employed, but the Newmark Method is perhaps the most common. Other methods, such as the Houbolt, Wilson-T, Hilber–Hughes–Taylor, Runge–Kutta, and Central Difference Methods, are also commonly used in finite-element solvers. Refer to Bathe (1995) for further details.

6.4.2 Finite-Element Analysis Methods

An FEA method is recommended when one or more of the following conditions exist:

1. More accurate damage estimate is required.
2. Ratio of a member's natural period to the natural period of the support system is in the range of 0.5 to 2.0, such that an uncoupled analysis approach

may yield significantly inaccurate results. In this case, either FEA or an MDOF model is recommended.

3. Time-varying support reactions or member forces are desired to evaluate the structure or its supports in greater detail in an effort to minimize the costs of new or retrofit construction.
4. Overall structural behavior is to be evaluated with regard to structural stability (e.g., frame buckling), gross displacements, and *P*-delta effects.
5. Structure has unusual features such as unsymmetrical or non-uniform mass and stiffness characteristics.
6. Structural components are expected to undergo large inelastic deformations or be subjected to secondary debris impacts.
7. There is a structural response benefit of sweeping the blast load along framing that is not normal to the blast.

Many commercial implicit and explicit formulation computer programs are available to perform nonlinear FEA with dynamic load application and may be suitable for use in blast-resistant design. These programs include ABAQUS, ADINA, ANSYS, LS-DYNA, and MSC/NASTRAN, and mostly favor the explicit formulation for blast analysis. This list is only representative and does not constitute an endorsement of the programs or the techniques used. New capabilities and material constitutive models are continually being developed and the analyst is encouraged to review the documentation for these programs to understand the techniques and limitations. Of particular interest to the analyst may be the solution formulation and the ability to resolve gaps, impacts, and contact surfaces within a specific FEA program.

Certain considerations should be made to achieve adequate results at a reasonable cost when using FEA methods. One item to consider is the appropriateness and practicality of the element type. The most suitable element types from the simplest to the most complex include spring elements, line (beam) elements, plate/shell elements, and solid elements.

Another important item is to consider how the FEA output data would be used to confirm compliance with acceptance criteria. For example, the traditional use of stress output data from plate or shell elements to evaluate a reinforced concrete slab or steel wide flange member is not very practical because strains are more indicative of damage in post-elastic behavior. Nonlinear FEA codes employ a yield criterion for plate and shell elements on the basis of stress resultants (i.e., forces and moments). Another potential difficulty arises when trying to determine relative displacements of a member so as to check its maximum deflection against the allowable deflection. These issues are discussed in detail in Section 6.4.4.

Finite-element analysis has become much more common than other methods for analyzing building structures in recent years. This is driven by the improved user-friendliness of programs and advances in computing power, which make the analysis of nonlinear response problems feasible on microcomputers. A greater familiarity with FEA tools is also leading to their increased use in more conventional projects. Still, it is important for the analyst to avoid a false sense of reliance on FEA results without an independent estimate of results. Despite the increasing use of FEA, SDOF methods remain widely used to solve blast problems, even though they can lead to conservative results. This is due in large part to their relative simplicity to implement on the analyst's end and to check assumptions and results on the reviewer's end. Still, the limitations of simplified methods must be recognized.

As with the use of SDOF and MDOF codes, the engineer should ensure that the FEA program (which addresses geometric nonlinearity) and material model (addressing material nonlinearity) has been vetted against experimental test data, and that a study of the literature surrounding dynamic FEA for structural response, including the relevant program manuals or guides, has been consulted.

6.4.3 Benefits of Multiple-Degree-of-Freedom and Finite-Element Analyses

Although the complexities of FEA require an experienced engineer to generate and interpret analysis results, the benefits of increased fidelity are realized in reduced support reactions, increased structural capacity, and visualization of the problem. The use of explicit (direct) time integration is identical with SDOF procedures. The increased complexity of material and geometric modeling (i.e., material failure and contact, respectively) allows for a response that is closer to a realistic response.

Single-Degree-of-Freedom analysis is both a simple and quick calculation for an engineer to quantify the structural behavior of components subjected to blast load effects. However, it is important for the blast engineer to be aware of the inherent conservatism when using SDOF methodologies. As described previously, SDOF analysis utilizes a single structural response mode (typically flexural) to dictate the total deflection or response of the component. Although this assumption may be acceptable for highly dominant singular modes of response, in reality, structural components respond to blast loads through a combination of low- and higher-order structural modes to dictate their overall behavior. SDOF analyses may also inherently overestimate support reactions. As the system approaches its ultimate resistance, R_u, energy is dissipated through plastic strain of the structural component (an increase in μ_d) instead of the plasticity being spread between the structural component and its supporting members.

For applications where connection design is crucial, a higher-fidelity analysis (i.e., an MDOF or FEA) can be performed to calculate more realistic values for support reactions and ductilities. Furthermore, material and geometric nonlinearity and higher-order deformations are limited in an SDOF with flexural response. In systems where the structural response cannot be simply determined by the engineer, a higher-fidelity analysis is warranted. Higher-fidelity models that capture higher-order structural behavior, as well as MDOF nonlinearity and combined inertial resistance, can lead to more cost-effective design solutions and retrofits. More realistic boundary conditions can also be employed, especially when the actual connections are also included in the model.

6.4.4 Acceptance Criteria for Use with Finite-Element Analyses

An important item is to consider how the FEA output data would be used to confirm compliance with acceptance criteria. As discussed in Chapter 5, the acceptance criteria for blast-resisting design for SDOF systems is based on peak dynamic deformations with respect to support rotation and ductility. Support rotations are determined by the relative displacement between a support point and the point of maximum displacement in the span, as illustrated in Chapter 5, Figure 5-9. Similar response limits can directly be applied for simple MDOF systems.

Finite-element analysis programs can output maximum/minimum displacement time histories at each node. Contour plots of nodal displacements are a practical way for evaluating the relative displacements between nodes. Meanwhile, the FEA results of element plastic strains can be used for evaluating member ductility. The ductility is usually calculated as the ratio of plastic strain to yield strain. FEA programs can output maximum plastic strains (among the entire time history) of each element. Contour plots of maximum plastic strains are the most effective way for evaluating member ductility as it relates to material yield strain. However, FEA can output plastic strain that can spread far beyond the regions of concentrated plastic hinges. For this reason, it is important to determine a reference length over which the element strain values are averaged to compute ductility in the plastic region. Several industry practice documents are available to provide guidance on this topic such as DNV-RP-C208 (DNV 2022) and UKOOA (2002).

With that stated, if the FEA response is elastic in both tension and compression, the damage response level would conservatively qualify as meeting a low damage response level. If the FEA results in tensile and compressive plastic strains near their ultimate strain limits, the damage response level would qualify as meeting a high damage response level. It is recommended that a high response should maintain at least a small margin of safety against complete failure. Defining a medium damage response level would require an experienced FEA analyst to have a broad background in inelastic behavior of ductile and brittle material responses in tension and compression under high strain rates as well as their constitutive material model capabilities and deficiencies. Conservatively, an estimate of medium response based on FEA results following the SDOF methodology limits for support rotation and ductility presented in Chapter 5 would be acceptable.

Figure 6-10 shows an example of a framed building FEA model consisting of beam elements where a sweeping load was applied to better assess the lateral resisting system under blast loading. It is recommended that support rotation criteria be used to assess damage when using beam elements. For this type of analysis, the use of end rotation or ductility limits as provided in Chapter 5 for SDOF analysis should be applied to establish low, medium, and high component responses.

Figure 6-11 demonstrates a discrete FEA calculation of a pre-engineered metal building wall panel section subjected to blast loading, where the metal

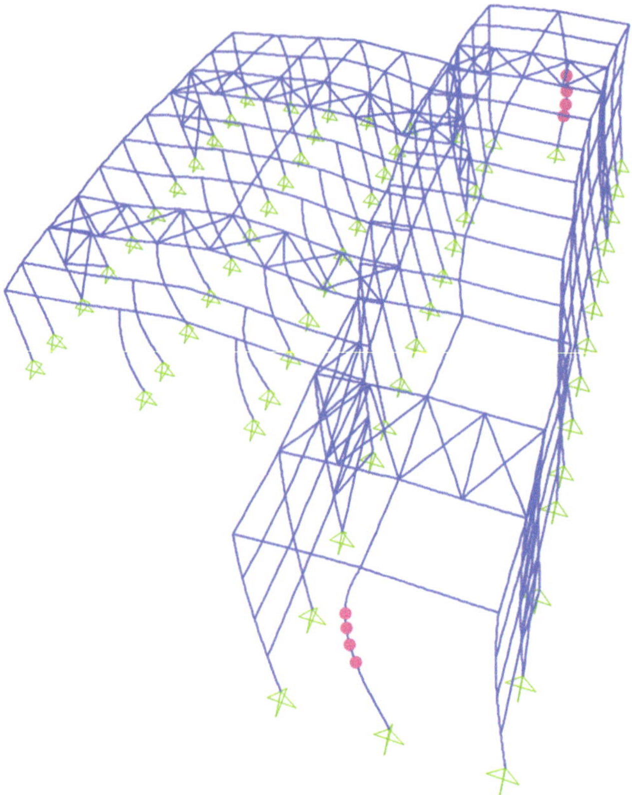

Figure 6-10. Beam element FEA model response of a building subjected to sweeping blast load.

Source: Thornton Tomasetti.

panels, girts, and column are broken into shell elements. The engineer can make informed design decisions in this case on the basis of quantifying both deflections and allowable plastic strain limits. Discrete FEA studies are usually more accurate than simplified analyses and could potentially provide owner-operators of petrochemical facilities with more economical structural solutions.

6.5 APPLICATIONS

Dynamic analysis approaches for some typical applications are described subsequently.

6.5.1 Shear Wall/Diaphragm–Type Structures

A common type of blast-resistant structure at petrochemical facilities consists of a reinforced concrete, steel, or masonry, single-story building with a rectangular

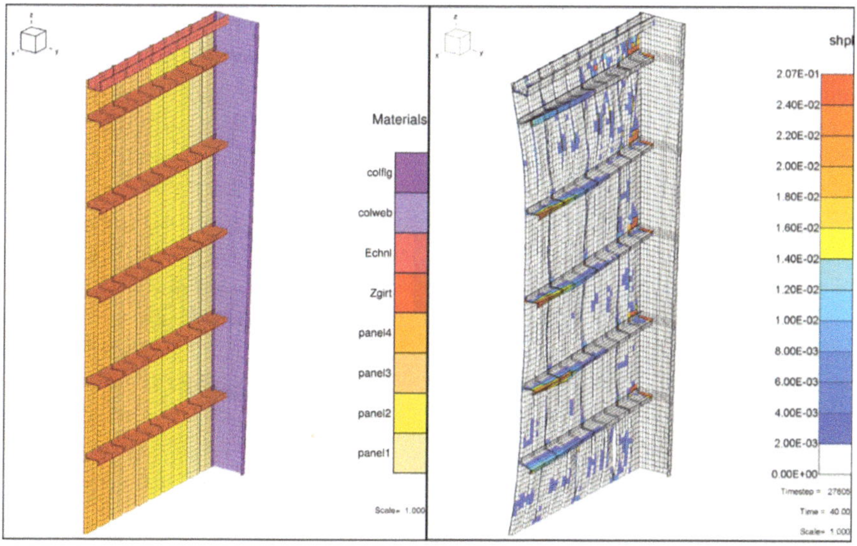

Figure 6-11. Typical pre-engineered metal building discrete FEA analysis visualizing plastic strain in the girts and metal panels.

Source: Thornton Tomasetti.

footprint. The usual approach for designing for lateral blast loads is to design the wall facing the blast as a flexural member spanning vertically between the roof and the foundation. The roof system is designed as a horizontal diaphragm spanning between the side walls of the building. The side walls are then designed as shear walls that carry the lateral loads as well as the overturning effects to the foundation. This concept is illustrated in Figure 6-1.

Several considerations are essential when analyzing this type of structure. First, the usual load path described previously may not be appropriate, which depends on the proportions of the building. The predominant mode of resisting lateral loads by a compact building may be through cantilever beam action as opposed to the shear wall/diaphragm action described previously.

Another consideration in the analysis of the shear wall/diaphragm systems is the effective width of the diaphragm flanges. Some portions of the front and rear walls can be expected to act as compression and tension flanges, respectively, of the horizontal diaphragm slab. The effective width of the flange is usually taken as approximately six times the wall thickness (Derecho et al. 1974). Because the dynamic response of a member is affected by its natural period and maximum resistance, the flanges should be considered when determining the diaphragm's stiffness and strength.

A similar situation exists for the side shear walls. Some portions of the connecting front and rear walls will act as beam flanges as in a C-shaped cross section in plan. Here again, an effective width of six times the flange wall thickness may be used.

6.5.2 Frame Structures

Modeling of frame-type structures usually involves the use of an MDOF or FEA approach owing to a simultaneous application of lateral and vertical blast loads on the frame. A simultaneous application of these forces usually results in combined axial and bending load conditions in the individual frame members, which significantly affect the member design. Otherwise, a conservative combination of the separate effects of each loading condition on the response of the frame must be used. Advantage can be taken of the fact that peak responses caused by the vertical and lateral loads do not usually occur simultaneously, especially when the analysis allows for the blast load to sweep over and to the back side of the frame based on the blast wave speed.

Another consideration for frame-type structures is whether to use a 2D- or 3D model. The appropriate choice depends on the symmetry of the structural resistance, mass, and the loading. If all three are symmetric, a 2D plane frame model may suffice. Also, the shear deformation needs to be considered to model stocky members.

Some studies of one- and two-story plane frames have examined the level of modeling detail required to obtain reasonable results, which are summarized by Baker (1983). These studies considered factors such as the number and spacing of joints, member loads versus joint loads, girder flexibility, sweeping roof loads, and mass distribution. Because of the large number of variables studied, the reader is encouraged to refer to the referenced document to obtain a clear understanding and appreciation of the results.

Selection of the material model is another important factor to be considered. Some programs allow the user to specify plastic moment–rotation curves for beam elements. However, a more rigorous and widely available method of defining nonlinear material properties is to specify the stress versus strain data that account for strain rate effects. Plastic behavior is approximated at the section level in the former method, whereas the latter method tracks plastic behavior at the individual integration points (fibers) through the thickness of the member. Each method has its advantages and disadvantages.

The plastic hinge nonlinear material model is easier to use but usually cannot consider axial load effects. Plastic hinge locations must usually be predetermined and are usually limited to the ends and/or midspan of the members. Analysis results that include displacements and plastic hinge rotations are directly comparable against acceptance criteria provided in Chapter 5.

Fiber beam models are simplified versions of the finite-element method that can also be used to model frame structures. Fiber beam models can use detailed geometry and material models to obtain an accurate representation of yielding and nonlinear behavior along the length of the member. These are more sophisticated than the plastic hinge model and less computationally demanding than the finite-element models. Plastic hinge models require the details of moment–rotation relationships at the beam ends, which are unlikely to be available to the designer. Details of the section and geometry of the structural component and properties of the materials are essential components to building the fiber models, which are usually available and simple to obtain. Fiber models are also computationally efficient because of their minimal storage and processing requirements.

The more rigorous stress/strain nonlinear material model, often referred to as the plastic zone method, is theoretically capable of handling any general cross section. Both isotropic and kinematic hardening rules are usually available. This method is most practical for homogeneous materials such as structural steel because of the complications involved with modeling composite materials such as reinforced concrete. Output results include stresses and strains at various locations along the length and through the thickness of each member. As with FEA methods, the analyst needs to be cognizant of what response fringe data plots are important for their response damage assessments.

6.5.3 Walls with Openings

Openings in walls and roof sections can create a complex structural element in terms of dynamic response calculations. Plate-type elements without openings, such as one-way span walls, can readily be represented by a simple SDOF system. Openings interrupt the distribution of stresses and reduce the ultimate resistance of a component. The resistance can be determined through a yield line analysis, which reflects the boundary conditions of the actual component including openings. The process consists of establishing a trial yield pattern, computing the resistance of each sector of the slab, and iterating the yield line locations until a convergence of the resistance in each segment is obtained. The reader can refer to UFC 3-340-02 (Park and Gamble 2000), and ASCE MOP 42 (ASCE 1961) for more detailed information.

Another approach is to use linear or nonlinear FEA methods to determine the resistance function. This method is more favorable when considering the effects of openings in panels. The resistance function can be used to develop an SDOF system for computing the dynamic response through numerical integration or to develop *P–I* diagrams. Alternatively, a nonlinear dynamic FEA approach can be used to determine the dynamic response of the component directly. In either case, it is important to include the effects of architectural elements that cover openings such as a door or window. If the architectural cover is relatively weak, as in the case of unimproved windows and doors, the opening can be treated as uncovered with little error because these elements have little mass and fail well before transferring any significant load to the perimeter of the opening. If blast-resistant windows and doors are used, their mass and ability to transfer the blast load to the panel need to be included as loaded area in the analyses.

In some cases, the opening is relatively small and the loss of resistance is minor. In other cases, the perimeter of the opening can be designed with additional reinforcing to act as a support for the door or window and transfer loads directly to their supports. This "framing" configuration eliminates the weakening effect of the opening on the structural panel component.

6.5.4 Empirical Methods

Empirical methods based on structural damage data collected from tests and actual explosions have been used in evaluating existing structures for blast loading, especially when coupled with SDOF, MDOF, and/or finite-element analyses to fill in the gaps between data. Future understanding of building and component response performance will continue to increase as more data are collected and evaluated.

As briefly mentioned in Section 6.3.3, the *P–I* diagram method is sometimes based on a combination of empirical and analytical relationships. Mathematical expressions of *P–I* damage curves can also be derived from test results. Refer to Baker (1983) and the CEDAW workbook (2008) for further details.

APPENDIX 6A. NUMERICAL INTEGRATION METHOD

The basic steps for numerical integration using the linear acceleration method and a bilinear resistance-deflection function for compression and tension are outlined subsequently. These steps are easily programmed for use with personal computers using spreadsheets or computer script. The following procedure is based on *Microcomputer-Aided Engineering: Structural Dynamics* (Paz 1986). Implementation examples are included in Chapters 10 and 11. Other numerical integration methods are based on a range of approaches such as the constant velocity approach, multilinear resistance-deflection functions, or other methods based on the Newmark beta method. Modest differences in the resulting response are not a cause for concern.

Initialize

a. Determine the stiffness, K, mass, M, positive resistance, R_u, rebound resistance, $R_{u\text{-}}$, damping value, C, time increment, Δt, static load F_s, reaction resistance coefficient, C_R, and reaction force coefficient, C_F.

   For blast design, damping is usually set to zero.

   The mass term in Equation (6-6) can be adjusted with an average of the elastic and plastic transformation factors over all time steps or can account for the applicable K_{LM} at each time step.

   Reaction resistance coefficients are needed only if dynamic reactions are to be calculated. C_R and C_F can be extracted from Tables 6-1 through 6-3.

b. At each time step (step = 0 to last), determine the value of the dynamic forcing function, F_0 to F_{last}.

c. For the initial time step (step = 0), initialize the displacement y, velocity v, acceleration a, positive and negative yield displacements y_e and y_{e-}, and resistance R.

$y_0 = F_s/K$

$v_0 = 0$

$a_0 = F_0/M$

$y_{e,0} = R_u/K$

$y_{e-,0} = R_{u-}/K$

$R_0 = F_s$

For a horizontal component, y_0 depends on the dead load, whereas $y_0 = 0$ for a vertical component.

d. Initialize the response indicator.

$\text{KEY}_0 = 0$

NOTE: Values for KEY represent the following:

KEY = 0 is elastic

KEY = +1 is positive plastic

KEY = −1 is rebound plastic

For each time step: (step = i, beginning with $i = 0$)

a. Calculate the effective stiffness

if $(\text{KEY}_i = 0)$, then $K_{e,i} = K + (6/\Delta t^2)\, M + (3/\Delta t)\, C$ (6A-1)

otherwise, $K_{e,i} = (6/\Delta t^2)\, M + (3/\Delta t)\, C$

b. Calculate the effective incremental force

$\Delta F_{e,i} = (F_{i+1} - F_i) + [(6/\Delta t)\, M + (3)\, C]\, v_i + [(3)\, M + (\Delta t/2)\, C]$ (6A-2)

c. Solve for the incremental displacement

$\Delta y_i = \Delta F_{e,i}/K_{e,i}$ (6A-3)

d. Calculate the incremental velocity

$\Delta v_i = (3/\Delta t)\, \Delta y_i - (3)\, v_i - (\Delta t/2)\, a_i$ (6A-4)

e. Calculate the displacement and velocity at the next time step (step = $i + 1$)

$y_{i+1} = y_i + \Delta y_i$ (6A-5)

$v_{i+1} = v_i + \Delta v_i$ (6A-6)

f. Determine the calculation case indicator, Z, for the next time step (Z is used as a switching mechanism in selecting the appropriate formulas for KEY, y_e, and y_{e-}.)

if $(\text{KEY}_i = 0)$ and $(y_{i+1} > y_{e,i})$, then $Z_{i+1} = 2$

if $(\text{KEY}_i > 0)$ and $(y_{i+1} > y_i)$, then $Z_{i+1} = 3$

if $(\text{KEY}_i > 0)$ and $(y_{i+1} \leq y_i)$, then $Z_{i+1} = 4$

if $(\text{KEY}_i = 0)$ and $(y_{i+1} < y_{e-,i})$, then $Z_{i+1} = 6$

$$\text{if } (\text{KEY}_i < 0) \text{ and } (y_{i+1} < y_i), \text{ then } Z_{i+1} = 3 \tag{6A-7}$$
$$\text{if } (\text{KEY}_i < 0) \text{ and } (y_{i+1} \geq y_i), \text{ then } Z_{i+1} = 8;$$
$$\text{otherwise, } Z_{i+1} = 1$$

Note: Values for Z represent the following:

$Z = 1$ is continued elastic deformation,

$Z = 2$ is the start of positive plastic deformation,

$Z = 3$ is increasing plastic deformation,

$Z = 4$ is the end of positive plastic deformation,

$Z = 6$ is the start of rebound plastic deformation, and

$Z = 8$ is the end of rebound plastic deformation.

Notes: This cycle progresses from $Z = 1$ through $Z = 8$, then returns to $Z = 1$ when plastic response is achieved in both inbound and rebound. The $Z = 5$ case is skipped as this elastic case is covered when $Z = 1$. The $Z = 7$ case is skipped as this plastic case is covered when $Z = 3$. Other cases may be skipped, for example, if only elastic deformations or only elastic and positive plastic deformations are achieved.

g. Determine the response indicator for the next time step

$$\text{if } (Z_{i+1} = 1 \text{ or } 3), \text{ then } \text{KEY}_{i+1} = \text{KEY}_i \tag{6A-8}$$
$$\text{if } (Z_{i+1} = 2), \text{ then } \text{KEY}_{i+1} = 1$$
$$\text{if } (Z_{i+1} = 6), \text{ then } \text{KEY}_{i+1} = -1$$
$$\text{otherwise, } \text{KEY}_{i+1} = 0$$

h. Determine the positive yield displacement at the next time step

$$\text{if } (Z_{i+1} = 4), \text{ then } y_{e,i+1} = y_i$$
$$\text{if } (Z_{i+1} = 8), \text{ then } y_{e,i+1} = y_i + (R_u - R_{u-})/K; \tag{6A-9}$$
$$\text{otherwise, } y_{e,i+1} = y_{e,i}$$

i. Determine the rebound yield displacement at the next time step

$$\text{if } (Z_{i+1} = 4), \text{ then } y_{e-,i+1} = y_i - (R_u - R_{u-})/K$$
$$\text{if } (Z_{i+1} = 8), \text{ then } y_{e-,i+1} = y_i; \tag{6A-10}$$
$$\text{otherwise, } y_{e-,i+1} = y_{e-,i}$$

j. Calculate the resistance at the next time step on the basis of the value of KEY

$$\text{if } (\text{KEY}_{i+1} = 0), \text{ then } R_{i+1} = R_u - (y_{e,i+1} - y_{i+1})\, K$$
$$\text{if } (\text{KEY}_{i+1} = 1), \text{ then } R_{i+1} = R_u; \tag{6A-11}$$
$$\text{otherwise, } R_{i+1} = R_{u-}$$

k. Calculate acceleration at the next time step

$$a_{i+1} = [F_s + F_{i+1} - (C)\, v_{i+1} - R_{i+1}]/M \tag{6A-12}$$

l. If desired, calculate the dynamic reaction

$$V_i = (C_R)\, R_i + (C_F)\, F_i \tag{6A-13}$$

m. Positive and rebound ductility ratios may be calculated as follows:

$$\mu_{d,i} = (y_i - y_{e,i} + y_{e,0})/y_{e,0} \tag{6A-14}$$

$$\mu_{d-,i} = (y_i - y_{e-,i} + y_{e-,0})/y_{e-,0} \tag{6A-15}$$

n. Repeat the loop until the desired deformations are reached.

Maximum ductility and support rotation values can then be determined on the basis of the maximum calculated deformations.

References

ASCE. 1961. *Design of structures to resist nuclear weapons effects*. MOP 42. Reston, VA: ASCE.

Baker, W. E. 1983. *Explosion hazards and evaluation*. New York: Elsevier.

Bathe, K. 1995. *Finite element procedures in engineering analysis*. Englewood Cliffs, NJ: Prentice-Hall.

Biggs, J. M. 1964. *Introduction to structural dynamics*. New York: McGraw-Hill.

CEDAW. 2008. *Component explosive damage assessment workbook (CEDAW)*. PDC TR-05-02. Omaha, NE: US Army Corps of Engineers, Protective Design Center. Accessed November 21, 2024. https://www.nwo.usace.army.mil/Portals/23/docs/PDC/PDC_TR-05-02_Component_Explosive_Damage_Assessment_Workbook_DistribA.pdf?ver=2018-11-01-090340-113.

Clough, R. W., and J. Penzien. 1993. *Dynamics of structures*. 2nd ed. New York: McGraw-Hill.

Derecho, A. T., D. M. Schultz, and M. Fintel. 1974. *Analysis and design of small reinforced concrete buildings for earthquake forces*. Engineering Bulletin No. EB004.05D. Skokie, IL: Portland Cement Association.

DNV (Det Norske Veritas). 2022. *Determination of structural capacity by non-linear finite element analysis methods*. DNV-RP-C208. Oslo, Norway: DNV AS.

Krauthammer, T., N. Bazeos, and T. J. Holmquist. 1986. "Modified SDOF analysis of RC box-type structures." *J. Struct. Eng.* 112 (4): 726–744.

Krauthammer, T., S. Shahriar, and H. M. Shanaa. 1990. "Response of reinforced concrete elements to severe impulsive loads." *J. Struct. Eng.* 116 (4): 1061–1079.

Mays, G. C., and P. D. Smith. 1995. *Blast effects on buildings—Design of buildings to optimize resistance to blast loading*. London: Thomas Tedford.

Park, R., and W. Gamble. 2000. *Reinforced concrete slabs*. 2nd ed. New York: Wiley.

Paz, M. 1986. *Microcomputer-aided engineering: Structural dynamics*. New York: Van Nostrand Reinhold.

Paz, M. 1991. *Structural dynamics: Theory and computation*. 3rd ed. New York: Van Nostrand Reinhold.

Stronge, W. J., and T. X. Yu. 1993. *Dynamic models for structural plasticity*. London: Springer.

UKOOA (UK Offshore Operators Association Limited). 2002. *Fire and explosion guidance*. Issue 2, Part 0: Fire and explosion hazard management. London: UKOOA.

USACE/NAVFAC/AFCESA (US Army Corp of Engineers/Naval Facilities Engineering Command/Air Force Civil Engineer Support Agency). 2014. *Structures to resist the effects of accidental explosions,* UFC 3-340-02, Change 2. Washington, DC: National Institute of Building Sciences.

US DoD (US Department of Defense). 2008. *Structures to resist the effects of accidental explosions*, UFC 3-340-02. Washington, DC: US DoD.

CHAPTER 7
Design Procedures

7.1 INTRODUCTION

The purpose of this chapter is to bring together all the subjects of the preceding chapters and to discuss design requirements for structural elements. The design approaches provided in the preceding chapters represent a performance-based design procedure. General blast design concepts that apply to all structures are discussed. Next, a design sequence is outlined. Finally, specific design methods for blast-resistant building construction are presented.

7.2 GENERAL DESIGN CONCEPTS

Several important concepts should be kept in mind while designing buildings for blast resistance. These concepts include energy absorption, safety factors, limit states, load combinations, resistance-deflection functions, structural performance considerations, and most important, redundancy.

Although typical structural design codes, for example, AISC 360 (AISC 2016), ACI 318 (ACI 2019), and TMS 402/602 (TMS 2016), do not specifically cover blast-resistant design, they remain the best commonly available design tools that are supplemented by these design recommendations.

7.2.1 Energy Absorption

The need for achieving ductile responses has been discussed previously in Chapter 5, "Dynamic Material Strength and Response Criteria." However, both strength and ductility are necessary to achieve high energy absorption. Energy absorption capacity equates to the area under the elastic-plastic load versus displacement diagram, or resistance-deflection function, of a member or overall structure (refer to Chapter 5, Figure 5-1). High energy absorption capacity is achieved through the use of appropriate structural materials and details. These details must accommodate relatively large deflections and rotations. High strength with little-to-no ductility is undesirable for conventional design and is even less desirable for blast-resistant design.

7.2.2 Safety Factors

Traditional definitions of safety factors in terms of strength requirements, such as load-resistance factors or allowable stresses, are not applicable in blast-resistant design. Safety factors are more appropriately measured in terms of strain energy demand versus strain energy absorption capacity. Allowable deformations are a practical method to quantify energy absorption capacity.

Margins of safety against structural failure are achieved through the use of allowable deformation criteria, as presented in Chapter 5. As long as the calculated deformations do not exceed the allowable values, a margin of safety against failure exists.

An additional method that has been used to achieve a margin of safety is to increase the design blast pressure loading. For example, UFC 3-340-02 (US DoD 2008b) recommends adding 20% to the weight of the charge. However, increasing the blast load is not common, and is not recommended, for petrochemical explosions because of the level of accuracy and conservatism associated with the methods used in their load prediction compared to known stored charge weights where equivalent high explosive charge weights can be more accurately determined.

7.2.3 Limit-State Design

Limit-state design methods are used in blast-resistant design. These methods provide a comprehensive, reliable, and realistic means of predicting damage, failure mechanisms, and structural capacities. Limit-state design specifications are available for structural steel, cold-formed steel, reinforced concrete, aluminum, and reinforced masonry. Details on the implementation of these methods are given in subsequent sections in this chapter for each class of material.

Each of the limit-state design specifications contains special provisions for high seismic conditions, which are commonly used for blast-resistant design. These provisions are intended to protect against nonductile failure modes, such as buckling or premature crushing of brittle materials, through the use of special detailing and design requirements.

7.2.4 Loading Combinations

Limit-state design specifications define the load factors and combinations of loads to be used for conventional loading conditions such as dead, live, wind, and earthquake. Blast loads are combined with only those loads that are expected to be present at the time of the explosion. Therefore, blast loads are not combined with earthquake or wind loads because the likelihood of a significant explosion during a design seismic or wind event is highly improbable.

The dead and live loads are applied as a static preload to the component prior to the application of the dynamic blast load. In the case of horizontal components, this accounts for the initial deflection under service conditions. In the case of vertical components, this accounts for the axial load present under service conditions. Note that although a portion of the live load is used to establish the initial loading conditions, the live load should not necessarily be included in the mass of the resisting structural system; refer to Chapter 6, Section 6.2.4, for more discussion on live load contribution to component mass.

The basic limit-state loading combination for all material types used in blast-resistant design is as follows:

$$1.0(\mathrm{DL}) + a(\mathrm{LL}) + 1.0(\mathrm{BL}) \tag{7-1}$$

where

DL = Dead load,

LL = Live load,

BL = Blast load, and

a = Load factor applied to conventional live loads to reflect the portion of live load expected to occur simultaneously with the blast load. The factor can be taken as 0 for cases where the live load is opposite the blast load.

7.2.5 Resistance-Deflection Functions

To determine the dynamic response of a system, one needs to develop generalized force versus elasto-plastic deflection relationships for the overall structure or each member. These force versus deflection relationships are usually nonlinear (because of ductile materials or geometry), which are called *resistance-deflection functions*. They are an essential input parameter for the analysis of equivalent single-degree-of-freedom (SDOF) systems. Resistance-deflection functions are not usually needed for analyses of multidegree-of-freedom (MDOF) systems, or more specifically, FEA models. Material models employing nonlinear stress versus strain data, as discussed in Chapter 5, are used in MDOF systems, and are included in constitutive FEA material models.

The first step in developing a resistance-deflection function is to determine the plastic section capacities, such as plastic moment, M_p, as shown in Figure 7-1. The next step is to determine the sequence of plastic hinge formation and the corresponding load and deformation values. This is done by incrementally applying loads until a collapse mechanism is formed, as illustrated in Figure 7-2 for a fixed end beam with a uniform load. Collapse loads for other common one-way members are included in Chapter 6, Tables 6-1 through 6-3. For a more complete treatment, refer to Chapter 5 of Biggs (1964).

The piecewise linear curve representing the resistance-deflection function shown in Figure 7-2b is an approximation made to simplify the analysis and design process. This approximation ignores the nonlinear effects listed below, which are typically included in advanced FEA constitutive material models.

- Concrete and masonry—softening due to cracking, initial yielding,
- Reinforcing steel and structural steel—strain hardening, and
- Structural steel—progressive yielding of fibers through the section thickness.

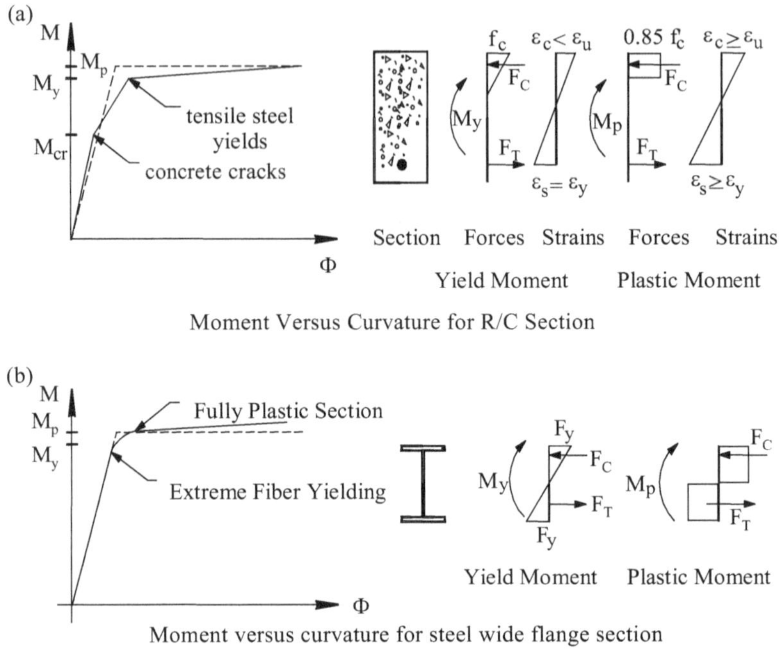

Figure 7-1. Typical moment versus curvature diagrams: (a) moment versus curvature for an R/C section, (b) moment versus curvature for a steel wide flange section.

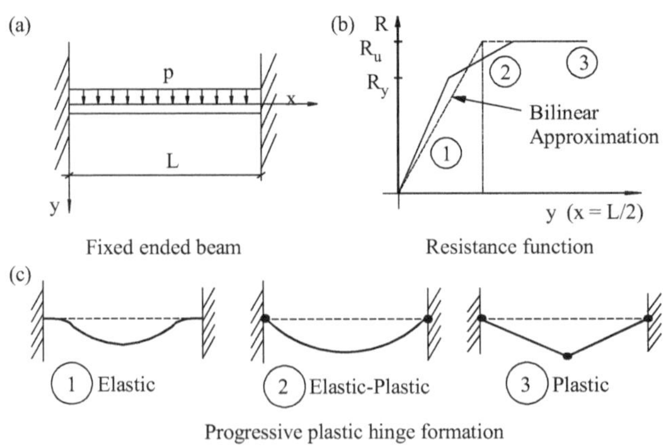

Figure 7-2. Resistance function for a member with sequential plastic hinges: (a) fixed ended beam, (b) resistance function, (c) progressive plastic hinge formation.

Preloads are sometimes considered in developing resistance-deflection functions. Preloads are any dead or live loads that cause a deformation in the member and thereby use up some of the available elastic strain energy. The effects of preload on equivalent SDOF system analyses are sometimes handled by reducing the calculated available resistance by the amount of the preload.

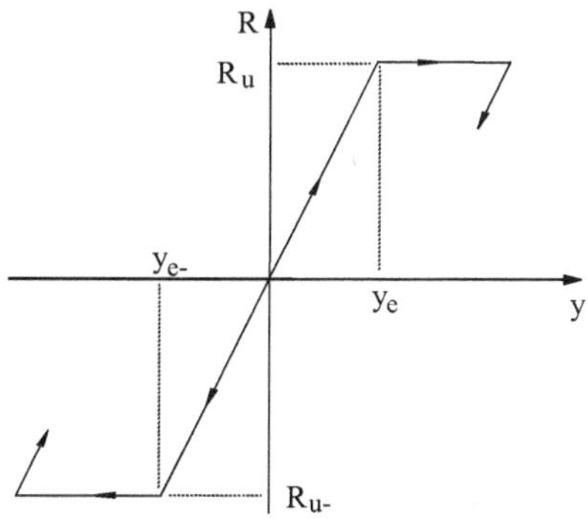

Figure 7-3. Typical simplified elasto-plastic resistance function.

Resistance-deflection functions can further be approximated by elastic, perfectly plastic bilinear functions that are used in the development of response charts and formulas. The approximation is made by maintaining maximum resistance and equating areas under the curve (strain energy) up to maximum resistance, R_u, as shown in Figure 7-2. Maximum resistance values may be different for the positive and negative loading directions. Strain-hardening effects can be considered; refer to Chapter 5, Section 5.5.5. A typical resistance-deflection function is illustrated in Figure 7-3.

The basic steps outlined previously for the design of flexural members can also be applied for shear-dominated response members with one major difference in the determination of the initial stiffness (slope) of the resistance-deflection function. For these types of members, shear deformations are as large or larger than flexural deformations and therefore cannot be ignored as is the case of flexural members. Maximum resistance is determined in accordance with the member shear strength without strength reduction factors (i.e., $\phi = 1.0$).

7.2.6 Structural Performance Considerations

Structural performance requirements for blast-resistant design include limits imposed on member deflections, story drifts, and damage tolerance levels. Conventional serviceability requirements are not applicable for one-time severe-event blast-loading conditions. Refer to Chapter 5 for additional information.

7.3 MEMBER DESIGN PROCESS

The following steps depict the design process for individual members. Descriptions of each individual step are given in the following sections.

- Step 1: Load determination,
- Step 2: Determination of member properties,

- Step 3: Model representation,
- Step 4: Trial member selection,
- Step 5: Dynamic analysis,
- Step 6: Deformation criteria check, and
- Step 7: Connection design.

These steps are described in the following sections.

7.3.1 Load Determination

For primary members (external walls, roof slabs, etc.), the load computation is performed in accordance with Chapter 3. Loads on supporting, or interior members, are determined either by the tributary area method or from a computed dynamic reaction. In the tributary area method, blast pressures are multiplied by the exterior surface area tributary to a support location. The resulting force is then applied to the supporting member. Dynamic reactions result from a numerical time history analysis (refer to Chapter 6, Section 6.4.5) and provide a more accurate time-varying load on the supporting member. It is recommended that the use of dynamic reactions in SDOF analyses be limited to one supported component and limit the load function to one positive and one negative phase of response. Use for a sequence of more than one supported member could result in erroneous results, whereas MDOF and FEA can accurately model the dynamic loading transfer to any number of supporting members. More specifically, dynamic interaction and inherent damping in an actual structure cannot be properly captured through the use of SDOF methods.

7.3.2 Determination of Member Properties

Member properties are determined in accordance with Chapter 5. Required properties usually include mass, stiffness, yield strength, and resistance. Additional properties include post-yield strength or membrane resistance.

7.3.3 Model Representation

Mathematical models for individual structural members will have to be developed using the SDOF method. The engineer will then decide the most appropriate structural representation, such as one-way versus two-way action, and loading distributions for each member. Individual members are usually idealized as simple one-way beams or two-way plates because these types of members can be analyzed as equivalent SDOF systems with minimal engineering effort. One-way members are the most common.

Boundary conditions need to be evaluated by considering the type of connections to be used for the member supports. The engineer must keep in mind that support details must provide sufficient strength, ductility, and stability to enable the member to develop the expected mechanism. Support capability to resist reaction forces for both the loading and the rebound phases of the response must be considered when evaluating boundary conditions.

7.3.4 Trial Member Selection

Unlike most static design procedures, dynamic design requires a trial and error approach. Only in the verification of shear capacities and in the design of support connections can member proportions be directly determined. For the dynamic analysis, the needed nonlinear response properties are determined from a trial section. The analysis results then indicate the adequacy of the trial section. Experience on the part of the engineer will help in reducing the number of iterations.

7.3.5 Dynamic Analysis

The dynamic analysis itself is then performed by one of a number of different methods ranging from simple chart or equation solutions to complex nonlinear finite-element analysis. Analysis methods are covered in Chapter 6. The purpose of this step is to compute member deformations and reactions.

7.3.6 Deformation Criteria Check

Single-degree-of-freedom and MDOF results will provide peak element deformations that should be compared to the allowable values given in Chapter 5. Deformations will be dealt with in terms of ductility ratios, support rotations, deflections, or as deflection-span ratios. If the allowable values are not met, then some changes to trial member sizes or to structural configurations must be made and the analysis repeated. Material-specific criteria are provided in Sections 7.4 through 7.6.

In addition, calculated peak effective plastic strains in FEA results can provide insight into the degree of damage, as shown in Figure 7-4. Peak effective plastic strain is commonly used for assessing blast damage to offshore structures in both the base material and the welds. Applicable references were presented in Chapter 6, Section 6.4.4.

7.3.7 Connection Sizing

Connections must be sized to transfer computed reaction forces and to ensure that plastic hinges can be maintained in the assumed locations using the SDOF method. For reinforced concrete design, splices and development lengths must account for the full yield capacities of reinforcing. For structural steel design, connections are designed for a capacity somewhat greater than that of its supported member. Further information is provided in later sections of this chapter.

It is essential that the design engineer recognize that the job is not complete until the structural system has been detailed in a manner that ensures the response will be consistent with the design intent. The development of details should also consider cost and constructability.

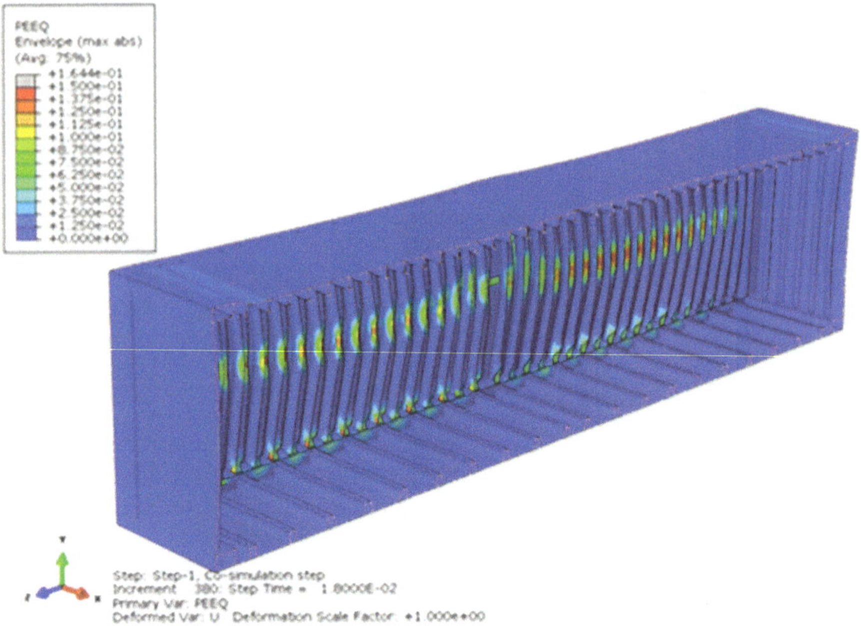

Figure 7-4. Plastic strain distribution under blast loading on a metal building.
Source: Courtesy of RedGuard, used with permission.

The details discussed or illustrated in this chapter are some that have been found to be cost-effective and constructible. The design details included are not intended to limit the use of alternate designs.

7.4 REINFORCED CONCRETE DESIGN

Reinforced concrete is sometimes used in petrochemical buildings for the exterior faces directly exposed to blast effects. The exterior faces may be cast-in-place or precast.

Wall and roof elements are usually made of reinforced concrete for projectile penetration resistance. Roof and side wall structural elements may also use the inherent in-plane strength of concrete to resist lateral blast forces. Being relatively thin flexural elements, walls and roofs should be designed for a considerable ductile response to absorb blast energy without transmitting it to the supporting elements. Construction preferences often indicate the need to eliminate shear reinforcing if at all possible to reduce field labor costs. The combination of these objectives leads to the need for higher-strength concrete.

Precast walls are used for two reasons: to reduce the cost of the building through decreased field labor and to shorten the schedule by constructing the walls and foundations simultaneously. The largest drawback for the use of precast structural elements is the design and detailing of connections. As in seismic design, special attention to ductility is required.

Foundations are almost always constructed of reinforced concrete. Blast-resistant buildings are typically supported on piled or soil-supported mats.

7.4.1 Design Principles

ACI 318 (ACI 2019) is used as the basis for structural properties that should be adjusted for blast-resistant design. The resistance of concrete elements is computed using the dynamic material strengths given in Chapter 5, Section 5.4. Strength reduction factors are not applied (i.e., $\phi = 1.0$) to load cases involving blast. The plastic response used in blast design is similar in concept to the moment redistribution provisions in ACI 318, Section 6.6.5, and the seismic criteria provided in ACI 318, Chapter 18. The more extensive seismic detailing provisions are applied to provide the necessary ductile response.

7.4.2 Supplementary Design Requirements

In addition to ACI 318 requirements, the following items should be considered for blast-resistant design.

1. Minimum Reinforcing: The minimum reinforcing provisions of ACI 318 apply; however, the option to use one-third more reinforcing than computed should not be considered. The moment capacity of underreinforced concrete members is controlled by the uncracked strength of the member. To prevent a premature ductile failure, reinforcing in excess of the cracking moment should be provided. In computing minimum reinforcing, the dynamic material strengths discussed in Chapter 5 should be used.
2. Maximum Reinforcing: Code provisions for maximum reinforcing are included to prevent crushing of concrete prior to yielding of steel. Code provisions also allow compression reinforcing to offset maximum tension reinforcing requirements. Because blast-resistant concrete members typically have reinforcing on each face to resist rebound forces, meeting the maximum reinforcing provisions should not be a problem. Note that panels with single reinforcement at middepth can be used for both inbound and rebound resistance, especially for low blast loads.
3. Substitution of Higher Grades of Reinforcing: The substitution of higher grades of reinforcing is typically not recommended. Stronger reinforcing tends to increase the moment capacity of a concrete section, while not affecting the concrete shear capacity. This could cause a ductile response to become nonductile. In addition, a higher moment capacity will tend to increase the dynamic reaction that the supporting member must resist. Because ASTM specifications provide minimum requirements, mill test reports should be reviewed for possible significant overstrength.
4. Development Lengths: Development lengths should not be reduced for excessive reinforcement. Because plastic hinges will cause overdesigned

reinforcing to yield, the full actual strength of reinforcing should be used in computing section capacities. The development of reinforcing should be computed accordingly.

5. Serviceability Requirements: Criteria intended to reduce cracking at service load levels need not be applied to load combinations including blast. Cracking, as well as permanent deformations resulting from a plastic range response, is an expected result of such an unusual type of load. The ductility limits of Chapter 5 are consistent with the performance requirements of the building under blast.
6. Precast Panel Connections: Precast panel connections usually consist of embedded steel plates and/or clip angles anchored and/or welded to the panel and the supporting component. Cast-in-place and/or post-installed anchors are typically utilized. In cases where the connection capacity is controlled by a brittle failure mode (e.g., anchor shear failure, concrete punching shear, or concrete embed pullout), the engineer should consider the peak calculated dynamic reaction, as discussed in Section 6.3.6, when designing the connection. According to Oswald and Bazan (2014), the peak dynamic reaction load is expected to surpass the equivalent static reaction load associated with the ultimate resistance of the panel if the peak applied blast pressure exceeds approximately 1.5 times the ultimate resistance of the panel.
7. Combined Forces: Some concrete elements are simultaneously subjected to out-of-plane bending loads in combination with in-plane shear loads. For example, side walls must resist side overpressures acting into the plane of the side wall. In addition, reactions from the roof diaphragm acting in the plane of the side wall must also be resisted. There are three means of dealing with this situation:
   a. Separate sets of reinforcing may be determined for each type of force to be resisted. For example, exterior reinforcing may be sized to resist bending, whereas a layer of center reinforcing may be used to resist in-plane shear. Care must be taken to make sure that hinge capacities are not changed as a result of reinforcing intended for other purposes.
   b. An interaction equation, based on criteria from ASCE MOP 42 (ASCE 1985), can be applied to determine acceptable behavior:

$$[\Delta_d / \Delta_a]_{ip}^2 + [\Delta_d / \Delta_a]_{op}^2 \leq 1.0 \qquad (7\text{-}2)$$

where

Δ_d = Computed deformation (ductility ratio or support rotation),
Δ_a = Allowable deformation (ductility ratio or support rotation),
ip = In-plane deformations (subscript), and
op = Out-of-plane deformations (subscript).

(c) The time phasing of in-plane shear and normal loads can be determined from a numerical integration. Provided the peak forces are reached at different times, these forces can be treated separately. Judgment must be used to make this determination.

7.4.3 Failure Mechanisms

The primary failure mechanisms encountered in reinforced concrete buildings are flexure, diagonal shear, and direct shear. Of these three mechanisms, flexure is preferred under blast loading because an extended plastic response is provided prior to failure. To ensure a ductile response, sections are designed so that the flexural capacity is less than the capacity of nonductile mechanisms.

Shear reinforcing is not commonly used in wall and roof elements even though reinforced elements can undergo an extended plastic response. Shear reinforcing increases the diagonal shear capacity of the member, but more important, it provides lateral restraint for the principal reinforcing. Such restraint is vital for large deformations where exterior protective concrete may spall and be unavailable for resistance during ongoing cycles of deformation.

Other failure mechanisms involve portions of structural elements or the transmissions of loads between elements. These other mechanisms must be sized so as not to control the overall structural response. Such failure mechanisms include failures of reinforcing development, precast connection, anchor bolt embedments, and door and connections. These types of failures involve reinforcing development and anchor bolt embedment. Nonductile failures are prevented by providing a concrete embedment strength greater than the material strength of the anchor bolt or reinforcing bar. Connection-type failures involving precast connections or door and window frame embedment are avoided by designing these connections so that the plastic hinge occurs away from these connections.

Situations will occur where a ductile bending mechanism is not attainable. Deep roof diaphragms and side walls resisting in-plane shear are two examples. For these cases, the response must be limited accordingly. Refer to Chapter 5 for these response limits.

7.4.4 Connection Details

Precast construction uses a steel or concrete frame and precast concrete wall panels. Many details have been developed for precast concrete walls. Details for precast walls should be in accordance with the seismic requirements of ACI 318, Chapter 18 (ACI 2019).

The precast details covered in this section can be grouped into two categories: conventional enhanced details and those that mimic cast-in-place details. Conventional enhanced details need to be strengthened for blast-resistant design. Figures 7-5 to 7-8 are examples of these details. One way to provide a reliable degree of strength and ductility is to mimic cast-in-place construction. This approach has been suggested for conventional precast construction in seismic areas. Figures 7-9 to 7-11 are examples of these details.

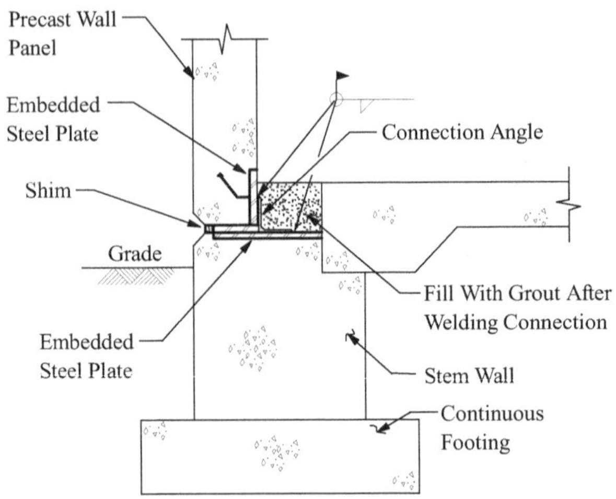

Figure 7-5. Precast panel connection to a foundation.

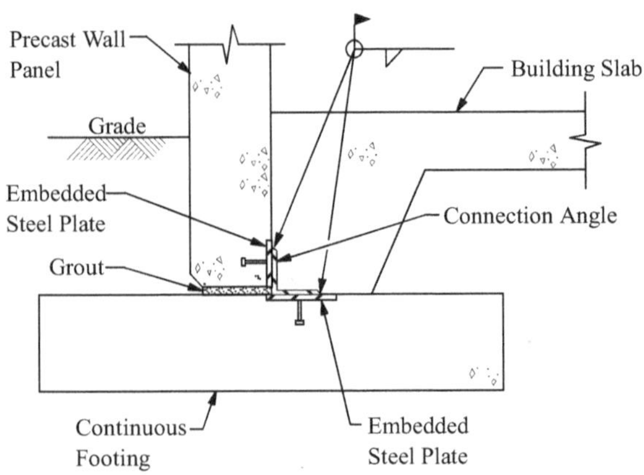

Figure 7-6. Precast panel connection to a foundation.

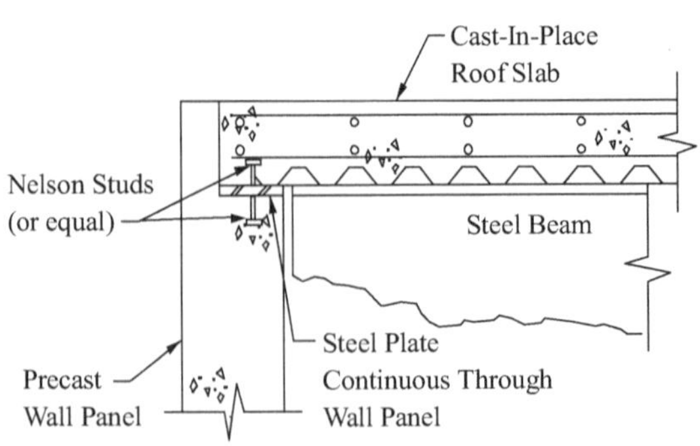

Figure 7-7. Precast panel connection to a roof slab.

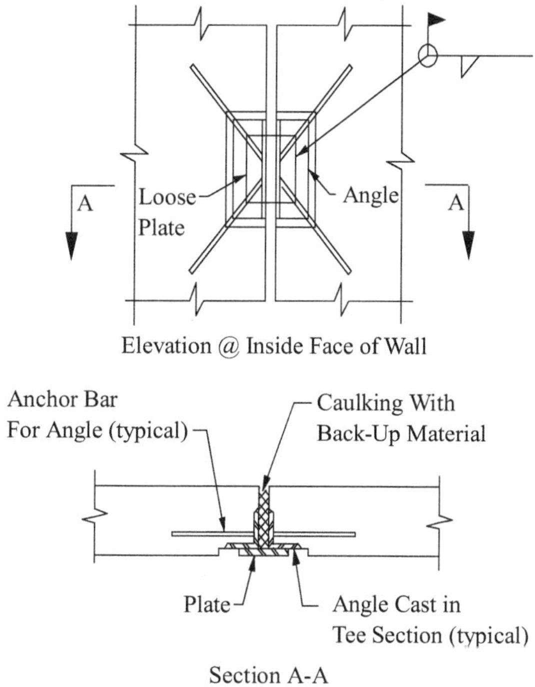

Figure 7-8. Precast panel vertical joint.

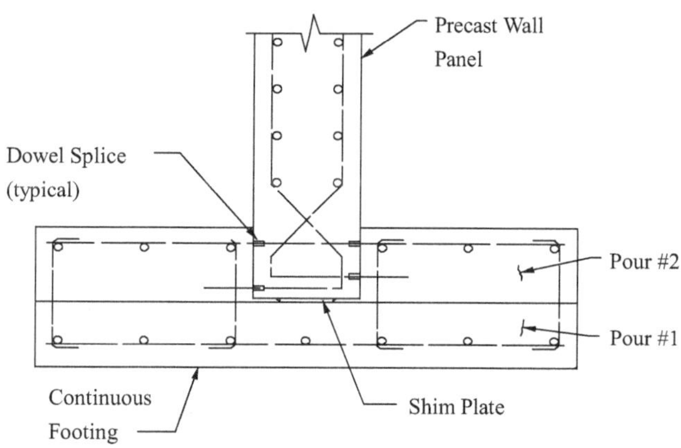

Figure 7-9. Precast panel connection to a foundation.

Cast-in-place construction may be totally reinforced concrete or may be a combination of concrete or steel frames with cast-in-place or precast walls. Shear wall details should be developed using the seismic provisions of ACI 318, Chapter 18. Figures 7-12 and 7-13 are typical cast-in-place details.

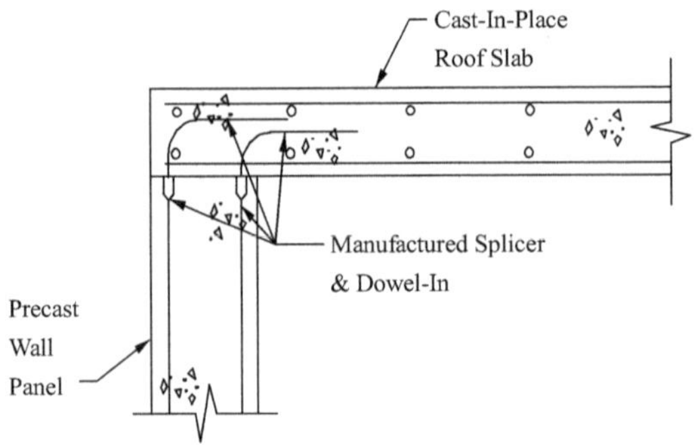

Figure 7-10. Precast panel connection to a roof slab.

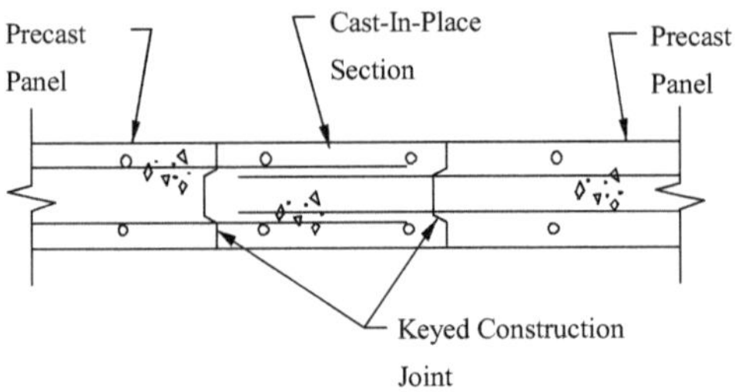

Figure 7-11. Precast panel vertical joint.

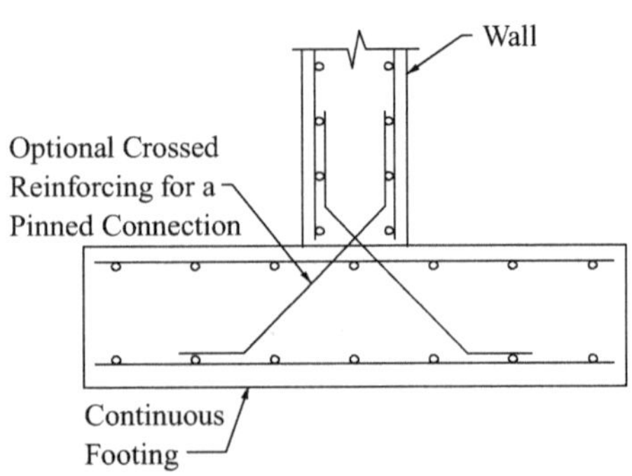

Figure 7-12. Cast-in-place wall to a foundation joint.

7.5 STEEL DESIGN

Applications for structural steel in blast-resistant design include beams and columns for the support of vertical loads, braced and rigid frames for the support of vertical and horizontal loads, and specialized elements such as doors, window frames, decking, and protection for duct openings. For lower blast loads, steel siding supported by steel girts can be used.

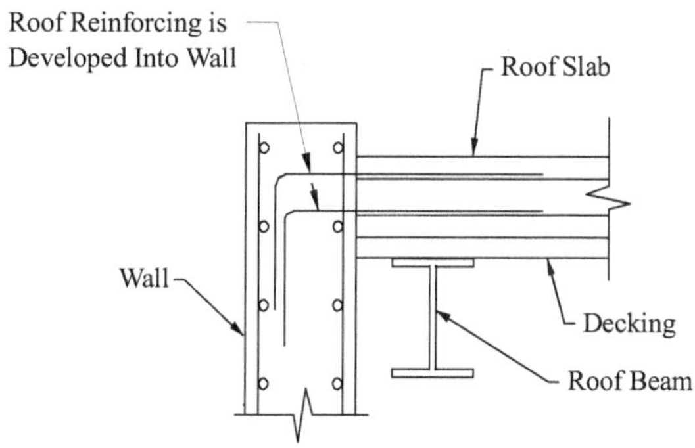

Figure 7-13. Cast-in-place wall to a roof slab joint.

Structural steel has the advantage of quick assembly at the jobsite. Specialized elements, such as doors, are usually delivered in one piece ready for installation into concrete formwork or into the building framing. Being a factory produced material, steel has well-controlled and predictable strength and post-yield properties. Unlike concrete, steel has good tensile as well as compressive strength.

There are several disadvantages of structural steel in blast design. The most significant is the inherent slenderness of steel and the possibility of premature local or general buckling. A less significant disadvantage is that steel siding has a lower resistance to projectile penetration. In addition, steel framing has less inertial resistance compared to reinforced concrete.

7.5.1 Design Principles

The *AISC Specification for Structural Steel Buildings* (AISC 360) (AISC 2016), is used as the basis for structural properties that should be adjusted for blast-resistant design. The resistance of structural steel elements is computed using the

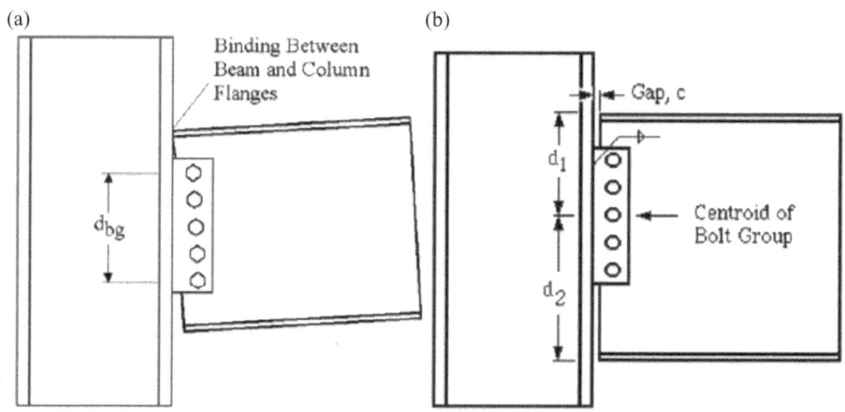

Figure 7-14. Single shear plate connection: (a) binding of a beam, (b) critical dimensions for binding.

dynamic material strengths given in Chapter 5, Section 5.4. Strength reduction factors are not applied (i.e., $\phi = 1.0$) to load cases involving blast. The resistance of structural steel elements is computed using plastic analysis techniques and seismic detailing provisions.

Slenderness considerations are of particular importance to the ductility of structural steel members. Steel, compared to other building materials used in blast design, is considerably thinner, in terms of both the overall structure and the components of a typical member cross section. As a result, the effect of overall and local instability upon the ultimate capacity is an important consideration. Width-thickness provisions must be applied not only to the extent that a full plastic capacity can be achieved, but to the extent that higher ductility ratios can also be safely reached. The limiting width-to-thickness ratios, from Table D1.1 of *Seismic Provisions for Structural Steel Buildings* (AISC 341) (AISC 2016), are used for this purpose.

7.5.2 Supplementary Design Requirements

In addition to AISC 360 (AISC 2016b) requirements, the following should be considered for blast-resistant design:

1. Substitution of Higher Grades of Steel: Substitutions of higher grades of steel is typically not recommended. Higher grades of steel possess less effective resistance-deflection curves, may alter the relationship between flexural and shear capacity, and tend to increase the dynamic reaction that supporting members must resist. Mild steels are preferable because they provide more ductility and are mass-produced.
2. Cold-Formed Steel: *North American Specification for the Design of Cold-Formed Steel Structural Members* (AISI S100) (AISI 2016b) can be used for blast design with several adjustments. *North American Standard for the Seismic Design of Cold-Formed Steel Structural Systems* (AISI S400) (AISI 2015) can also be used for blast-resistant design.
3. Diaphragms: In the design of walls to resist blast pressure loads, it is usually assumed that the walls are supported at opposite sides for one-way slab design or supported at four sides for two-way slab design. Therefore, the roofs or the floors should be designed adequately as diaphragms to resist the in-plane loads and transmit them to the resisting shear walls or other lateral resisting systems.

   The most common type of steel diaphragm is a cold-formed, corrugated floor or roof deck that transfers lateral loads to shear walls or braced frames. Very little published technical guidance exists pertaining to the design of diaphragms for severe loading conditions. The recommended procedure is to design metal diaphragms elastically using conventional design methods outlined in AISI S310 (AISI 2016a). Refer to AISI D310 (AISI 2017) for examples on the design of steel diaphragms.

Special considerations for attaching roof decking to the structural frame are listed in UFC 3-340-02, Sections 6-17 through 6-22 (US DoD 2008b), and in NFEC (1986), Section 5.4, for blast loading conditions. These considerations include items such as material specifications, minimum recommended rib depth and sheet metal gauge, side lap requirements, and fastener details. The emphasis is on providing connections having adequate strength to secure the roof deck under combined in-plane and normal loads.

In addition to in-plane loads, roof diaphragms are also subjected to normal positive overpressures and, to a less severe extent, normal negative pressures. The time lag between in-plane and normal loads may be taken into account in the design. Although Equation (7-2) could be used for this load interaction, separate structural bracing members are normally added to transfer lateral wall reactions. Refer to AISI S310 (AISI 2017) for the design of profiled steel diaphragm panels.

4. Connection Design: To maximize the plastic response, the connection must not control the capacity of the member. Preferably, a moment connection will force a plastic hinge away from the connection and into the member. Connection strength is determined through AISC 360 (AISC 2016) design methods. Ductility requirements are implemented through the use of appropriate connection details.

   Both welded and bolted-type connections are used in rigid and semirigid constructions. There is no particular advantage of using one type over the other with regard to joint performance under blast loading conditions. Because plastic hinges are likely to be formed at member connections, special connection details require careful consideration of the effects of possible stress concentrations. Sharp corners and weld details prone to undercutting should be avoided. AISC 360 fatigue criteria should be consulted for additional information.

   Some insight on what types of details should be used or avoided can be obtained by referring to AISC 341 (AISC 2016). Additional discussions of the basic design concepts for structural steel connections are given in Chapter 6 of TR 4837 (US Army Picatinny Arsenal 1975) and Chapter 5 of UFC 3-340-02 (US DoD 2008b).

   Detailed evaluations of connection ductility are usually not performed. However, in some special cases, it may be necessary to evaluate moment versus rotation characteristics. Theoretical methods for predicting connection behavior, as well an electronic database of actual test data, are available from Chen and Toma (1994). Useful information on moment versus rotation relationships for various types of connections can also be obtained from Committee 43 (1993) and ASCE 41 (ASCE 2006).

   For secondary beams with shear connections, the support rotation limit for high damage response is 12 degrees, as given in Chapter 5, Table 5B-2. The engineer should consider the rotational capacities of the shear connection and ensure that the shear connection can accommodate the expected rotational limit of the support. From Mander et al. (2014), it is learnt that deep single shear tab connections could be prone to binding at support rotations less than the high damage response limit of 12 degrees. In addition, large tension forces could develop within the connection under large rotations (Figure 7-14).

5. Cladding: Cold-formed light gauge sheet metal panels are a common cladding material used in petrochemical buildings. Prefabricated buildings with metal siding and roof deck panels are quite common in petrochemical facilities. These are used only in low blast pressure applications because of premature buckling of the relatively thin webs. Reducing girt and purlin spacing is common to increase the blast capacity of these panels.

   AISI S100 (AISI 2016b) and AISI S400 (AISI 2015) are used as the basis for blast-resistant design. Strength reduction factors are not applied (i.e., $\phi = 1.0$) to load cases involving blast. ASTM A653 (ASTM 2022) is the widely used material by cold-formed steel fabricators. Properties of steel panels can be found from manufacturer catalogs. It is also to be noted that the section properties of cold-formed steel panels will change with the increase of load intensity. As the load increases beyond the level of local buckling, properties like area and moment of inertia decrease and deflection increases. For large deflections, the steel panel acts as a membrane in tension. Therefore, care must be exercised in selecting the proper section for the anticipated load. End-panel connection detailing is important to develop end-span membrane action.

   The resistance of a cold-formed steel panel is computed using dynamic increase factors given in Chapter 5. Chapter 5 also suggests a factor of 0.9 in computing resistance in flexure and provides the necessary equations.

   The primary failure mechanisms encountered in cold-formed steel panels are bending and shear. Care must be exercised to preclude shear failure by increasing the span length, etc. Because cold-formed steel panels will usually have thin webs, web crippling must be prevented by providing sufficient bearing area.

   Response limits are given in Chapter 5. Use low response limits when tension membrane action is not present. Use high response limits when tension membrane action is permitted and steel panel end connections are properly designed.

7.5.3 Failure Mechanisms

Ductility limits for structural steel members are established such that gross member collapse caused by the failure of the member itself or its connections is precluded. It is presumed that local and gross member instabilities are prevented by providing adequate bracing and stiffeners. Shear failure modes are also to be precluded by design. Determination of failure mechanisms and corresponding capacities for flexural members and beam columns are adequately covered by AISC 360 (AISC 2016).

Connections of structural steel members are usually designed to develop the full strength of the member. However, for certain members using the full strength of the member to analyze the connection may be overly conservative if the peak response is below the elastic limit. In these cases, the engineer may alternatively use the calculated dynamic reaction with a sufficient factor of safety in lieu of the reaction from the full strength of the member. With regard to ductility evaluations for connections, explicit checks are usually not made. It is presumed that satisfaction of the gross member displacement ductility criteria ensures the integrity of the member connections.

7.5.4 Connection Details

The enhancement of prefab metal buildings may be achieved by using closer spacing for the building frames and girts and combining sections of the standard AISI cold-formed shapes to achieve symmetry.

Oversized washers are typically used to secure the cladding to the frames to minimize tearing under the effects of blast or rebound structural response and loading (i.e., negative phase). Figure 7-15 illustrates the use of oversized washers. An alternative is to use conventional plug or puddle welds at spacings required to meet the load conditions.

Most details for metal clad construction are not uniquely influenced by blast-resistant design. For steel frame buildings, appropriate AISC steel details used for plastic design methods should be used. The attachment of the siding and roofing requires special attention and the details shown in Figure 7-13 are applicable.

7.6 REINFORCED MASONRY DESIGN

Masonry, both reinforced and unreinforced, is a common construction material in petrochemical facilities. However, unreinforced masonry is inappropriate in blast-resistant design because of its limited strength and its nonductile failure mechanisms, especially when used as load-bearing walls. Reinforced masonry walls with independent structural framing for vertical loads are commonly used in blast-resistant design, but they may be used as load-bearing walls if *P*-delta effects are considered.

The blast capacity and ductility of reinforced masonry walls are much lower than can be achieved with reinforced concrete of comparable dimensions. The lower capacities are attributed to the limited available space for placing steel reinforcing, low compressive strength of the masonry, and limited mortar bond strength.

7.6.1 Design Principles

TMS 402/602 (TMS 2016) is used as the basis for structural properties that should be adjusted for blast-resistant design. The resistance of masonry elements is computed using the dynamic material strengths given in Chapter 5, Section 5.4. Strength reduction factors are not applied (i.e., $\phi = 1.0$) to load cases involving

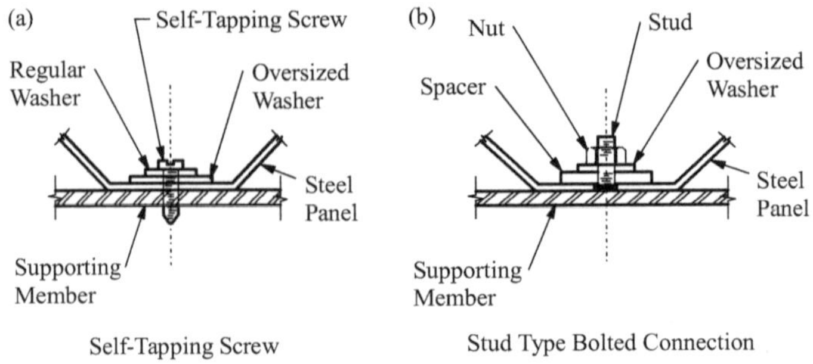

Figure 7-15. Use of oversize washers to connect cladding to framing.

blast. In addition, strength design principles for reinforced masonry are well-documented in many texts such as Schneider and Dickey (1987). Ductility is achieved by adhering to TMS 402/602 detailing provisions for high seismic zones.

7.6.2 Supplementary Design Requirements

Design requirements corresponding to TMS 402/602 seismic criteria are used in blast-resistant design of masonry structures, with some minor adjustments:

1. Interaction: Interaction between in-plane and out-of-plane loading effects is considered by using Equation (7-2).
2. Shear Walls: As is the case for reinforced concrete, it is not usually practical to achieve flexural failure modes for reinforced masonry shear walls. However, the use of shear reinforcing in the form of horizontal joint reinforcing does provide some limited ductility.

7.6.3 Failure Mechanisms

The failure mechanisms of interest in reinforced masonry wall elements include flexural, transverse shear, in-plane shear, and in some cases, combined axial compression and flexure. Buckling failure modes of compression elements and connection failures are to be avoided.

7.6.4 Connection Details

Masonry should always be reinforced, and details typically used for reinforced masonry construction are available.

However, one additional requirement for blast-resistant design should be considered. The presence of negative pressures and rebound forces require that wall-to-frame connections be provided to ensure proper transfer of these outward acting forces. Figure 7-16 shows an application of anchor straps to handle rebound forces.

7.7 FOUNDATION DESIGN

The lack of validation in blast design of foundations, coupled with an absence of any real known examples of direct blast loading causing foundation failure,

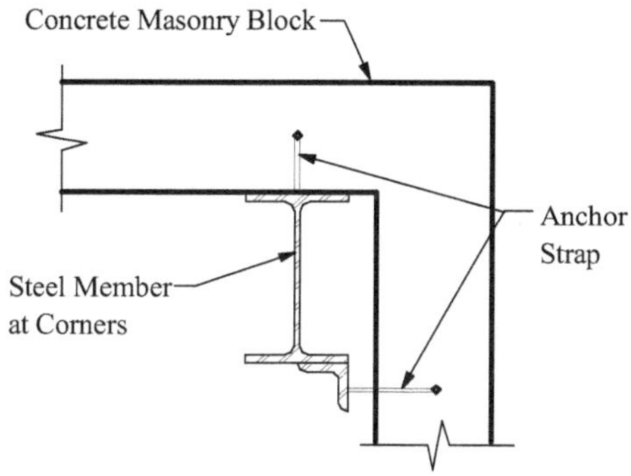

Figure 7-16. Masonry anchors.

indicates that blast loading typically does not govern a conventional foundation design. In many design cases, the duration of blast loads is too short to activate the soil movement that would result in structural failure. Typical building foundations designed for environmental loads are expected to provide satisfactory performance in most cases.

The engineer should consider the foundation when designing or analyzing a building for blast response. However, it is typically only under unique structural configurations or conditions that blast design of foundations should be considered, including (but not limited to) the following:

- Near field blast loads.
- When evaluating concentrated load from a buttress, where performance directly relies on the adequate support of the foundation during blast loading.
- Elevated structures or buildings, where the overall structural stability could be compromised by excessive foundation deformation or settlement.
- Basement-supported structures, where there are similar stability concerns as elevated structures. These should include blast surcharge loads.
- Multistory buildings and blast-resistant modules (BRMs)resisting high blast loads.

A detailed blast-resistant design of foundations may not be warranted for several reasons. First, there are no known instances of foundation failures in actual blast events. Second, estimating geotechnical parameters under very short duration dynamic loading is very difficult and carries large uncertainties. Third, the actual response in the foundation may be minimized from the large amount of inertial mass activated and the absorption of blast energy through inelastic deformations in the superstructure, depending on the foundation type.

Foundation design for conventional loading results in adequate performance in most cases. For information on foundation analysis for blast loading, the reader is referred to the previous edition of this report.

A suitable connection between the foundation and the superstructure shall always be provided to ensure that the foundation is an integral part of the blast load–resistant system. This includes the design of components such as anchor bolts and shear keys for lateral and uplift blast forces as well as the design of reinforcing for embedment into the concrete foundation. Furthermore, the foundation support conditions should match the assumptions made regarding the component being analyzed. For example, if a fixed connection was assumed at the bottom of a wall, then any rotation of the foundation at the support would result in larger wall deformations than the design predicted.

7.8 DESIGN AGAINST PROJECTILES

If a building design incorporates the "neutral risk" philosophy (Chapter 5, Section 5.6), then external projectiles need not be considered.

If a building is required to resist a projectile impact, a safe standoff distance is important. There are two types of projectiles, primary and secondary. Primary projectiles are caused by the disruptive failure of pressurized equipment or rotating machinery. Secondary projectiles are defined as accelerated unsecured objects that have been picked up by the blast from an explosion. Projectile impact effects on a structural component are usually divided into two categories: local effects and overall response. Local effects are largely independent of the dynamic characteristics of the structure, whereas the overall response primarily depends on the dynamic characteristics of the structure. Both local effect and overall response require the mass and velocity of the projectile. Appropriate methods given in SCI-P-112 (Fire and Blast Information Group 1993), UFC 3-340-02 (US DoD 2008b), UFC 4-023-07 (US DoD 2008a), DOE/TIC-11268 (USDOE 1992), ASCE MOP 58 (ASCE 1980), Kennedy (1976), and Baker (1983) may be used to calculate the speeds of projectiles as a function of blast overpressure.

There is no closed-form solution to calculate the damage from a projectile impact. However, projectile impact can be possibly simulated by numerical techniques with valid material models, for example, the use of FEA methods. It should be noted that if finite-element analysis methods are used to predict the complex and highly nonlinear interaction between a projectile and its target, the results should be carefully reviewed and bounding calculations using empirical methods should be considered. Fragment impact and perforation effects can also be evaluated using semi-empirical techniques, such as

- Design for impact load can be found in Chapter 6 of Baker (1983), "Fragmentation and Missile Effects."
- Reinforced concrete wall design for impact load can be found in Section 8.4 of ACI 376, "Requirements for Concrete Structures for Containment of Refrigerated Liquid Gases" (ACI 2011).
- Design of concrete structures for impulsive and impactive loading can be found in Appendix F of ACI 349, "Requirements for Nuclear Safety-Related Concrete Structures" (ACI 2014).

7.9 BLAST-RESISTANT MODULAR BUILDINGS

Blast-resistant modular (BRM) buildings are used at petrochemical facilities because of their portability, customization, and robust blast resistance. Most commercially available BRMs are metal-framed structures with fully welded steel cladding. The following sections discuss design considerations associated with this construction type. However, other material types may be used to construct BRMs if they are properly designed.

7.9.1 Blast Rating of BRM Buildings

The blast rating of a BRM building should include the blast free-field overpressure, blast positive duration, and response level (damage level). For example, 5 psi–200 ms with a medium level rating indicates that maximum free-field overpressure of 5 psi and positive blast duration of 200 ms are considered in the design of the BRM building and that the acceptance level is chosen to be a Medium Damage Level. It should be noted that reflected overpressure values should be calculated to design the structural wall members and any other members that are exposed to reflected overpressure. Because of their versatility in orientation over time, it is typical to have all sides of a BRM building designed against reflected overpressures.

7.9.2 Design Approach

With fully welded connections between the exterior steel crimped or flat wall panels and the structural frame members, as well as the frame member-to-member connections, modular buildings improve their blast capacity by providing a high level of continuity. Design of welded connections between exterior panels and structural frame members should account for rebound structural response. The engineer should account for response/failure modes in the analysis, including tension membrane effects and plastic strain limitations. Also refer to Section 7.3.

Figure 7-17 shows a finite-element model for a stacked BRM configuration. Doors and other openings can be explicitly modeled as shown.

Although there is an increasing interest of oil and gas production industries and modular buildings into the Arctic region, currently, there are no specific industry standards or guidelines addressing blast-resistant design of BRM buildings under low-temperature conditions. Sari and Paradkar (2014) provided various recommendations for blast assessments of both existing and proposed BRM buildings in the Arctic, or low service temperatures. They also presented a comprehensive overview of the low-temperature effects on the engineering properties of materials that are used, or that can be potentially be used.

7.9.3 Anchorage Considerations

Some BRMs are designed to be permanent installations, with their anchorage designed to resist the total anticipated blast loads.

However, some owners have taken the approach that the foundations and anchorages need only be designed for conventional design loads (e.g., seismic,

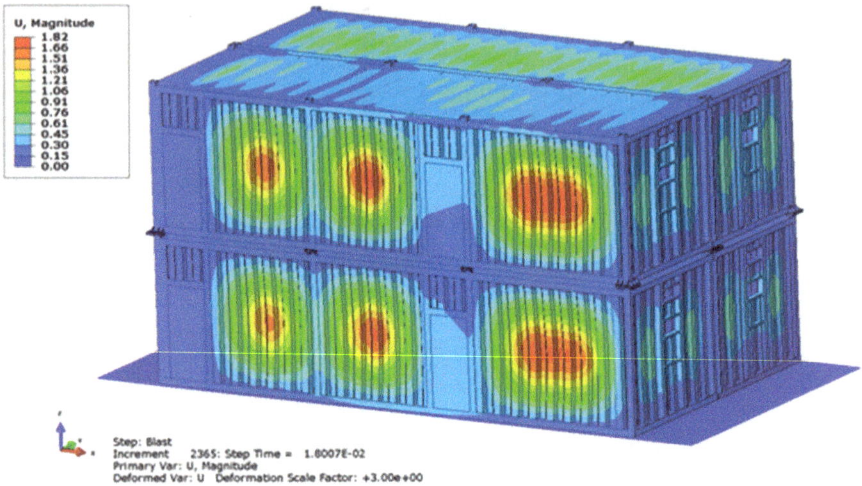

Figure 7-17. Example deformed shape of a stacked BRM configuration.
Source: Courtesy of RedGuard, used with permission.

wind). In this case, the building's anchorages are permitted to fail during a blast event but are designed to remain intact under conventional design loads. For this case, the breakaway anchorage should not induce a tipping response in the building during structural response to blast loads. Alternatively, the building can be designed to be completely unanchored (free to slide) for both blast and other loading effects, subject to the local building official's anchorage requirements for conventional loading. Whether a building should be anchored for blast, anchored for other loads (but not for blast), or completely unanchored (and free to slide) depends on the anticipated use of the building, whether potential down time is acceptable following a blast event, the amount of flexibility in the utility connections (power, water, wastewater, gas) and, most important, the owner's tolerance to occupant risk. Owners should make this decision on the basis of many factors, including risk, safety, cost, magnitude, and probability of blast.

If the building is unanchored (free to slide) for blast loading, or only anchored for conventional loads, the maximum sliding displacement, velocity, and acceleration of the building can be estimated using impulse-momentum first-principles, simplified numerical integration methods or FEA methods. The contents and personnel within the building should be evaluated for these actions. In this case, the structural movement may result in impact/damage to attached utilities and building contents, as well as the possibility of injury to personnel, because of interaction with the structure, fixed equipment, and internal moving objects (typically unrestrained and falling objects). In such interactions, the critical components of motion can be local accelerations, velocities, and displacements that govern local forces and energies of impact, including the propensity to topple over and fall. Permanent fixtures and equipment should be designed to withstand the calculated local building motions as a result of blast loadings. Anchorage and restraint techniques for nonstructural items typically follow those used for earthquake design [FEMA 412 (FEMA 2002), FEMA 413 (FEMA 2004b), FEMA 414 (FEMA 2004a), SMACNA (2008)]. Attached utilities should also be designed to accommodate expected movements or fail in a safe manner.

As stated, the decision as to whether these displacements, velocities, and accelerations are acceptable to an owner depends on the anticipated use of the building, whether potential down time is acceptable following an event, flexibility in the utility connections and, most important, the owner's tolerance to risk. Because the building will act as an external pressure barrier, the design of internals needs only consider the effects of movement. Definitive evaluation criteria for interactions with personnel are not available, but criteria do exist (Baker 1983). Additional criteria for projectile impact (such as falling objects) are also available (Committee for the Prevention of Disasters 1992). Note that architectural and nonstructural components may become debris hazards. UFC 3-340-02 (US DoD 2008b) provides some guidance on the tolerance of mechanical and electrical equipment as well as personnel. For sensitive and critical equipment that must function during and after the event, verification by shock testing with the induced motions consistent with expected structural motions may be warranted.

In other words, BRMs can be designed to be free to slide if a sliding analysis shows that the risk is acceptable by the owner. Sari et al. (2016) discussed the individual risk calculation methods on the basis of various key response parameters, including building floor acceleration, head acceleration, jerk, and impact velocities. UFC 3-340-02 suggests a maximum horizontal acceleration of 0.5 g to provide protection against explosion ground-shock effects for nonrestrained personnel (standing, sitting, or reclining). However, as stated previously, this is an owner decision and is specific to the building being designed. In all cases, buildings that are not anchored for blast must have a high margin against overturning, but the propensity to uplift should be verified in stacked configurations.

7.9.4 Installation Considerations

For a modular blast-resistant building to be fully effective, proper installation and use is required. Even with appropriately designed structures, improperly installed BRMs may not fully protect occupants. For example, modular blast-resistant buildings should always be placed on a level, flat surface. Dunnage or leveling blocks should not be used for permanent or temporary BRM installations. Also, BRMs should not span drainage ditches, or low areas, as this provides blast leakage pressures under the BRMs, thereby requiring blast design of the floor framing. This, in turn, increases the chance of sliding, uplift, and overturning.

When available space is limited, BRMs can be stacked. However, certain considerations must be made when stacking buildings.

- Connections: Connections between levels need to be properly designed by a qualified blast engineer.

- Gaps between Levels: It is recommended that closure plates be provided for potential gaps between levels to avoid blast leakage pressures under the top-level BRMs.
- Height: It is recommended that stacking be limited to two stories.
- Equal Module Numbers: An equal number of module sections should be used on the first and second stories (e.g., 2×2 configuration for a four-module BRM building) to limit amplified reflective pressures in re-entry corners.
- Module Widths: Stacked configurations should be a minimum of two-module sections in width to limit overturning demands.

7.9.5 Connection Design

If a multimodule building is used, connections between adjacent modules are a critical aspect to consider. Racking of adjacent modules should be avoided.

7.9.6 Transportation and Lifting Analysis

Interior modules of multimodule complexes tend to be flexible and, if this is the case, transportation and lifting analyses may be warranted to prevent possible damage to interior nonstructural components during these phases of the life cycle. Temporary bracing may be required during transportation and lifting operations.

7.9.7 Temporary Buildings

BRMs may be used as temporary structures and may be moved from site to site over their lifetime. If this is the case, care should be taken to ensure that accumulated damage to the building's primary structural framing caused by rugged service conditions (e.g., corrosion) and repeated transportation does not result in a compromised blast-resisting system. Therefore, routine inspections should be performed to this end.

References

ACI (American Concrete Institute). 2011. *Code reqs for design and construction of concrete structures for containment of refrigerated liquefied gases*, ACI 376. Farmington Hills, MI: ACI.

ACI. 2014. *Code requirements for nuclear safety-related concrete structures and commentary*, ACI 349. Farmington Hills, MI: ACI.

ACI. 2019. *Building code requirements for structural concrete and commentary*, ACI 318-19 and ACI 318R-19. Farmington Hills, MI: ACI.

AISC (American Institute of Steel Construction). 2016. *Specification for structural steel buildings*, AISC 360. Chicago: AISC.

AISC. 2016. *Seismic provisions for structural steel buildings*, AISC 341. Chicago: AISC.

AISI (American Iron and Steel Institute). 2015. *North American standard for seismic design of cold-formed steel structural systems*, AISI S400. Washington, DC: AISI.

AISI. 2016a. *North American standard for the design of profiled steel diaphragm panels*, AISI S310. Washington, DC: AISI.

AISI. 2016b. *North American specification for the design of cold-formed steel structural members*, AISI S100. Washington, DC: AISI.

AISI. 2017. *Design examples based on AISI S310-16*, AISI D310-17. Washington, DC: AISI.

ASCE. 1980. *Structural analysis and design of nuclear plant facilities*, ASCE MOP 58. Reston, VA: ASCE.

ASCE. 1985. *Design of structures to resist nuclear weapons effects*, ASCE MOP 42. Reston, VA: ASCE.

ASCE. 2006. *Seismic rehabilitation of existing buildings*, ASCE 41. Reston, VA: ASCE.

ASTM. 2022. *Standard specification for steel sheet, zinc-coated (galvanized) or zinc-iron alloy-coated (galvannealed) by the hot-dip process*, ASTM A653/A653M-20. West Conshohocken, PA: ASTM.

Baker, W. E. 1983. *Explosion hazards and evaluation*. New York: Elsevier Scientific.

Biggs, J. M. 1964. *Introduction to structural dynamics*. New York: McGraw-Hill.

Chen, W. F., and S. Toma, eds. 1994. *Advanced analysis of steel frames; theory, software and applications*. Boca Raton, FL: CRC Press.

Committee 43 (Council on Tall Buildings and Urban Habitat Committee 43). 1993. *Semi-rigid connections in steel frames*. New York: McGraw-Hill.

Committee for the Prevention of Disasters. 1992. *Methods for the determination of possible damage to people and objects resulting from releases of hazardous materials (Green Book)*. Den Haag, The Netherlands: Committee for the Prevention of Disasters.

FEMA (Federal Emergency Management Agency). 2002. *Installing seismic restraints for mechanical equipment*, FEMA 412. Washington, DC: FEMA.

FEMA. 2004a. *Installing seismic restraints for duct and pipe*, FEMA 414. Washington, DC: FEMA.

FEMA. 2004b. *Installing seismic restraints for electrical equipment*, FEMA 413. Washington, DC: FEMA.

Fire and Blast Information Group. 1993. *Interim guidance notes for the design and protection of topsides of structures against explosions and fire*, SCI-P-112. Ascot Berkshire, UK: Fire and Blast Information Group.

Kennedy, P. P. 1976. "A review of procedures for the analysis and design of concrete structures to resist missile impact effects." *Nucl. Eng. Des.* 37 (2): 183–203.

Mander, T. J., L. Magenes, and T. Anderson. 2014. "Steel shear tab connection performance at large rotations during blast loading." In *Proc., ASCE/SEI Structures Congress 2014*. https://ascelibrary.org/doi/10.1061/9780784413357.023

NFEC (Naval Facilities Engineering Command). 1986. *Blast resistant structures, Naval Facilities Engineering Command, design manual 2.08*. Alexandria, VA: NFEC.

Oswald, C. J., and M. Bazan. 2014. "Performance and blast design for non-load bearing precast concrete panels blast." In *Proc., ASCE/SEI Structures Congress 2014*. https://ascelibrary.org/doi/10.1061/9780784413357.014.

Sari, A., U. Azimov, and W. Sahasakkul. 2016. "Risk-based blast design for safer occupied buildings in industrial facilities." In *Proc., 66th Canadian Chemical Engineering Conf.* (CSChE), Quebec City, PQ: Chemical Institute of Canada.

Sari, A., and M. Paradkar. 2014. *Blast resistant design at low temperatures*. Tech. Newsletter 064. Aberdeen, UK: FABIG.

Schneider, R. R., and W. L. Dickey. 1987. *Reinforced masonry design*. 2nd ed. Englewood Cliffs, NJ: Prentice-Hall.

SMACNA (Sheet Metal and Air Conditioning Contractor's National Association). 2008. *Seismic restraint manual guidelines for mechanical systems*. Chantilly, VA: SMACNA.

TMS (The Masonry Society). 2016. *Building code requirements and specification for masonry structures*, TMS402/602-16. Longmont, CO: TMS.

US Army Picatinny Arsenal. 1975. *Design of steel structures to resist the effects of the explosions*, TR 4837. Dover, NJ: US Army Picatinny Arsenal.

US DoD (US Department of Defense). 2008a. *Design to resist direct fire weapons effects*, UFC 4-023-07. Washington, DC: US DoD.

US DoD. 2008b. *Structures to resist the effects of accidental explosions*, UFC 3-340-02 (formerly TM 5-1300). Washington, DC: US DoD.

USDOE (US Department of Energy). 1992. *Blast and fragment loadings on structures*, DOE/TIC-11268. Washington, DC: USDOE.

CHAPTER 8

Ancillary and Architectural Considerations

8.1 INTRODUCTION

This chapter addresses blast-resistant considerations for doors, windows, and utility openings and also deals with special exterior and interior requirements. These considerations should be jointly addressed by the different discipline members of the building design team. A guiding principle is that the number of size of openings should be minimized. In cases where an opening is necessary, the covering (glazing, door panel, louver) should be designed to resist the applied load and be properly supported by the structure. In certain situations, it may be acceptable to allow the opening covering to fail but to remain attached to the structure so that it does not become debris. Propagation of the blast wave into the building should be closely evaluated for this approach to ensure that adequate protection for building occupants is provided.

8.2 GENERAL CONSIDERATIONS

When there is an opening in the blast-resistant envelope that is not designed for the applied blast load, the blast wave may propagate inside the structure and increase in the interior pressure. The amplitude of the pressure increase is a function of the opening size and the peak pressure and duration of the blast load impinging on the structure. UFC 3-340-02 (USACE/NAVFAC/AFCESA 2014) illustrates a method for calculating the change in pressure inside a building. The pressure increase should be evaluated for what is tolerable for people and nonstructural items, and all interior components should be designed accordingly.

If a maximum permissible interior pressure is not specified in the building's design criteria, the design team must ensure that each opening either completely blocks the interior propagation or that the effects are suitably mitigated.

The design of the various devices used to protect building openings is a very specialized field. Normally, the detailed design of the different components of doors, window glass and frames, blast valves, and attenuators is performed by the manufacturer on the basis of design criteria provided by the engineer. Typically, a performance specification provided by the engineer defines the applied blast load and the performance criteria. This specification will also define the methodology for demonstrating acceptable performance in the form of calculations and test data. It is important that the specification define the permissible variation of blast load or component size with test articles. Less often, the engineer will develop the detailed design and prescriptive specification.

8.3 DOORS

This section deals with door design for resistance against an accidental explosion. The types and applications of blast-resistant doors are discussed, and design approaches are provided.

Conventional (non-blast-resistant) doors, due to their functional requirements and associated hardware limitations, are a potential weak link in blast-resistant design. Because doors are likely to be the largest opening into a building, they can create the most significant potential source of blast wave propagation into the structure. This is especially true for buildings that have rollup doors such as shops, warehouses, and similar facilities. The engineer should develop the performance specification to meet the required protection levels for building occupants. It is important to consider emergency egress and potential shelter-in-place requirements when establishing the performance specifications. The protection requirements must be balanced with operability requirements. If doors are not designed with functionality considerations, building occupants may take measures to circumvent protection features such as closers and interlocks. Ergonomic aspects of doors must also be considered. Doors that are much heavier than conventional doors may require automatic operators to prevent injury and ensure that doors are properly closed.

Blast doors are available from a limited number of manufacturers with a wide range of blast capacity. Low-range blast doors (refer to Section 8.3.1) are similar to conventional hollow metal doors, but have special detailing to increase the blast capacity of the doors. Mid-range and high-range blast doors are designed to provide increased levels of blast load as the names imply. In general, blast-resistant doors are heavier than conventional hollow metal doors, and the higher the blast capacity, the heavier the door will be. The additional weight increases inertial resistance to movement and causes these doors to move more slowly in manual operation than personnel are likely expecting, making them more difficult to operate. Therefore, it is important to choose an appropriate door for a given situation and avoid using heavier doors except when absolutely required. An alternative is the use of lightweight blast doors, which are principally composed of structural fiberglass. These doors can have significantly greater deflection and lower blast capacity than commonly available steel doors but may be appropriate for low blast demand situations.

It may be appropriate to achieve a balance in functionality and protection through building layout. Limiting personnel occupancy near door openings and providing additional exits may be viable strategies if ingress of pressures can be accommodated through interior design.

Rollup doors are a particular challenge for blast resistance. These doors typically have very little blast resistance and limited restraint. Consequently, they can create significant debris hazards to personnel, although observations after accidental explosions show that rollup doors tend to swing open without failing. This response may be acceptable for blast design of areas with low occupancy. The use of high-wind and hurricane-rated rollup doors can provide some resistance to blast loads and offer better protection than conventional doors. Increased resistance can be created by using retainers and guides are designed to allow the slats or blades to act as a membrane. The performance should be properly verified by calculation, simulation, and/or testing. Blast-mitigating and blast-resistant rollup doors have been under development and are available in the marketplace. These designs use wind-rated door systems with improved connections to enhance ductility and blast performance.

8.3.1 Definitions

Throughout this section, the terms *low range, midrange, and high range* are used to distinguish varying levels of blast pressures applied to blast-resistant doors. The ranges come from the research of product literature of several blast-resistant door manufacturers. The terms are loosely defined as follows:

- Low-Range Door: A door designed to withstand an equivalent static pressure that is less than 3 psi (21 kPa).
- Mid-Range Door: A door designed to withstand an equivalent static pressure in the range of 3 to 10 psi (21 to 69 kPa).
- High-Range Door: A door designed to withstand an equivalent static pressure that exceeds 10 psi (69 kPa).

For elastic behavior, an applied static force is twice that of an applied dynamic force of infinitely long duration.

It is typical for manufacturers to have several models in each range. The doors may vary significantly in terms of material, thickness, restraining hardware, frame profile, and anchorage.

It is important to understand that the static capacity of a blast door is not equal to the peak blast pressure that the door can resist. The static capacity is only important to show the relative strength of the door. A dynamic analysis or blast testing is necessary to determine whether a given blast door can resist a design blast load.

8.3.2 Performance Limitations of Commercial Industrial Doors

The average industrial personnel door is a hollow steel or composite door typically 1-3/4 in. (44 mm) thick with an 18-gauge steel facing. A composite door consists of a center sound-deadening noncombustible core, usually of polyurethane foam or slab. Light-gauge vertical reinforcement channels are used in hollow metal panels to add strength and rigidity. Heavier doors (14-gauge and 16-gauge) can provide some additional blast resistance.

These doors may be acceptable for withstanding applied blast overpressures in the 1.0 psi (7 kPa) to 2.0 psi (14 kPa) range. Limited testing has indicated that the actual performance is better than expected by analysis; however, most of the test data available have been conducted in the impulsive region of the *P–I* diagram, and performance in the pressure-sensitive region is less certain. PDC TR 11-01, *SDOF Resistance Function Methodology Study-Door Study* (USACE 2011), discusses static and shock tube test data on conventional 16-gauge and 14-gauge doors.

A *P–I* diagram for 18-gauge steel doors developed from test data is shown in Figure 8-1. *P–I* diagrams for 16-gauge and 14-gauge doors, respectively, are shown in Figures 8-2 and 8-3. The *P–I* diagrams in these figures are applicable for 3 × 7 ft personnel doors that are seated into the frame by the applied blast load. The data show similar performance for 14- and 16-gauge doors for the impulsive regime in which the tests were performed. The performance-level IV/V impulse asymptote is shown slightly lower for the 14-gauge door, but the two test data points would support a higher-impulse asymptote of 90 to 100 psi·ms for the 14-guage door. A more significant difference in performance for loads in the pressure regime might be expected. Testing has shown that inadequate connections between conventional door frames and the supporting structure with adequate capacity will fail and allow the whole door and frame to be thrown into the protected space of the building. The door frame connections to the supporting structural component should have a total load capacity equal to the ultimate resistance of the structural member multiplied by the supported door area and increased by a dynamic increase factor of 1.2 for blast loading (refer to Chapter 5, Section 5.5.4). The connections of conventional door frames to supporting wood studs failed in numerous shock tube tests that produced the data for the *P–I* diagrams; therefore, conventional 14- and 16-gauge doors that are expected to resist blast loads should not be supported by wood studs.

When the initial direction of the blast wave tends to seat the door into the frame, these doors are susceptible to localized deformations or component failure that could render the door inoperable. Increasing the door rebate can improve the door performance when significant deflection of the door occurs. If the magnitude of the blast is significant enough, catastrophic failure of the entire door assembly could occur. Rebound forces can also create concern. These doors are equipped with standard builder's hardware. This type of hardware has severe limitations for withstanding forces resulting from a blast. The forces created by the blast often exceed the load ratings of the most commonly used latch sets and hinges. Knowing this, one realizes that there is little, if any, factor of safety when an untested common door is accepted as a suitable alternate to a certified blast-resistant door.

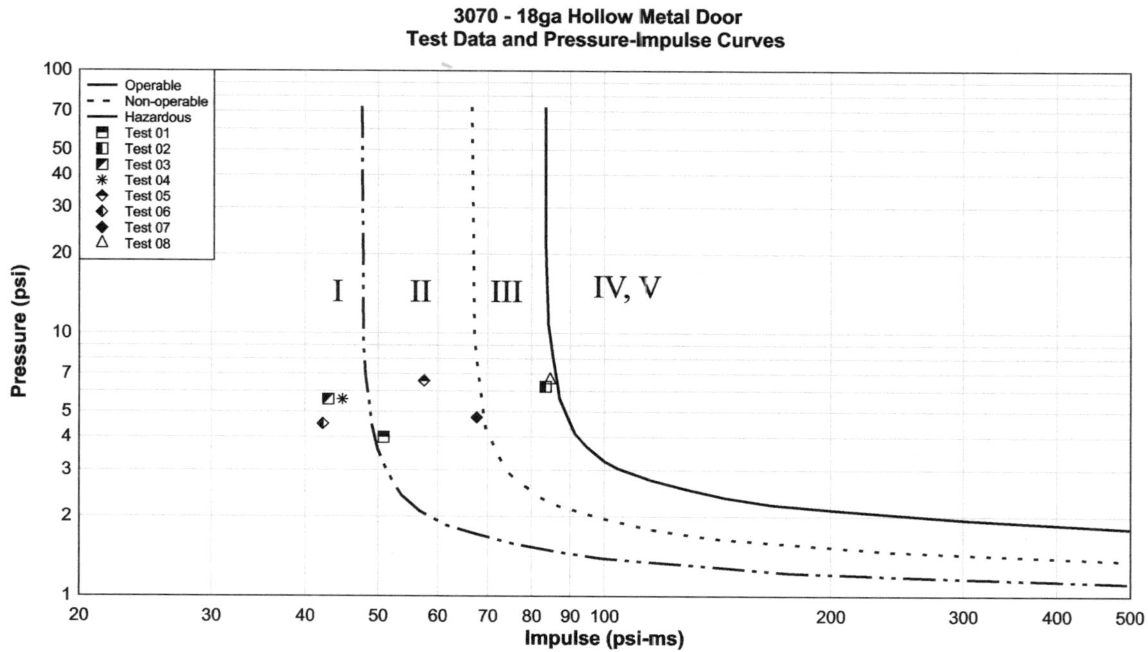

Figure 8-1. P–I diagram for 18 Ga hollow metal doors.

Source: Hurd et al. (2014) with permission from ABS.

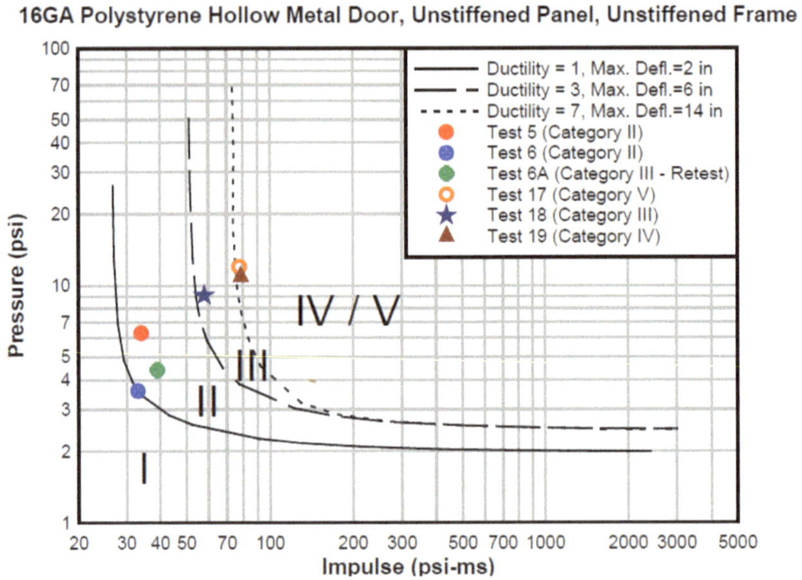

Figure 8-2. P–I diagram for 16 Ga hollow metal doors.

Source: Lowak et al. (2011) with permission from Baker Engineering and Risk Consultants.

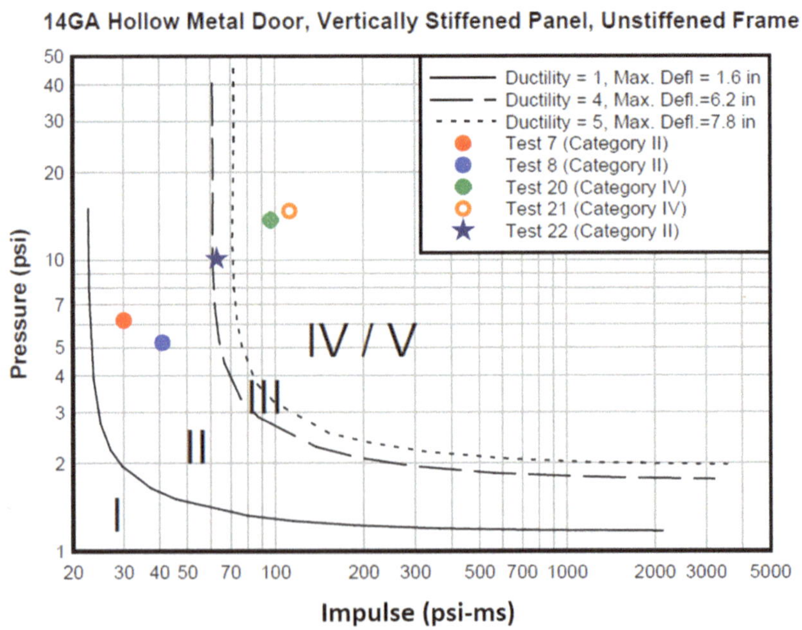

Figure 8-3. P–I diagram for 14 Ga hollow metal doors.

Source: Lowak et al. (2011) with permission from Baker Engineering and Risk Consultants.

8.3.3 Guidelines for Blast-Resistant Door Design

Depending on the desired end-use of the door, guidelines for acceptance are classified into categories. The following categorization developed by the Process Industry Practices (PIP), has been used by owner/operators of petrochemical facilities.

Category I—(Operable) The door is to be operable after the loading event and pre-established design criteria for stress, deflection, and the limitation of permanent deformation have not been exceeded. A ductility ratio of 1.0 or less (elastic range) and a door edge rotation of 1.2 degrees should be specified. This category should be specified when the door may be required to withstand repeated blasts or when entrapment of personnel is of concern and the door is a primary exit to the building. Design of the door section, hardware, and support framing must include provisions for rebound forces determined from a dynamic analysis of response. If continued occupancy after an explosion is required, the door may need to remain fire resistant after the application of the blast load. Because this door is to remain operable, performance under blast loading should be demonstrated by conducting tests.

Category II—(Operable) The door is to be operable after the loading event, but significant permanent deformation to the door is permitted. A ductility ratio in the range of 2 to 3 and a door edge rotation of 2.0 degrees is recommended. The door must remain operable and this category should be specified when entrapment of personnel is a concern. Design of the door section, hardware, and support framing must include provisions for rebound forces determined from a dynamic analysis of response. Similar to Category I, this door type may be required to be fire resistant following the blast loading. Because this door is to remain operable, performance under blast loading should be demonstrated by conducting tests.

Category III—(Inoperable) Noncatastrophic failure is permitted. The door assembly remains in the opening. No major structural failure occurs in the door panel structure, the restraining hardware system, the frame, or the frame anchorage that would prevent the door assembly from providing a barrier to blast wave propagation. However, the door will be rendered inoperable. A ductility ratio in the range of 5 to 10 and a door edge rotation of no greater than 8 degrees is recommended. This category should be specified only when entrapment of personnel is not a possibility.

Category IV—(Inoperable) Outward rebound force and resulting hardware failure is acceptable. A ductility ratio in the range of 10 to 20 and a door edge rotation of no greater than 12 degrees is recommended. The design objective is to prevent the door from becoming hazardous debris.

At least one door must remain operational after the VCE event.

Although the positive-phase overpressure load usually controls the design, negative-phase loads should be evaluated to ensure adequate performance. The door support conditions will be quite different in the negative-phase-load direction. In addition, rebound forces may act in conjunction with negative-phase loads and create unexpected demand on the door and frame.

8.3.4 Coordinating Efforts with a Blast-Resistant Door Manufacturer

Because blast door designs interfaces with other structural components of a facility, it is wise to approach the preliminary design of the blast-resistant door system early in the design stage of the project. As a minimum, door manufacturers will need the following information to furnish pricing and complete detail design of the doors:

1. Blast-resistant door frames may be anchored into surrounding walls by several methods. They may be cast in place in new concrete, bolted in with concrete expansion anchors, welded to an existing steel embed or structure, or bolted to an existing structure. What method of anchorage will be used?
2. What is the wall's rough opening size and the door's jamb opening size?
3. Furnish the same information relative to peak incident overpressure, peak reflected pressure, and blast load duration that has been used for the structural components.
4. Does the direction of the blast force act to seat the door into the frame or unseat the door from the frame?
5. Must the door's material remain in the elastic or elasto-plastic range? Is permanent deformation permitted?
6. What is the limit for the ductility ratio?
7. What is the total permissible deflection at the midspan of the door panel or the end rotation of the door panel?
8. Must the door be operable after the blast?
9. Furnish information about such architectural requirements as hardware functions, door closers, door opening assists, paint and finish, fire labeling requirements, etc.

The PIP consortium has developed a common specification for blast doors (PIP ARS08390) (PIP 2002) covering design, fabrication, and installation. Data sheets for specifying principal requirements for blast doors are included in PIP STC01018 (PIP 2014).

8.3.5 Testing and Structural Analysis Methods

Most blast door manufacturers opt to perform static load tests on prototype assemblies of low-range blast doors to demonstrate that the assembly will resist the blast overpressure specified. Static tests should be accepted only if the dynamic structural response and dynamic load factors have been considered and the door, frame, and restraining hardware are manufactured using the same materials, dimensions, and tolerances as those in the prototype static test.

It is common practice among manufacturers to substantiate the structural integrity of mid- and high-range blast-resistant doors by design calculations. Calculations supporting the ability of the door to meet performance criteria under the specified blast loading should be supplied to the specifier for review before manufacturing of the door proceeds. The calculations must cover the initial response of the door, rebound, and all secondary items such as stresses in welds and fasteners, local buckling and web crippling in structural members, the structural capacity of the hinges and latches, and frame anchorage to the surrounding structure.

Testing may be used to validate a calculation method to confirm door performance. Alternatively, door performance may be demonstrated solely by testing. Given that doors are typically common sizes, it is feasible to have standard load and performance ranges for standard sizes. However, often, standard designs are tested for blast loads associated with high-explosive hazards. In these cases, the peak pressure can be significantly higher and blast load duration substantially shorter than for petrochemical blast hazards. This makes it difficult to ascertain how a door will perform for lower-pressure, longer-duration loads, and a calculation-based approach may be required. Test data from short-duration load tests can be used to validation the calculation method or model.

ASTM has published ASTM F2247-18, *Standard Test Method for Metal Doors Used in Blast Resistant Applications (Equivalent Static Load Method)* (ASTM 2018). This document describes a method of testing a door under static uniform pressure and provides guidance on the determination of an appropriate equivalent static load for a given design blast load.

8.3.6 Fire Labels and Fire Label Construction

Many blast-resistant door manufacturers can offer 3 h "A" and 1-1/2 h "B" fire labels on low-range and mid-range doors that certify that the construction of the door has been fire-tested by an agency such as Underwriters Laboratories. Few manufacturers offer a fire label on high-range doors. When a door design conflicts with a manufacturer's fire label procedure, often the manufacturer will offer a letter certifying that the doors are fabricated from fire-resistant materials that will not contribute to flame spread. Often, this method is accepted by fire protection authorities on the project; however, the specifier should consult the authorities early in the design stages of the project to verify acceptance.

8.3.7 Delivery Lead Times

Blast-resistant doors are not "off the shelf" items. They are built to order, and manufacturers usually require 6 to 8 weeks after notification of approval of the shop drawings and design data to schedule and fabricate low-range doors, 10 to 12 weeks for mid-range doors, and 12 weeks or more for high-range doors.

8.4 WINDOWS

Historically, ordinary glass windows are not adequate for blast overpressures as low as 0.2 psi (1.4 kPa). Many injuries in explosion accidents result from glass fragments.

Annealed glass is of relatively low strength when compared to tempered glass and tends to fracture into angular fragments. Wire glass is an annealed glass with an embedded layer of wire mesh used as a fire-resistant barrier. Although the wire helps bind fragments, wire glass should be avoided unless required by NFPA considerations.

When windows are required to be blast resistant or blast mitigating, they should be designed to withstand the same blast loads as the walls containing the windows. The engineer should define the structural design criteria and coordinate with the building's architect to ensure the manufacturer's correct interpretation. Unless the design can be adequately justified, large windows should be discouraged for high blast pressures with long durations. There are higher-strength-type glass and glazing materials such as laminated glass, polycarbonate, and plastic interlayer that may be considered acceptable depending on the design overpressures. These materials may be used either by themselves or as components in a composite glazing layup.

The glazing rebate or bite is important for the performance of ductile systems such as laminated glass or polycarbonate. As the glazing lites deform under blast loading, the mode of failure for these ductile window systems involves the release of the lite from the frame. The addition of stops or similar components to increase the bite can substantially increase the blast capacity of the glazing.

Chapter 6 of ASCE's *Structural Design for Physical Security*, MOP 142(ASCE 2021) addresses the various types of glazing materials and structural components of window frames and can be referenced for additional discussion of the topic.

Significant development of blast-resistant windows has increased the options available to mitigate glass fragment hazards for new and retrofit designs. The US Department of Defense (US DoD) and US General Services Administration (GSA) have developed minimum standards for protection of personnel, which have resulted in the development of products based on these requirements. Because the GSA and DoD are concerned with high-explosive terrorist attacks with usually short-duration loads, available products may not have been evaluated for longer-duration loads anticipated in petrochemical facilities. However, these products can provide substantial protection from glass fragments from blast loads caused by vapor cloud explosion, BLEVES, and vessel burst hazards. A typical minimum blast load standard used by the GSA for which many vendors have developed products is a short-duration blast load of 4 psi (28 kPa) with an impulse of 28 psi·ms (193 kPa·ms) and a load duration of 14 ms. A second common load specification for DoD facilities is a peak pressure of 6 psi (41 kPa) with an impulse of 42 psi·ms (290 kPa·ms) and a load duration of 14 ms. Because of the longer duration of vapor cloud explosions, these products may be applicable at lower peak pressures.

Products developed to mitigate glass fragment hazards include laminated glass, blast shades, blast curtains, and safety film in daylight, mechanically anchored, and wet-glazed configurations. In general, laminated glass is used for new blast-resistant window design or replacement window designs. Anchored safety film, daylight safety film with catch bars, and laminated glass replacement are usually used for retrofit situations. The specific blast capacity of these products is dependent on the load duration and the reader is urged to investigate product claims closely, because often, these products have been tested only for short-duration loads representative of high-explosive threats.

For new blast-resistant windows, laminated glass is preferred as the protective layer (the inner lite in an insulating glass window). Laminated glass is composed of two or more layers of monolithic glass bonded with a thin but tough sheet interlayer material, most commonly polyvinyl butyral (PVB). The interlayer material is available in thicknesses ranging from 0.015 to 0.090 in. The interlayer enables composite action and provides ductility and some tension membrane capacity, which significantly enhances the performance of glass alone. When laminated glass fails, the laminate interlayer tends to retain the glass fragments, therefore significantly reducing the hazardous fragments entering inhabited areas.

Blast-resistant laminated glass is typically attached to the frame with structural silicone. If it is not attached with structural silicone, then a large bite on the order of 0.75 in. is often necessary or the window will deflect out of the frame after the glass cracks. Blast-resistant polycarbonate windows also typically require a large bite to prevent the window pane from deflecting out of the frame.

8.4.1 Performance Criteria for Window Glazing

The GSA, DoD, and ASTM have developed performance criteria for windows to standardize specification of blast requirements. These criteria categorize glass fragment debris hazards in terms of throw distance. The reader is referred to PBS P100 (GSA 2022) for GSA criteria and UFC 4-010-01 (US DoD 2024) for DoD criteria. ASTM F2912-17 (ASTM 2017a) contains a matrix of performance levels with descriptions of glazing and debris flight distance and impact in a test configuration.

Glazing performance criteria given in Table 8-1 are extracted from ASTM F2912. These criteria are used to categorize the performance of glazing systems during testing and can be used for specifying window glazing performance at petrochemical facilities.

8.4.2 Testing Methods

Window systems may be dynamically tested to demonstrate performance equivalent to or better than the hazard rating associated with the applicable level of protection. ASTM F1642/F1642M-17 (ASTM 2017c), *Standard Test Method for Glazing and Glazing Systems Subject to Airblast Loadings*, is available for evaluating blast-resistant windows to any prescribed blast load using open-air testing or shock tube testing. Testing should include the entire window system, including connections, and should be in accordance with ASTM F1642 with hazard ratings in accordance with ASTM F2912-17 (ASTM 2017a) or GSA-TS01-2003 (GSA 2003).

The structural supporting system used in the test for fastener attachment should be representative of the fielded application. Any deviations in field application of the connections or the connected elements from the test should be demonstrated by calculation to provide the applicable level of protection for the specific application.

Table 8-1. Glazing hazard levels.

Hazard Rating Description	*Description of Fenestration Glazing Response*	*Hazard Level*
No Break	The glazing is observed not to fracture and there is no visible damage to the glazing system or glazing retrofit system.	H1
No Hazard	The glazing is observed to fracture but is fully retained in the facility test frame or glazing system frame and the rear surface (the side opposite the airblast loaded side of the specimen) is unbroken.	H1
Minimal Hazard	The glazing is observed to fracture and the total length of tears in the glazing plus the total length of glazing pullout from the frame is less than 50% of the glazing sight perimeter. Also, there are three or less perforations or indents anywhere in a vertical witness panel located 3 m (120 in.) from the interior face of the specimen and there are fragments with a total mass of 15 g (0.033 lbm) or less on the floor of the witness area between 0 to 1 m (0 to 40 in.) from the interior face of the specimen and no fragments in the witness area greater than 1 m (40 in.) from the interior face of the specimen.	H2
Very Low Hazard	The glazing is observed to fracture and fragments are located 1 m (40 in.) or less from the original interior surface of the specimen. Also, there are six or less perforations anywhere in a vertical witness panel located 3 m (120 in.) from the interior face of the specimen and there are fragments with a total mass of 15 g (0.033 lbm) or less on the floor of the witness area between 1 and 3 m (40 and 120 in.) from the interior face of the specimen.	H3
Low Hazard	The glazing is observed to fracture, and fragments generally fall between 1 and 3 m (40 and 120 in.) of the interior face of the specimen. Also there are 25 or less perforations 50 cm (20 in.) or less above the floor of a vertical witness panel located 3 m (120 in.) from the interior face of the specimen and 10 or less perforation high than 50 cm (20 in.) in a vertical witness panel located 3 m (120 in.) from the interior face of the specimen. For perforations in the vertical witness panel, up to 5 of the perforations may penetrate through the full thickness of the foil backed insulation board layer of the witness panel as defined in Test Method F1642.	H4

(Continued)

Table 8-1. Glazing hazard levels. (Continued)

Hazard Rating Description	*Description of Fenestration Glazing Response*	*Hazard Level*
Moderate Hazard	The glazing is observed to fracture, and fragments generally fall between 1 and 3 m (40 and 120 in.) of the interior face of the specimen with numerous perforations and penetrations lower than 50 cm (20 in.) in a vertical witness panel located 3 m (120 in.) from the interior face of the specimen. Also there are 25 or less perforations 50 cm (20 in.) or higher above the floor of a vertical witness panel located 3 m (120 in.) from the interior face of the specimen. For perforations, in the vertical witness panel, up to 5 of the perforations may penetrate through the full thickness of the foil backed insulation board layer of the witness panel as defined in Test Method F1642.	H5
High Hazard	Glazing is observed to fracture and there are more than 25 perforations in the area of a vertical witness panel located 3 m (120 in.) from the interior face of the specimen and higher than 50 cm (20 in.) above the floor or there are more than 5 perforations in the same witness panel area with fragment penetration through the full thickness of the foil backed insulation board layer of the witness panel as defined in Test Method F1642.	N/A

Source: Used with permission of ASTM International, from *Standard Specification for Glazing and Glazing Systems Subject to Airblast Loadings*, Subcommittee F12.10, Vol. 15.08, 2017; permission conveyed through Copyright Clearance Center, Inc.

It should be noted that testing for windows is typically for blast loads associated with high explosive hazards with significantly shorter durations compared to vapor cloud explosions at petrochemical facilities. This makes it difficult to ascertain how a window will perform for lower-pressure, longer-duration loads. Engineering judgment would be required for proper justification if the tested windows are to be used at petrochemical facilities.

8.4.3 Analysis and Design Methods

Glazing performance may be determined through analytical methods or testing. Analytical methods typically involve the use of specialized SDOF software codes that are developed using test data for validation. US government agencies have developed specialized software for analysis and design of windows subjected to blast loads; however, much of the software is restricted distribution. It is important that these software codes be used for glazing within the limits of data from which they were developed. Finite-element Analysis methods for glazing analysis are available but require validation through test data.

To ensure that glazing and framing systems do not fail and become hazardous debris, anchorage should be designed to develop the maximum resistance of the glazing. A load factor on the reaction loads should be applied to compensate for underprediction. Because of the brittle nature of glass and the large variability in flexural strength, glazing resistance is determined on the basis of material properties that are a function of the probability of failure. This probability is expressed in the form of number of breaks per 1,000 static bending tests. For example, one type of glass may have a resistance value associated with 500 breaks per 1,000 and a different, lower-resistance value expressed for 8 breaks per 1,000. It is understood that values with a lower number of breaks per 1,000 will be more conservative and result in a higher safety margin. Depending on whether it is a design of a new glazing system or an evaluation of an existing window, different resistance values can be selected for calculations. It may be appropriate to use a lower resistance value to determine glazing performance and a higher resistance value to determine reaction loads on the window framing. When selecting material properties for resistance calculations, it is important to note that the material strength of glass decreases with exposure to environmental conditions.

The structural system supporting the glazing should be designed for the blast loads applied to the wall of the structure rather than the reaction loads from the glazing. Experience has shown that design for reaction loads is not necessary to achieve adequate structural performance.

Protective glazing is typically designed for the positive phase of the blast load because negative-phase loads tend to reduce the demand on the supporting system. Similarly, rebound loads are often ignored because the principal concern is fragments being projected into the structure. For the case of a building used for shelter-in-place, a high protection level is often necessary to prevent ingress of toxics. This necessitates a consideration of negative-phase loads and rebound of the glazing and framing.

A dynamic analysis of framing for glazing systems can be accomplished with SDOF or FEA methods. Performance criteria for window framing are provided in Chapter 5, Table 5B-3.

A static analysis approach may also be used for glazing design. ASTM F2248-19, *Standard Practice for Specifying and Equivalent 3-Second Duration Design Loading for Blast Resistant Glazing with Laminated Glass* (ASTM 2019) is available to provide a method of determining an equivalent static load for use in standard glass design charts provided in ASTM E1300 *Practice for Determining Load Resistance of Glass in Buildings* (ASTM 2024).

8.5 UTILITY OPENINGS

Blast-resistant buildings require the same openings for air intake, exhaust, power and control cables, and service piping as conventional buildings. For blast-resistant buildings, it may be necessary to provide protection at the openings. Manufacturers of protective devices for these openings normally provide the detailed design.

Electrical and pipe penetrations may be brought into the building underground. Based on economy and design, this type of entry may be preferred.

Many building owners accept the risk of the blast entering an occupied blast-resistant building through utility openings and mitigate this risk by using designs developed for earthquake-resistant construction to prevent the failure of overhead interior items and interior walls.

8.5.1 Blast Dampers

Heating, ventilation, and air-conditioning blast dampers are devices with mechanical elements that close within milliseconds of the blast wave arrival. Blast dampers are available that will remain closed or that will reopen after pressures return to normal. Blast dampers are furnished in frames that require attachment to properly designed structural elements.

Because of the need to close within milliseconds, open blast dampers create a significant operating pressure loss. Therefore, the resulting blast damper opening is usually much larger than a normal duct penetration. This must be considered in the building opening layout.

8.5.2 Blast Attenuators

HVAC blast attenuators are similar to blast dampers, except that they do not have any moving parts. They typically consist of a metal perimeter frame with fixed or adjustable blades that provide weather protection and can lessen the blast wave effects by reducing the interior increase in pressure. The blades must be designed such that they do not become debris inside the building. The blades must be properly secured to perimeter frame which, in turn, must be secured to the wall system. Blast attenuators are typically designed for short blast durations and may not be suitable for long-duration loads from vapor cloud explosions. Manufacturers will provide the necessary design information.

8.5.3 Cable and Conduit Penetrations

Large concentrations of unprotected cable or conduit penetrations can result in a significant entry of blast pressures. Through the use of proprietary devices, the annular space around the cable or conduit can be completely sealed. Alternatively, custom-designed closure plates may also be used.

8.6 INTERIOR DESIGN CONSIDERATIONS

Consideration should be made to certain interior items. Functional or decorative objects should not be mounted on the interior surface of an exterior wall. Rapid inward movement of the wall may dislodge objects, causing injury to people or damage equipment. For the same reason, file cabinets and other furnishings should not be placed closer to the interior surface of a wall than the maximum predicted deflection of the wall. If there are objects that need to be mounted to the wall, they should be positively attached to the structure and their anchorage should properly account for the dynamic forces caused by the blast event. These objects should not have components that could become projectiles as a result of the wall motion. The best practice is still to avoid mounting objects to the exterior wall.

Suspended ceiling components are particularly susceptible to being dislodged during a blast. Ceiling lighting fixtures, diffusers, and others should be supported independently of the suspended ceiling. Although there are not any current requirements on how to attach suspended ceiling components inside blast-resistant buildings, there are references on how to deal with this same issue in high-seismic zones. Chapter 13 of ASCE 7-16, *Minimum Design Loads and Associated Criteria for Buildings and Other Structures* (ASCE 2017); ASTM E580/E580M-22, *Standard Practice for Installation of Ceiling Suspension Systems for Acoustical Tile and Lay-in Panels in Areas Subject to Earthquake Ground Motions* (ASTM 2024b), and FEMA E-74, *Reducing the Risks of Nonstructural Earthquake Damage—A Practical Guide* (FEMA 2012), all provide valuable information on how to mitigate the dislodgement of ceiling components. A similar approach can be adopted for blast events. The calculation of the force for the anchorage of these components is left to the structural engineer.

8.6.1 Power Control Equipment

Power control equipment includes low-voltage and medium-voltage switchgear, motor control centers (MCCs), battery racks, and other electrical equipment typically found in a power control building in a process facility.

In the event of a sudden loss of electrical control, protection, and monitoring capabilities of the power control equipment, the controlled process could become unstable. The vulnerability of a process to loss of electrical control depends on the type of process being controlled. Connected equipment may no longer be protected from the effects of overcurrent heating because of process system overload or overpressures.

During the 2005 Vapor Cloud Explosion at the Texas City refinery, a power control modular building, which had not been designed for blast loading, was involved in the event. The investigation report published by the US Chemical Safety Board reported that the side-on overpressure at the power control building location was approximately 0.8 psi (Magenes and Burse 2018). Records indicate that, although the power control room underwent medium-to-high structural damage, the enclosed power control equipment was not serviced or replaced after the event. The building contained medium- and low-voltage conventional switchgear and MCC (non-arc-rated). Battery racks, battery charger, and a relay panel were also found inside the building.

Although very limited data are available on the effect of blast-induced accelerations onto equipment, the effect of high lateral and vertical accelerations on electrical equipment has been witnessed extensively during seismic events such as the San Fernando earthquake in 1971.

No specific guidance is available to evaluate the vulnerability of power control equipment under blast loading; however, analytical tools are available for estimating the performance of electrical equipment under seismic accelerations. The Seismic Qualification Utility Group (SQUG) is an association formed by utility and power companies that collects and evaluates test data and earthquake experience data to facilitate their use as a basis for seismic compliance of power equipment. The SQUG database defines seismic response capacity spectra for 20 classes of electrical and mechanical equipment:

1. Motor control centers,
2. Low-voltage switchgear,
3. Medium-voltage switchgear,
4. Transformers,
5. Horizontal pumps,
6. Vertical pumps,
7. Fluid-operated valves,
8. Motor-operated valves,
9. Solenoid-operated valves,
10. Fans,
11. Chillers,
12. Air handlers,
13. Air compressors,
14. Distribution panels,

15. Batteries on racks,
16. Battery chargers and inverters,
17. Engine generators,
18. Instruments on racks,
19. Temperature sensors, and
20. Instrumentation control panels and cabinets.

A conservative approach to evaluate equipment under blast load excitation is to use these capacity spectra and compare them to the blast-induced accelerations. Although this approach can give a general idea of the equipment vulnerability to a blast event, it should be noted that seismic capacity spectra do not cover the same frequency domain as a blast event; thus, the approach could be largely conservative. Overall, more research is needed on the performance of power and control equipment subject to blast-induced shocks.

8.7 EXTERIOR CONSIDERATIONS

The peak reflected blast loading is calculated assuming that the air can move around the structure and efficiently relieve the pressure. Buildings should be configured to prevent trapping of the blast wave and therein increasing the load above those specified. Items such as reentrant corners and set back doors can experience loadings that are considerably higher than the peak reflected overpressures and should be accounted for in the design.

The building design should not contribute to the likelihood of flying debris. Canopies and vestibules should be avoided because they frequently become dislodged and could block critical means of egress.

References

ASCE. 2017. *Minimum design loads and associated criteria for buildings and other structures*, ASCE 7-16. Reston, VA: ASCE.

ASCE. 2021. *Structural design for physical security: State of the practice*, ASCE MOP 142. Reston, VA: ASCE.

ASTM International. 2017a. *Standard specification for glazing and glazing systems subject to airblast loadings*, ASTM F2912-17. West Conshohocken, PA: ASTM.

ASTM. 2017c. *Standard test method for glazing and glazing systems subject to airblast loadings*, ASTM F1642/F1642M-17. West Conshohocken, PA: ASTM.

ASTM. 2018. *Standard test method for metal doors used in blast resistant applications (equivalent static load method)*, ASTM F2247-18. West Conshohocken, PA: ASTM.

ASTM. 2019. *Standard practice for specifying an equivalent 3-second duration design loading for blast resistant glazing fabricated with laminated glass*, ASTM F2248-19. West Conshohocken, PA: ASTM.

ASTM. 2024a. *Standard practice for determining load resistance of glass in buildings*, ASTM E1300-16. West Conshohocken, PA: ASTM.

ASTM. 2024b. *Standard practice for installation of ceiling suspension systems for acoustical tile and lay-in panels in areas subject to earthquake ground motions*, ASTM E580/E580M-22. West Conshohocken, PA: ASTM.

FEMA (Federal Emergency Management Agency). 2012. *Reducing the risks of nonstructural earthquake damage—A practical guide*, FEMA E-74. Washington, DC: FEMA.

GSA (General Services Administration). 2003. *Standard test method for glazing and window systems subject to dynamic overpressure loadings*, GSA-TS01-2003. Washington, DC: NIBS.

GSA. 2022. *Facilities standard for the public buildings service, with 2022 Addendum*, PBS P100. Washington, DC: NIBS.

Hurd, C., J. Collinsworth, and M. Kraemer. 2014. "Determination of conventional hollow metal doors resistance to shock loading with dynamic testing and FEA modeling." In *Proc., 85th Shock and Vibration Symp.* Spring, TX: ABS Consulting.

Lowak, M., J. Idriss, and J. Wesevich. 2011. "Testing and analytical evaluation of doors." In *Proc., Structures Congress*, 1375–1410. https://ascelibrary.org/doi/10.1061/41171%28401%29121

Magenes, L., and T. Burse. 2018. "Design considerations for electrical enclosures subject to accidental vapor cloud explosions." In *Proc. Structures Congress*. https://ascelibrary.org/doi/10.1061/9780784481349.044

PIP (Process Industry Practices). 2002. *Blast resistant doors and frames*, PIP ARS08390. Austin, TX: University of Texas, Construction Industry Institute.

PIP. 2014. *Blast resistant building design criteria*, PIP STC01018. Austin, TX: University of Texas, Construction Industry Institute.

USACE (US Army Corps of Engineers). 2011. "*SDOF resistance function methodology study—Door study.*" *USACE PDC Tech. Rep. 11-01.* Washington, DC: USACE.

USACE/NAVFAC/AFCESA (US Army Corps of Engineers/Naval Facilities Engineering Command/Air Force Civil Engineer Support Agency). 2014. *Structures to resist the effects of accidental explosions*, UFC 3-340-02, Change 2. Washington, DC: NIBS.

US DoD (US Department of Defense). 2024. *Minimum antiterrorism standards for buildings, with change 3*, UFC 4-010-01. Washington, DC: NIBS.

CHAPTER 9

Evaluation and Upgrade of Existing Buildings

9.1 INTRODUCTION

This chapter discusses structural evaluation strategies and upgrade options for buildings at petrochemical plants subjected to potential blast hazards. If buildings do not have adequate blast resistance, their blast capacity can be increased with proven structural retrofits or upgrades that have been validated by open air and shock tube testing and analysis. Blast-resistant upgrades for existing structures are discussed in numerous public domain documents. In many cases, upgrades to increase seismic capacity are also applicable for blast design or can be adapted to increase the blast resistance of a structural component. These retrofits are discussed in a number of documents, including FEMA 356 (ASCE/FEMA 2000), ASCE 41-23 (ASCE 2023), and SAC (SAC 1995). However, there are also significant differences between seismic and blast design, and therefore, seismic upgrades should only be used as a basis for blast resistant upgrades by engineers with blast design experience.

Blast retrofits of existing buildings can range from minimal upgrades, such as window modifications, to very significant, such as providing a robust concrete shell or "cocoon" around the existing building. It is important to remember that costs involve not only construction but also potential downtime due to the interruption of operations that may be necessary to implement the upgrades. These costs can increase significantly if the upgrades are installed during building operations and involve internal building strengthening as opposed to external building strengthening.

9.2 EVALUATION STRATEGIES

Typically, a blast protection study for existing buildings involves the following steps (refer to Figure 9-1):

1. *Hazard Identification and Quantification*: Determine the location and size of potential explosions and establish the free-field overpressure and impulse at each building location.

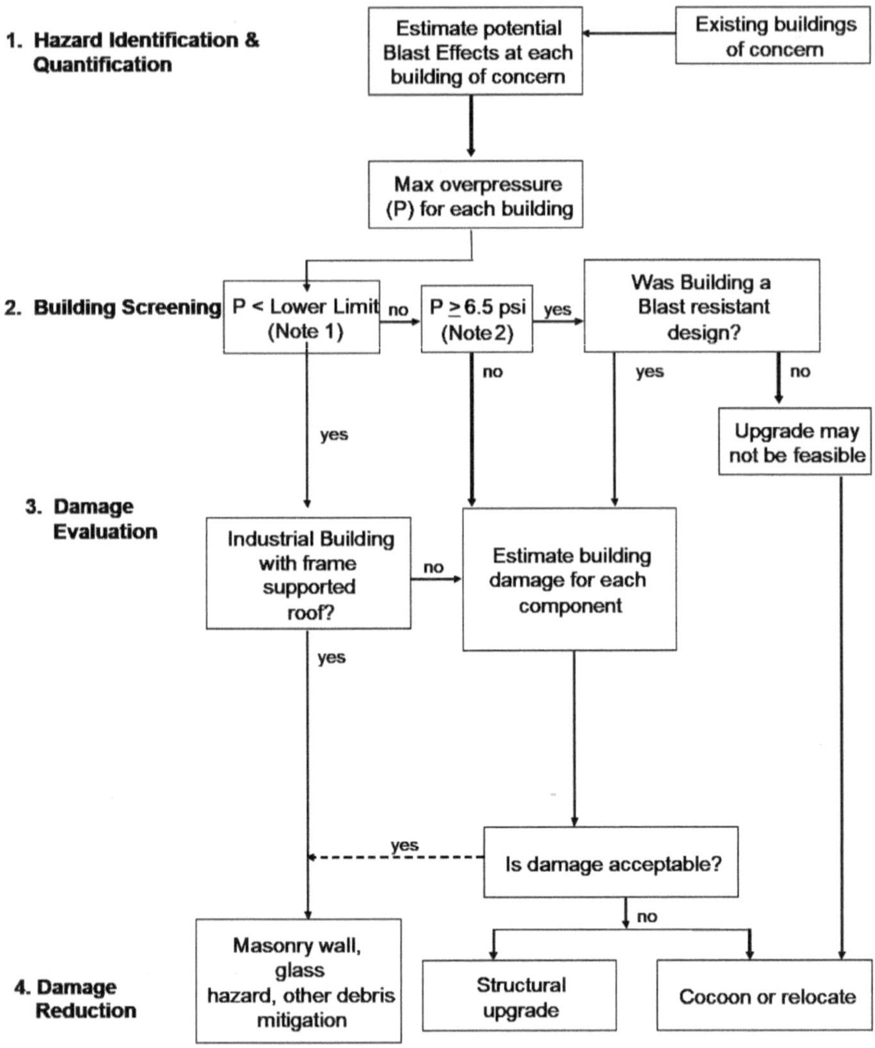

Figure 9-1. Methodology flowchart for existing building evaluations.

Notes: 1. Individual companies may have different lower-limit blast loads. These limit blast loads may depend on building type and acceptable building damage level. 2. At pressures > 6 psi (41 kPa), it may no longer be feasible or economical to upgrade a conventional building. In this situation, a reasonable approach would be to consider a blast-resistant shell around the existing structure.

2. *Define Acceptable Damage*: It is the owner's responsibility to define the acceptable damage for buildings. Typically, the acceptable damage to a building is limited such that the vulnerability of the occupants and/or the critical equipment inside the building remains at an acceptable level. Therefore, it can be acceptable in some cases for buildings to sustain damage

that is unrepairable, or even relatively severe. The most common method used to define acceptable building damage is in terms of the building damage levels shown in Chapter 5, Table 5B-1a.

3. *Building Screening*: To make the evaluation process efficient and focus the engineering efforts, buildings should be screened into the following categories:

    a. Buildings Not Requiring Structural Upgrade

       This category represents buildings subjected to a blast overpressure below a limiting threshold requiring structural upgrades. Individual companies may set this threshold on the basis of company building construction standards, blast overpressure, analysis or data, injury tolerance, and performance requirements. Note that threshold blast loads should be consistent with the owner's definition of acceptable damage.

    b. Buildings Requiring Evaluation to Determine the Need for Structural Upgrades

       Buildings in this category have blast loads exceeding the threshold values. They require further evaluation or analysis to determine whether calculated damage from potential explosions exceeds the acceptable damage level. If so, a structural upgrade should be defined that will reduce the damage to an acceptable level, as discussed in this chapter.

    c. Buildings Requiring Replacement or a Structural Shell

       At overpressures in excess of approximately 6 psi (45 kPa), it may not be reasonable to attempt an upgrade of conventional construction. In this situation, a new blast-resistant exterior structure that envelopes the existing building or a new robust building replacement may be the most feasible mitigation plan.

4. *Damage Evaluation*: For those buildings requiring further evaluation, a complete engineering evaluation should be conducted to determine the damage levels per Chapter 5, Appendix 5B, to each typical structural component from the applied blast loading. Damage to windows and doors is also an important factor. The calculated component damage levels, along with information defining the areas of the building that are occupied, can be used to determine whether the building requires an upgrade. If a risk-based approach is used to make this determination, the building component damage levels must be used to calculate vulnerability to building occupants (CCPS 2012).

5. *Damage Reduction to Nonstructural Components*: Upgrades to the building should reduce the expected damage to nonstructural components to tolerable levels. Studies have shown that failed nonstructural components can be as hazardous to occupants as failed structural components with similar mass and rigidity. Possible upgrades are discussed in the following sections of this

chapter. At a minimum, even for those buildings not requiring structural modifications, steps should be taken to mitigate the consequences of any calculated door failure or glass breakage and those arising from interior debris such as failed interior masonry walls, collapse of nonstructural components in the ceiling, and debris hazards caused by large dynamic displacements of the roof and wall components from blast loading. Methods to limit hazards from interior nonstructural debris during earthquakes are usually applicable to limit such hazards during blast loading.

9.3 BLAST-RESISTANT UPGRADE OPTIONS

A number of options are available to upgrade the blast capacity of an existing building. The choice of an upgrade option is usually based on the applied blast loads, characteristics of the existing structural component makeup, cost considerations, implementation approach around current operations, and building functionality requirements. The design of blast-resistant retrofits to existing buildings can be subject to many limiting factors not typically applicable for the design of new blast-resistant buildings. The engineer must consider the capacity of existing framing components to resist increased reaction loads from retrofitted wall and roof components, limitations on the achievable connection strengths between retrofitted and existing components, limits on available space for retrofits and accessibility for construction equipment, aesthetic considerations, and potential loss of building operational capability during and after construction. Retrofits may also increase the dead load on the building foundation and may need to resist loads transferred from attached components such as retrofitted windows and doors.

There are four basic approaches for blast-resistant upgrades to existing building components: (1) strengthening existing components and/or connections, (2) allowing existing components to fail into a catch system that protects building occupants from debris, (3) shielding existing components from blast loads with a new exterior blast-resistant component, and (4) using energy-absorbing connections that limit damage to the blast-loaded component and/or supporting components. Table 9-1 summarizes alternative concepts for these approaches, which are discussed in the following sections within the context of upgrades to different building systems and component types. This table is not meant to be a complete list of all available upgrade options. Any upgrade can be used that produces a structural system that meets an objective for limiting building blast damage and/or protecting building occupants against injury from applied blast loads. Approximate information is provided in Table 9-1 for the additional blast capacity that can be provided by typical applications of the concept. The actual increase in blast capacity can vary significantly depending on the size, material strength, and other properties of the upgrades.

Nonstructural upgrades to reduce debris hazards should also be considered in addition to structural blast-resistant upgrades. These upgrades include modifying

Table 9-1. Summary of Blast-Resistant Retrofit Concepts for Building Components.

Retrofit concept	*Wall or roof retrofit*	*Retrofit type**	*Additional blast capacity†*
Strengthen the connections of light metal components	Wall or roof	Wall or roof strengthening	Moderate
Weld or attach a steel plate to existing steel and reinforced concrete framing members	Wall or roof	Wall or roof strengthening	Moderate
Add additional beams between existing beams to reduce the cladding span and/or increase the overall capacity of the beams	Wall or roof	Wall or roof strengthening	Moderate
Attach high-strength material to a wall surface (i.e., Kevlar, E-glass, or carbon fiber)	Wall	Wall strengthening	Significant
Attach ductile polymer material to a wall surface	Wall	Wall strengthening	Significant
Attach new steel posts or heavy gage corrugated steel panels to masonry or a concrete wall	Wall	Wall strengthening	Very significant
Increase concrete wall thickness with a new reinforced concrete section attached to a wall with dowels	Wall	Wall strengthening	Very significant
Place reinforced concrete panels outside buildings attached to the foundation and a building lateral load–resisting system	Wall	Shield system	Very significant
Place a steel beam and a heavy gage corrugated steel panel system above the existing roof	Roof	Shield system	Very significant
Build a blast-resistant shell around the existing building	Wall and roof	Shield system	Very significant
Place a steel beam and grating below the existing roof	Roof	Catch system	Significant
Place strengthened metal stud walls inside the existing wall	Roof	Catch system	Significant
Place a geotextile well attached to building framing or floor systems over the inside surface of a wall	Roof	Catch system	Significant
Add energy-absorbing connections to the load path of blast-loaded components	Wall and roof	Energy-absorbing system	Significant

*Catch systems not applicable for buildings with load-bearing walls.

†Increase factors for blast capacity of conventional building components: moderate (100% to 150%), significant (150% to 300%), and very significant (>300%).

or eliminating architectural sections of the building having heavy material that could fall, providing positive attachment for all overhead equipment, lights, and false ceiling to roof structural members, attaching book shelves to walls, eliminating exterior enclosures around doors that could fail and block doors, and replacing interior unreinforced concrete block partitions with lightweight gypsum board metal stud walls or reinforced concrete block that are well attached to the floor and roof systems. In general, all nonstructural upgrades recommended for buildings subjected to earthquake loads are also applicable for blast-resistant design. Although these methods do not increase the blast resistance of the building directly, they are effective in reducing the potential injury to the personnel inside the building.

9.4 UPGRADES FOR STRUCTURAL MEMBER CONNECTIONS

Strengthening of structural connections can be the most cost-effective upgrade for existing buildings if it does not require removal of existing interior walls and equipment and if this upgrade is sufficient to resist the blast loads. This can be the case for members that have sufficient flexural capacity to resist the applied blast load, but their load capacity is limited by their connection capacity. For a member to absorb blast energy and be structurally efficient, it must develop its full plastic flexural capacity before reaching its shear and connection capacities. This can require a substantial increase in the capacity of connectors or surrounding material that is in tension, compression, bearing, or local shear.

A typical shear connection for a wall girt consists of a shear tab with two bolts. As a blast load is applied to the girt, connection failure may occur because of bolt failure, failure of the girt web around the bolts due to bearing failure or block tear-out, or failure of the shear tab due to insufficient weld capacity. Any connection failure will not allow a component to develop its plastic moment capacity and therefore will prevent the member to reach its maximum blast resistance. An upgrade for this type of connection can consist of additional web material welded or bolted to the girt or additional weld on the shear tab connection to the column. Also, lower-strength ASTM A325-14 (ASTM 2016) and ASTM A307 bolts (ASTM 2014, 2023) can be replaced with higher-strength ASTM A490 bolts (ASTM 2014). Bolt holes can also be enlarged for larger-diameter bolts.

Steel framing components often have moment-resisting connections that are designed to resist elastic stresses based on their calculated conventional load demands rather than the maximum tension, compression, and/or shear loads delivered by the connected members when these members develop their ultimate moment capacity under blast loads. Also, standard moment-resisting frame connections are based on the elastic distribution of stress rather than plastic distribution through the section. Connections for blast-resistant steel framing components need to transfer the full dynamic yield strength of the beam flanges into the columns, and horizontal column stiffeners are often required to transfer the tension and compression forces from the beam flanges through the column.

Stiffeners or doubler plates may also be required to resist high shear demand in the webs of perimeter columns of moment-resisting frames or to prevent shear buckling of beams and columns with noncompact sections. In addition, connections for blast-loaded members need to have sufficient rotation capacity to deform under the applied blast load without failing. This is particularly a problem for moment-resisting steel connections. A connection may have sufficient strength to resist the applied load; however, when significant deformation of the member occurs, this capacity may be reduced because of the buckling of stiffeners or flanges. Figure 9-2 is an example of a steel frame connection designed for large rotation capacity. FEMA 356 (ASCE/FEMA 2000) and ASCE 41-23 (ASCE 2023) describe a number of upgrades to increase the strength and ductility of steel moment–resisting frames subjected to earthquake forces that are also commonly applicable for blast-resistant design.

Connection and reinforcing steel splice details in blast-resistant structures can also be designed to provide an alternative response mode if they allow tension

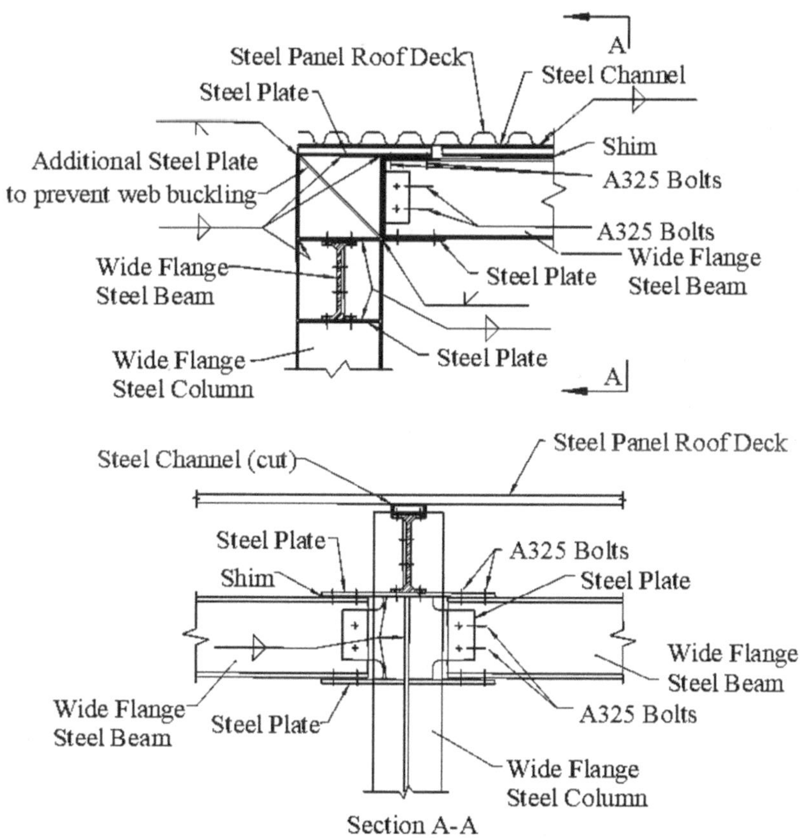

Figure 9-2. Large rotation connection detail.

Source: USACE/NAVFAC/AFCESA (2014).

membrane or catenary action to develop at large deflections. These response modes can increase the blast capacity of components that initially respond in flexure to blast loads. Also, steel and steel-reinforced concrete and masonry components that respond in a brittle response mode, such as lateral torsional buckling, local buckling of noncompact steel section, or shear failure, can continue to resist load in some cases if they are adequately attached to supporting components to develop tension membrane response and have continuous steel reinforcement along their span and all splices are full tension splices. The use of tension membrane response as an alternative response mode is discussed in UFC 4-023-02 (US DoD 2005).

Load reversal is typically not considered in the design of connections for conventional loads. As discussed in Chapter 7, rebound forces produced in a member's response can be quite high. These forces are a function of the mass and stiffness of member as well as the ratio of blast load to peak resistance. Connections that provide adequate support during a member's positive phase, or inward response to blast load, must also be analyzed for the rebound load. If the member fails during rebound because of inadequate connections, it may fall into the building if it is an overhead member or it may cause progressive collapse if it is a load-bearing member.

Design of connections for reaction forces during rebound can be a particular concern with many connections used for roof components and precast concrete wall components. Headed studs are normally used to secure a roof slab to a structural steel framing system for rebound loads in new construction. These studs are usually not found in existing structures unless the roof girders were originally designed to be composite with the concrete deck. It may be necessary to provide through bolts to the structural frame with a backing plate on the top side of the slab if the composite metal deck is found not to be adequately attached to the roof framing elements. The addition of knee bracing with bridging is sometimes required on the bottom flange of roof beams to reduce the unbraced compression flange length in rebound.

Precast concrete components that receive support from direct bearing during the positive phase response to blast load, such as the top of a wall slab that bears against a concrete roof deck, typically depend on the connection shear or tensile strength to resist rebound reaction forces that are much higher than those from conventional suction pressure loads. Typical wall panel connections with welded plates or bolts can often be used for blast-resistant design if the size and/or number of connections are increased.

It is often desirable to utilize the in-plane capacity of precast panels to function as shear walls that resist lateral loads. The connections typically provided between adjacent precast members are often inadequate to develop the required in-plane capacity to resist the high lateral loads transmitted into the building by blast-resistant wall components. In some cases, rectangular steel plates or steel ledger angles can be bolted to precast wall and roof deck panels to increase the in-plane shear connection capacity.

9.5 UPGRADES FOR STRUCTURAL FRAMING MEMBERS

Framing members can be upgraded by increasing their cross section. This can take on many forms, including welding cover plates onto steel members, bolting steel plates to the tension side of concrete beams, and welding or bolting steel angles, rods, rebar, or channels to the webs of steel beams near the flanges. Also, *knee struts* or *haunch members* can be added at corners where beams and columns frame together to reduce the effective frame member spans and add moment capacity to a simply supported connection. The compression capacity of steel I-beam columns members can be increased by welding plates onto the column flanges to create a box section. The capacity of concrete columns can be increased by jacketing the column with steel plates or additional reinforced concrete.

A particular problem can be pre-engineered frames with bolted moment connections that do not have sufficient moment capacity. A patented steel beam-to-column connection system that improves seismic resistance can be used to increase the flexural capacity and ductility of a steel moment connection (SidePlate Systems 1998). It consists of vertical steel plates placed along both sides of the bolted connection that are welded to wide cover plates attached to both flanges of the connected beams.

In some cases, steel framing members can be upgraded by adding lateral bracing. Many roof beams, roof purlins, and wall girts in existing petrochemical structures are not capable of developing their full plastic moment capacity because of inadequate lateral support of the compression flange for both inbound and rebound responses. These elements may have only adequate bracing of their compression flanges for conventional gravity loads and very limited bracing for wind loads in the direction for blast rebound loads.

New lateral bracing can be lightweight structural steel shapes welded or bolted to the compression flanges of existing load-carrying members. Material costs are very low. However, construction costs and plant operation interference usually prove to be deciding factors in whether a bracing upgrade can be feasible. In control rooms and laboratory facilities, the roof beams and purlins are typically below the primary roof and above an interior or dropped ceiling. The construction effort would have to be directed from inside the building and may interfere with building operations.

Taking advantage of existing structural members, lateral bracing may include tying the flanges of the steel members to the elements that they support, such as prestressed concrete decks, concrete decks on metal forms, and metal or fiber-reinforced plastic panels. Note that this bracing usually increases only the moment capacity for downward blast loads. If an analysis shows that upward rebound occurs, the bottom flanges may also need new bracing. Refer to Figure 9-3 for a typical bracing configuration. Bracing rods can also be used to tie the bottom flange to the roof beams (Figure 9-4).

For plate girders, the addition of transverse intermediate stiffeners at various spacings along the span length will increase the web buckling strength,

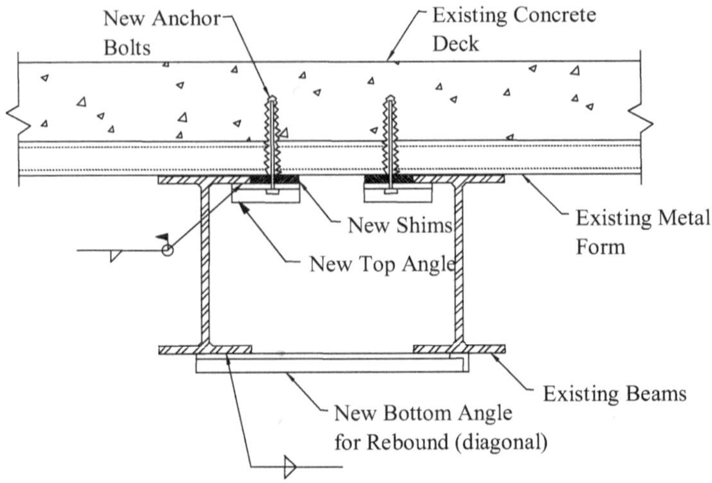

Figure 9-3. Typical bracing details.

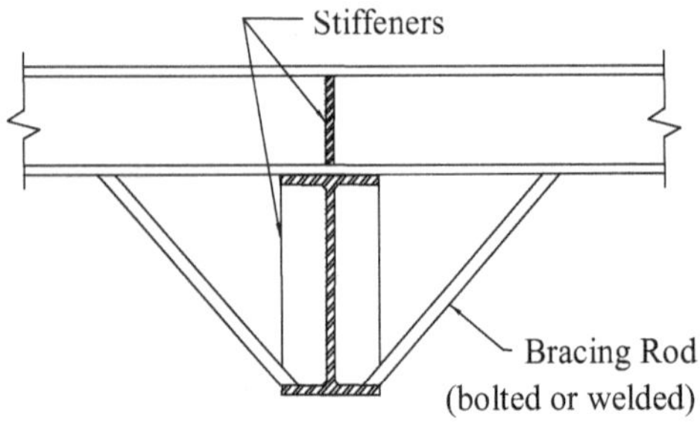

Figure 9-4. Typical bottom flange bracing.

Notes: Only one brace is needed, unless a rod is used. Top stiffeners are to prevent web crippling and buckling. Bottom stiffeners are to prevent web buckling and LTB. Typical web stiffening and bottom flange bracing are shown.

thereby increasing the web resistance to shear and moment. For deep web plates, longitudinal web stiffeners will also increase section capacity.

Upgrading lateral bracing is effective only if the full moment capacity of the component is sufficient to resist the applied blast loads. Existing components that are not well braced can be analyzed initially using a reduced flexural strength consistent with the maximum compressive strength that can be developed by existing lateral bracing and tension membrane resistance, depending on the existing connection strength, and including any effects from dead load. This analysis may show that the component is adequate to resist applied blast loads without failing, particularly for rebound response. The use of stiffeners to resist lateral torsional buckling should be considered for steel beams and girders with inadequate lateral bracing.

9.6 UPGRADES FOR METAL PANEL WALL AND ROOF SYSTEMS

Metal panel wall and roof systems consisting of corrugated metal panels and cold-formed wall girts and roof purlins are commonly used as exterior cladding. Metal panels are constructed of thin-gauge material that buckles in the compression flange at maximum moment regions before the panel cross section develops its ultimate plastic moment capacity. Resistance provided after buckling occurs in the maximum region is attributed to tension membrane response, which is characterized by stretching of the panel rather than flexure. This is described in Chapter 5, Section 5.4.4.

To achieve this type of response, it is necessary to restrain the ends of the panel to provide the required in-plane reaction. A typical conventional design utilizes small self-drilling/tapping screws attached to base angles and wall girts to secure the panels in place. These screws are typically sufficient to develop enough membrane capacity to maintain the flexural resistance up to the maximum allowable deflection for low or medium damage. At larger deflections, the connections tend to fail because of tearout through the panel ends as well as pullout over the head of the screws. The use of a flexible support will also limit the magnitude of load occurring at the ends of the member. This can be accomplished by developing a support that deforms in flexure and limits the end reaction. This type of connection is shown in Figure 9-5.

Reducing the span increases the capacity of metal panels in flexure. Because resistance is a function of the square of the span length, the addition of intermittent supporting members can be very effective in increasing the blast capacity of panels. This can be accomplished by adding wall girts or roof purlins to the structure. This upgrade also increases the blast resistance of the supporting members because their tributary-supported width is reduced. The cost for this

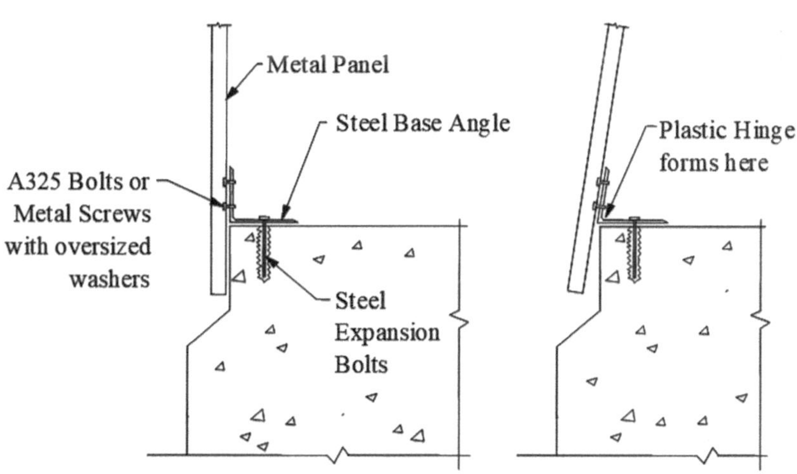

Figure 9-5. Base angle detail for a flexible connection.

upgrade can be quite high if the interior of the structural system is not easily accessible or if the construction requires interruption of operations.

When strengthening of existing panels is not feasible, panels can be replaced with heavier gauge metal panels, two nested conventional panels, or specially designed blast-resistant panels. Commercially available blast-resistant panels that have been developed for protection against terrorist attacks can be considered for protection against accidental explosions as well. Girts and purlins can also be upgraded by replacing them with heavier members. These components can also be overlapped and connected so that they act as a continuous component across interior supports. Typically, the girt or purlin connection must also be upgraded in these cases. Metal wall systems can also be upgraded with a blast-resistant shield wall, as discussed in Section 9.8.

In some cases, large metal buildings with very low occupancy can be designed to resist blast loads using controlled release panels (Oswald 2002). Controlled release panels are conventional corrugated steel panels that are continuous over only one supporting member with limited strength connections to supporting members at the panel edges and a strong connection to the intermediate supporting member, so that they fail by wrapping around the intermediate supporting member. Alternatively, light steel panels intended to fail can be tethered to framing members with steel cables. These supporting members and the building framing can be designed for a lower blast load if controlled release panels are used because the failing panel transfers less blast load into the supporting members. This has been demonstrated through shock tube test results. Attention must be devoted to securing interior building components and objects that can become hazardous debris because of interior blast loads. This upgrade with controlled release panels should be considered only when the free-field pressures will not directly cause serious injury to building occupants or cause failure to surrounding nonstructural components that could become debris and cause injury.

Vent panels designed according to NFPA 68 (NFPA 2023) can be used to reduce blast loads from interior explosions of flammable materials. These panels can have a restraint system (i.e., steel cable tethers) attached to the building framing so that they do not become missile hazards.

9.7 UPGRADES FOR MASONRY AND CONCRETE WALLS

Many petrochemical structures include masonry walls, including concrete masonry unit (CMU) walls with little or no steel reinforcement. This type of construction lacks ductility and has low resistance to blast loads. There are multiple ways to retrofit masonry walls to increase their blast resistance, as discussed in this section. In all these retrofit approaches, the upgraded masonry or concrete walls perpendicular to the blast loads can also serve as shear walls. Shear walls should be analyzed for in-plane shear and out-of-plane bending according to the procedure outlined in Chapter 7. Connections between shear walls and diaphragms and the foundation must also be evaluated. Most of the retrofit approaches in this section can also be used for concrete buildings, although they will net a smaller relative capacity increase than for CMU buildings. The upgrades are explained here primarily for CMU walls because these types of walls require upgrading more frequently.

9.7.1 Upgrade with Steel Posts

An upgrade to masonry and concrete walls with attached steel posts is a common type of retrofit approach used for CMU buildings on industrial sites. Figure 9-6 shows a cross section through a masonry wall retrofitted with vertical steel W-section posts. Vertical tube sections with welded angles can alternatively be used, as shown in Figure 9-7. In either case, the posts are assumed to not act compositely with the wall. This upgrade approach requires minimal available wall space compared to most other retrofits. The posts, which span between the floor and the roof diaphragms, are usually placed at 3 to 6 ft (1.2 to 2.1 m) spacing depending on the capacity of the masonry wall to span horizontally between posts. They can be placed on the inside or outside of the wall. Typically, posts are placed on the building exterior to avoid interior operational disruption and fixtures attached to the wall. In this case, the connections between the posts and the wall are more critical and the compression flange of the post is laterally unsupported during inward response to the blast load. These posts can be notched at low moment regions or doglegged around existing conduits that typically run horizontally along the wall exterior. Galvanized steel may be considered for externally mounted posts.

Vertical post retrofits can be designed for relatively high blast capacities depending on the post size and spacing; however, capacity may need to be limited as the lateral load–resisting system of the building must resist the peak lateral load transferred by the upgraded walls. The posts can resist blast-loading effects during both inward and rebound response, as required for load-bearing walls. The posts can also be positioned to support reaction forces from adjacent blast-resistant windows and doors. Interior posts can be covered with light architectural panels for aesthetic reasons, but these panels must be well attached to any interior posts so that they are not thrown into the room by the sudden acceleration of the posts under blast loading. Exterior posts can also be covered by architectural cladding designed for conventional loads.

The maximum size of post retrofits is often limited by the available strength of the connections of the post to the wall or the connections between the ends of the posts and the foundation and floor or roof diaphragm of the building. Through bolts with plate washers can be used to attach posts to ungrouted CMU walls. Drilled anchor bolts that are recommended for dynamic loading can be installed into grouted CMU blocks and solid masonry walls. Placement of grout into ungrouted CMU walls requires attention to details. Often, the face shells must be removed, and the cells must be cleaned of any mortar to ensure placement of solid grout that acts together with the CMU block to resist pullout of anchor

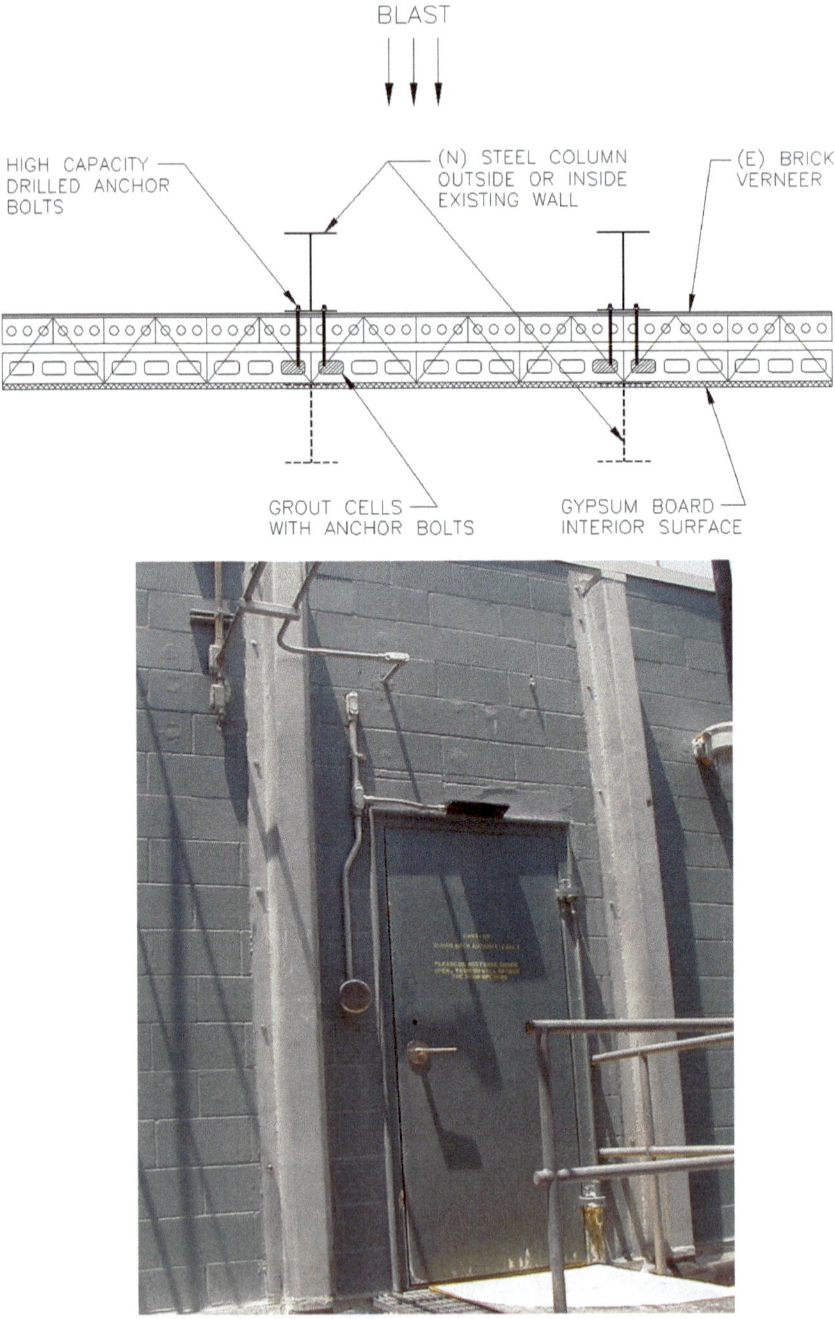

Figure 9-6. Masonry wall retrofit with vertical steel posts.

Source: BakerRisk.

Figure 9-7. Exterior tube section posts bolted to a masonry wall.

Source: BakerRisk.

bolts connecting the posts to the wall. The summed tensile capacity of the anchor bolts should be greater than the ultimate dynamic load resistance of the post and should be spaced no more than 24 in. on-center along the span of the post.

In addition, bolts may need to be spaced closer together at the ends of the post to properly transfer reaction forces into the floor/roof diaphragms. The attachments are also critical for interior posts on a load-bearing wall, because the posts must also support the wall during rebound response to the blast load. The posts are usually designed for low to medium damage because there is some concern regarding the ability of the drilled anchors on exterior posts to remain well attached to the wall at large post deflections. Figure 9-8 shows the through bolts and plate washers on the inside face of the wall. Note that the vertical strap along the wall in Figure 9-8 is unrelated to the wall upgrade.

Horizontal beams can be used rather than vertical posts, but such beams would typically need to transfer their lateral reaction loads into columns, which must also resist axial loads (as opposed to transferring reactions directly into floor and roof diaphragms). Therefore, this is not typically recommended. However, horizontal beams can be used with vertical posts to frame around wide equipment mounted on a wall. Horizontal beams must be well attached to the supported CMU wall in the same manner as vertical posts, and to the supporting vertical posts.

Another upgrade approach consists of using heavy corrugated steel panels, which function in a similar manner as closely spaced vertical posts, spanning vertically on the interior or exterior side (i.e., blast-loaded side) of an unreinforced,

Figure 9-8. Through bolts on the inside surface of a retrofitted wall.

Figure 9-9. Heavy steel panel on the blast side of a retrofitted wall.

ungrouted CMU wall. Figure 9-9 shows an example of this upgrade approach. Because the wall is ungrouted, through bolts were used to connect the corrugated steel panel to the wall.

The connections are critical for this upgrade. Typically, the structural upgrade component (e.g., steel post or heavy steel deck) will be attached to the upgraded wall with post-installed anchors. The design capacity for connections of the upgrade component to concrete supporting members (e.g., a concrete foundation stem wall beneath an upgraded wall) should be calculated using an LRFD-based approach. Typically, the material reduction factor is included in the calculation for the design of connections for upgrades. The manufacturer's values for the capacity of post-installed anchors into masonry are often provided in terms of allowable capacity. These values can be used with a 1.7 increase factor, which converts these to a factored nominal capacity for blast design.

Structural upgrade components attached to fully grouted CMUs with post-installed anchors using this approach have been validated through blast and shock tube tests and have performed adequately. Attachments of structural upgrade components into hollow CMUs, or other nonsolid masonry, with post-installed anchors should be used only if validated by blast or shock tube test data. The attachment of upgrade components with through bolts and plate washers to ungrouted CMUs, as previously discussed, has been validated in shock tube tests. Typically, the cells of all CMUs attached to a structural upgrade component are fully grouted as part of the upgrade or the upgrade component is attached with through bolts and plate washers.

9.7.2 Upgrades with Fiber-Reinforced Polymers Bonded to Walls

High-strength fiber strips or mats made from aramid fiber (e.g., Kevlar), carbon fiber, or glass fiber (e.g., E-glass) can be applied to the inside (i.e., non-blast-loaded side) or to both sides of reinforced and unreinforced masonry and reinforced concrete walls to significantly increase the wall blast capacity. These fiber-reinforced polymer (FRP) systems consist of parallel, very closely spaced high-strength fibers encased by the manufacturer in thin resin mats or strips that are bonded to the wall with the fibers oriented in the span direction. The fibers act compositely with the masonry (or concrete) similarly to steel reinforcement, where the tensile strength of the fibers and compression strength of the masonry form a resisting couple at the maximum moment regions. The fiber types have different characteristics that should be considered when choosing an appropriate FRP retrofit. For example, glass fibers are usually the most economical type, but they provide the least strength and stiffness, while being prone to degradation in acidic or alkaline environments. In any case, a proper application of the bonding agent and the high-strength fiber mats or strips to the wall is critical and must be performed by a trained applicator.

The ultimate tensile strength of the fibers per unit width is provided by the manufacturer, but typically only a fraction of the strength can be counted on for design because of debonding and/or environmental degradation factors. Also, high-strength fibers cannot be counted on to significantly increase the compression or shear strength of the masonry. The maximum blast resistance of a retrofit with high-strength fibers is usually limited by crushing of the masonry block in the maximum moment region, shear failure of the masonry near the supports, or fiber tension failure (often initiated by debonding), which are all brittle, nonductile, failure mechanisms. Therefore, this type of retrofit is usually designed to resist the blast load without yielding with a safety factor, on the basis of available blast test data, manufacturer's recommendations, and/or previous design experience.

Because it is typically designed not to yield, a wall retrofitted with FRP will have a relatively small deflection under the design blast load compared to the other, more ductile retrofits discussed in this chapter. This can be an advantage for load-bearing walls to limit the secondary moment caused by axial load as the wall deflects from blast load. It can also be a disadvantage, because nonductile construction must usually have a higher ultimate load capacity to resist a given blast load compared to ductile construction, and therefore have higher design loads for connections and for the lateral load–resisting system. For load-bearing walls, high-strength fibers should be attached to both sides of the wall to prevent wall failure during rebound. In this case, exterior exposures will warrant the use of protection applications against moisture and ultraviolet (UV) light.

Figure 9-10. E-glass retrofit to a concrete masonry unit and a brick masonry wall.

The primary advantages of an FRP retrofit are that it does not significantly affect the aesthetics or foundation load of the building, and it does not occupy any additional footprint area on the inside or outside of the building. Figure 9-10 shows that FRP retrofits cause only minimal change to the building surface, where the FRP layers only slightly protrude from the wall. The use of strips rather than a continuous mat can allow the placement of a retrofit around equipment attached to the wall, as shown in Figure 9-11. Additional high-strength fibers strips can

Figure 9-11. Fiber-reinforced polymer strips on walls and around existing interfaces.

be placed around window and door openings to resist loads transferred by blast-resistant window and door replacements.

The FRP upgrade is (at minimum) required to be applied to the interior wall surface. This can be a major drawback because fiber attachment to the wall interior will present construction challenges if there is significant interior equipment mounted to, or adjacent to, exterior walls, and interior construction may temporarily limit building operations. Also, hazardous decomposition and/or combustion products can be released by FRP in case of heat or fire exposure. Consult with the manufacturer and applicable material safety data sheets (MSDSs) for more specific information regarding this potential hazard.

CMU walls strengthened with high-strength fibers may need to be grouted so that the overall wall blast capacity is not controlled by the relatively low shear strength of ungrouted masonry. Grouting also improves the strength of any steel connections to the wall, as discussed previously. Ungrouted walls can be grouted by removing exterior face shells at various locations, cleaning out any mortar from the cells, and pumping grout into the voids of the CMU blocks.

Refer to Bazan (2012) for current-practice design methodology and response criteria for FRP retrofits to concrete and masonry walls based on shock tube testing. Refer to ACI 440.2R-17 (ACI 2017) for additional guidance on FRP retrofits of concrete and masonry walls.

9.7.3 Upgrade with Ductile Polymer Bonded to Masonry Wall

Masonry walls can be upgraded by applying a ductile polymer that bonds to the inside surface of the wall, so that it acts as ductile tensile reinforcement for unreinforced masonry and thereby greatly improves the ductility of the wall against blast loads. This is similar to the previously discussed retrofit with FRP, except that the polymer has a much lower strength and is more ductile than FRP. Elastomer such as polyurea is capable of elongating 100%, or more, with a typical yield strength of 1,500 psi minimum.

Figure 9-12 shows a spray-on application of a polymer to a CMU wall. The polymer coating is sprayed over the interior (nonloaded) surface of the exterior wall, with the typical thickness ranging from 1/4 to 3/8 in. Theoretically, the resistance provided by the polymer increases with the thickness, but it is practically difficult to apply the bonded polymer at a thickness > 0.5 in. It can be difficult to apply a uniform thickness $< 1/4$ in. A coating of a 3/8 in. thick polyurea applied to an unreinforced, ungrouted, 8 in. CMU wall with a 12 ft span is efficient for applied blast loads with a long-duration loading and applied peak pressure up to 1.5 psi (10.3 kPa) and short-duration loads (e.g., 12 ms) and applied impulses up to 65 psi·ms (450 kPa·ms). The coating should be applied to both sides of the wall if there is a requirement to improve the rebound response of the wall. However, this upgrade is not recommended for load-bearing walls because of the large deflections that may occur during blast loading. Anchorage by bonding of the coating to the supporting components shall be provided; it can consist of extending the polymer over the bottom and top supports or by mechanical anchoring such as steel angles. Figure 9-13 shows examples of these connections.

Figure 9-12. Application of a spray-on bonded polymer of a concrete masonry unit wall.

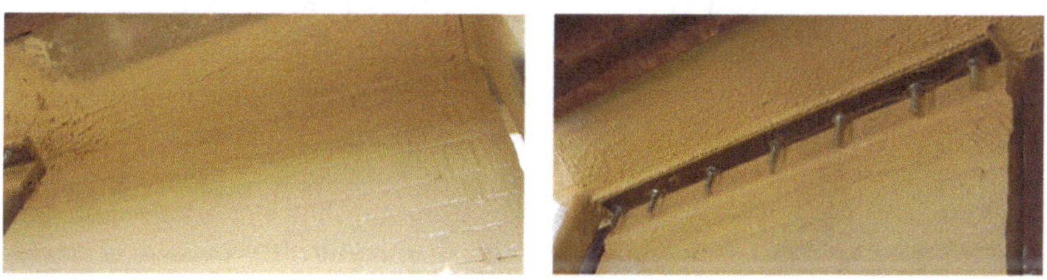

Figure 9-13. Bonded polymer attachment to supports with overlap and mechanical fasteners.

Source: MABS (Bazan and Conrath 2022).

Because the polymer has a much lower tensile strength than FRP, this upgrade will cause the upgraded wall to have lower reaction loads, and therefore transmit lower loads to the lateral load–resisting system for the building, and cause less potential shear and connection issues compared to a masonry wall upgraded with FRP. In addition to strengthening the wall, the polymer can act as a membrane after the masonry crushes in the maximum moment region of the upgraded wall and the wall no longer has any flexural capacity. Sufficient attachments to the lateral load–resisting system of the building are necessary for the polymer to act in membrane action as a catch system to prevent masonry debris throw into the building. Similarly, sufficient attachments to the main load–resisting system of the building are necessary for the polymer to act in membrane action as a catch system to prevent masonry debris throw into the building. Therefore, the capacity of the anchors should exceed the tensile capacity, if possible, or at least the conservatively calculated tensile force that develops in the geotextile. The main load–resisting system may need to be reinforced so that it can resist the reaction forces associated with the peak tensile force that develops in the polymer. The mechanical connection in Figure 9-13 is an example of the connection needed for the polymer to develop membrane action. Catch systems are discussed in Section 9.10. More details on the design of upgrades with this approach are discussed by Bazan et al. (2012).

Except as previously discussed, this retrofit generally has the same advantages and disadvantages as the FRP retrofit discussed in the Section 9.7.2. This includes possible hazardous effects from fire. Once cured, the coating is nontoxic under normal pressure. However, the installation operations must follow standard Occupational Safety and Health Administration (OSHA) safety procedures for hazardous materials. Some polymer retrofits may present fire and smoke hazards, which can be reduced by using fire-resistant coatings and covering with drywall. Some polymers work well within normal temperature ranges but may start to break down under extreme cold or hot temperatures. Nevertheless, different commercial polymer materials are available for different types of environments and applications, including fast curing, high temperature stability, and abrasion resistance. Also, Chapter 26 of the International Building Code (IBC) provides requirements for foam plastic related to surface burning characteristics (ICC 2021).

9.7.4 Increasing Wall Thickness with New Layer of Reinforced Concrete

Reinforced concrete panels can be upgraded by increasing the wall thickness on the exterior side with a layer of reinforced concrete that is well connected to the existing wall to create composite action between the new and existing wall sections. This upgrade can be installed by doweling into the existing concrete panel exterior and hanging a reinforcement grid parallel to the wall. A nozzle-applied shotcrete or cast-in-place concrete can then be placed to increase the overall wall thickness and flexural capacity. If shotcrete is used, the concrete surface must be prepared, and existing conduits and piping should be shielded against overspray of the concrete. It is not recommended that existing conduits along the panel exterior be cast within the new wall thickness. This process is viable if the required thickness increase is only a few inches and the additional wall weight does not require modification of the existing foundation. This upgrade also allows for casting concrete around existing wall penetrations.

9.8 UPGRADE WITH A BLAST-RESISTANT SHIELD WALL

Almost any wall system can be retrofitted by placing a blast-resistant shield wall outside the existing wall. The shield wall can consist of a blast-resistant precast concrete panel, as shown in Figure 9-14. It can also consist of a blast-resistant girt/steel cladding system, as shown in Figure 9-15. Some vendors have developed prefabricated blast-resistant wall systems that may be appropriate for a given upgrade depending on available test data validating the system for the given applied blast load. A third-party review of a vendor's design by an engineer familiar with blast testing and design is usually recommended.

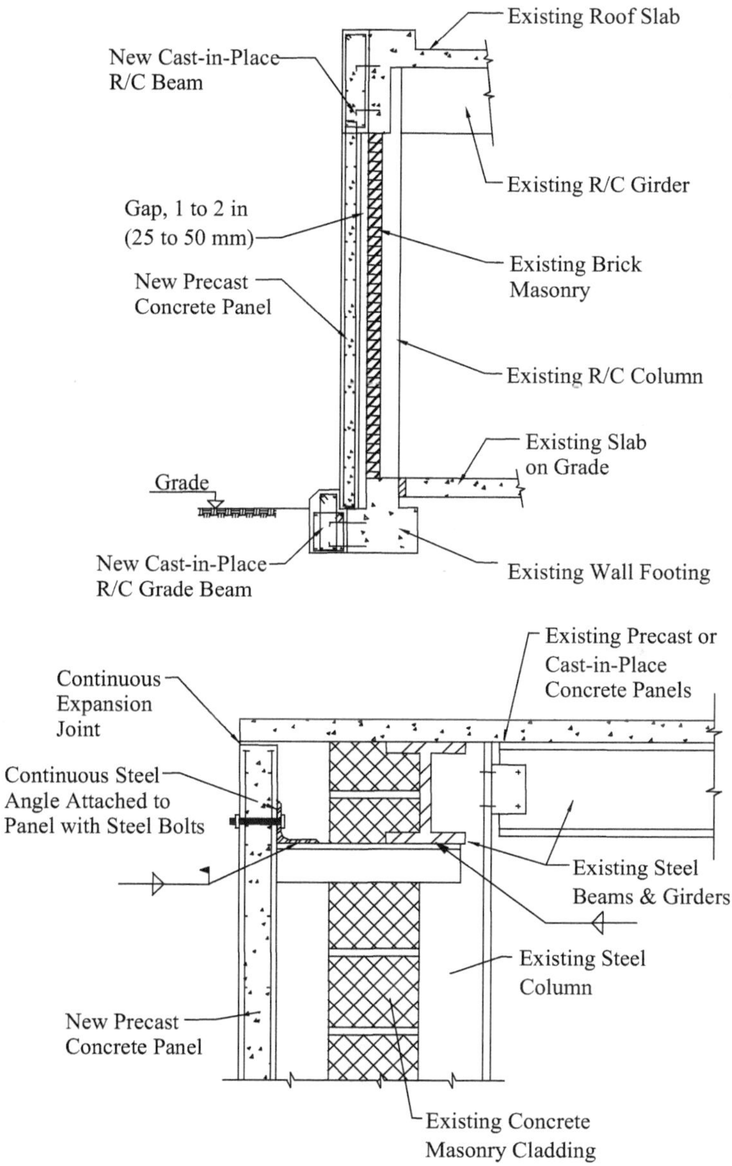

Figure 9-14. Wall upgrade with a blast-resistant precast concrete shield wall.

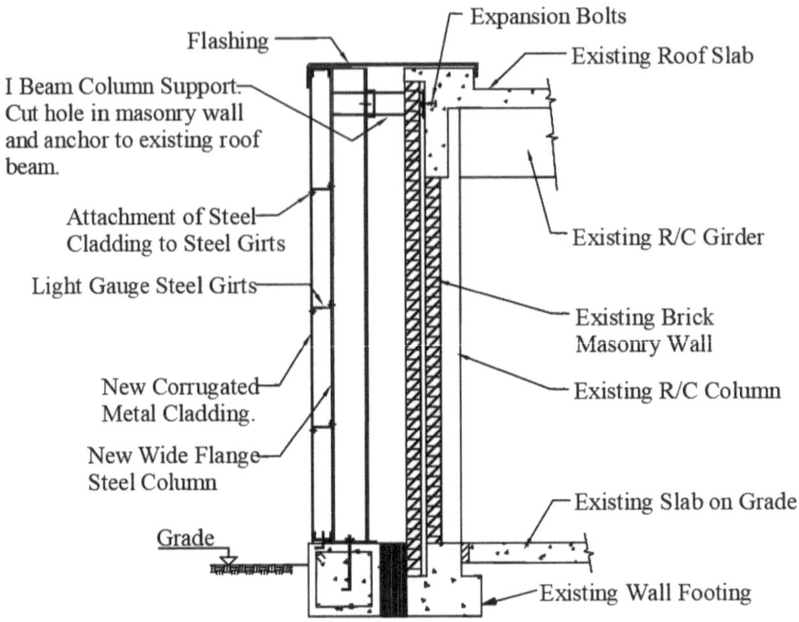

Figure 9-15. Wall upgrade with a blast-resistant steel panel and a grit shield wall system.

A gap that is greater than the predicted shield wall deflection should be maintained between the shield wall and the existing building wall. This will prevent the shield wall from deflecting into the existing wall, and possibly failing it, as it responds to the blast load. The engineer should also verify that the sudden deflection of the shield wall does not decrease the void volume between the shield wall and the existing wall such that there is an increased air pressure that can fail the existing wall. Small vent openings in the existing wall, combined with adequate shield wall standoff from the existing wall, on the order of several feet, can be configured to prevent this potential problem from arising. These vent openings can be covered with low-strength, lightweight material if necessary.

The blast-resistant shield wall panels can be supported at the base by an extension of the existing building foundation, or by a separate foundation, and at the top by the lateral load–resisting system of the building. This retrofit can be designed to resist very large blast loads, depending on the strength of the lateral load–resisting system of the building. It has the advantage of not requiring any construction work inside the building, but the building exterior must be accessible. The shield wall panels can be precast or prefabricated in a remote yard or on site. Blast-resistant doors and windows can be designed into the shield wall panels to line up with doors and windows in the existing wall. Any windows covered up by blast-resistant shield walls should be removed to ensure that any pressure buildup between the shield walls and the windows as the shield wall deflects does not fail the windows and throw glass into the building.

If a building without moment-resisting frames has a shield wall, either the existing wall or the shield wall can be used as the shear wall. If the shield wall is used as a shear wall, the connections of the shield wall to the roof and floor diaphragms and foundation must be designed accordingly.

9.9 UPGRADES FOR ROOF SYSTEMS

Roofs are often particularly difficult to upgrade and usually represent a significant portion of a building upgrade cost. Limited upgrade options can be applied outside the building, and interior upgrades often require removing the drop ceiling, HVAC ducting, and other wiring systems, which results in significant disruption to building operations. If an interior upgrade is an option, applicable wall upgrade options described previously can be applied to the roof system. In the case where an external upgrade is required, the most practical option is often the placement of a new blast-resistant shield roof system over the existing roof. Barring roof-mounted HVAC equipment, the shield roof upgrade is non-intrusive to the building, and therefore, it does not require any penetrations into the existing roof, removal of the interior false ceiling and ducting, transport of construction materials into the building, or overhead welding work. If air handling units are mounted on the existing roof, temporary conditioned air must be supplied to the building until the new roof and HVAC system are installed. The air intake stack may also need to be addressed for new roof shields over existing control rooms.

Figure 9-16 shows an exterior roof retrofit, where a new blast-resistant roof system is placed over the existing roof, which remains in place as an environmental barrier and structural diaphragm. The new roof system resists the blast load, shielding the existing roof from blast load. The spacer beams in the figure are sized to prevent beam and panel deformation into the existing roof. The spacer beams also act with the existing girders to transfer reaction loads from the shield roof into the columns. The columns may need to be upgraded for these increased reaction loads. The effect of the additional dead load from the new roof on the foundation will also need to be considered. Blast-resistant panels should be used to prevent any blast pressure from entering the space between the new shield roof and the existing roof.

As with the shield walls, the engineer should verify that sudden deflection of the shield roof retrofit will not create a significant dynamic pressure buildup within the void between the shield roof and the existing roof. This is usually not a problem if the shield roof system is significantly stiffer than the existing roof system because the existing roof is flexible enough to deflect an amount equal to the relatively small deflections of the overlying shield roof system and alleviate a pressure buildup. In some cases, both the beams and the girders in the shield roof can be placed directly over existing framing members, so that both the new and the existing framing members deflect together and resist the blast load together as a superimposed structural system.

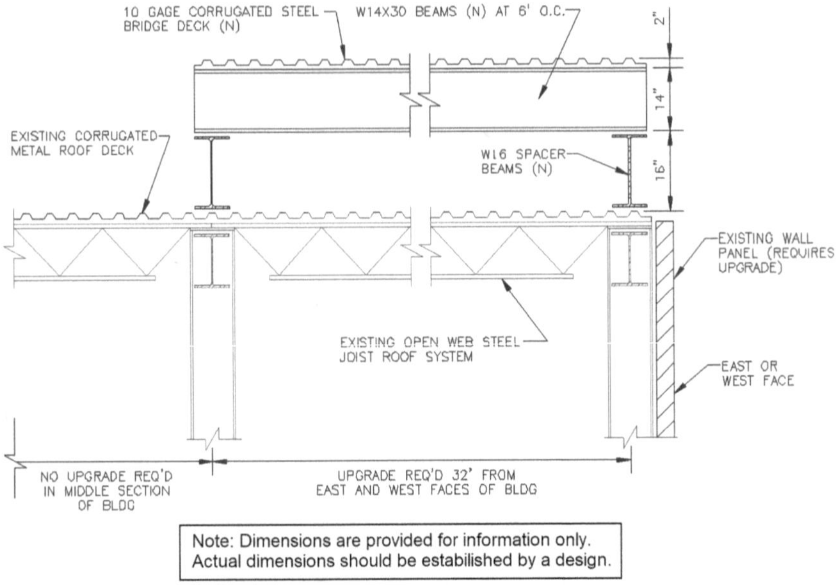

Figure 9-16. Shield roof over a part of an existing building.

9.10 WALL AND ROOF CATCH SYSTEM UPGRADES

Various "catch wall" and "catch roof" systems can be placed inside the building that stop debris from failed wall and roof systems before injuring building occupants. These systems are typically used in small, heavily occupied areas, where the building framing can resist the blast load, but the overhead roof or adjacent wall panels cannot be upgraded to resist the applied blast load. Catch wall retrofit concepts include building an internal steel frame with heavy steel grating around the occupied area, placing an interior blast-resistant wall system in front of occupied areas that is well attached to blast-resistant framing, and installing a geotextile fabric behind a wall that is well attached to blast-resistant framing or components.

This latter retrofit is illustrated in Figure 9-17. The geotextile fabric is designed to resist the applied blast load in membrane action with the mass of the wall and the strain is limited to less than the material failure strain with a factor of safety. Fabrics with failure strains in the range of 7% to 12% are recommended. The capacity of the anchors should exceed the tensile capacity of the geotextile. Elastomeric-based spray and continuous E-glass sheets can also be applied on the wall's interior and anchored to the floor and roof to serve as catch systems.

Additionally, Figure 9-18 shows a retrofit that consists of an interior steel roof catch system. The retrofit consists of new tube steel beams that support heavy steel

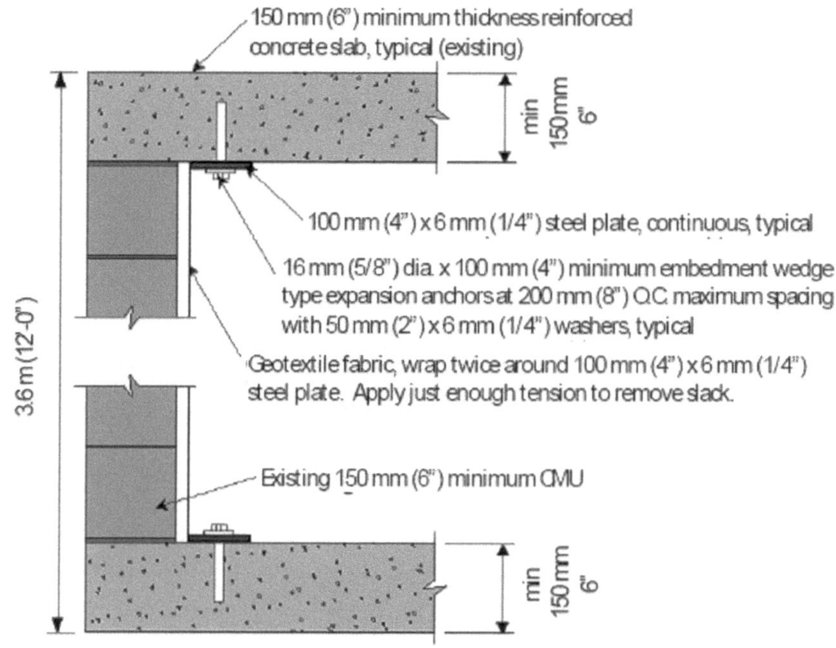

Note: Dimensions are provided for information only.
Actual dimensions should be estabilished by a design.

Figure 9-17. Geotextile retrofit behind a masonry non-load-bearing wall.

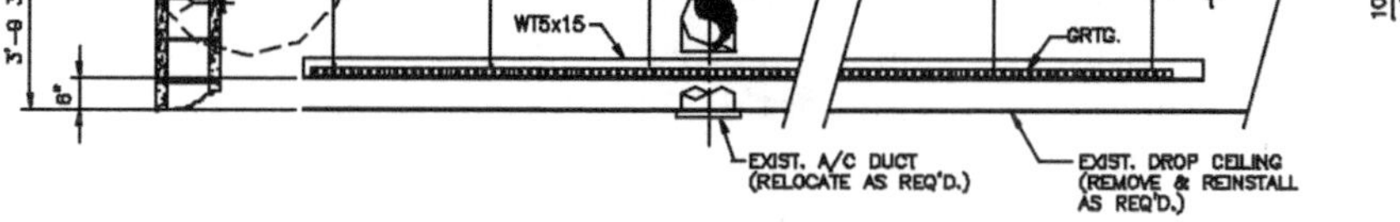

Note: Dimensions are provided for information only.
Actual dimensions should be estabilished by a design.

Figure 9-18. Elevation section of an interior roof retrofit with TS steel beams and steel grating.

grating, which hangs by steel cables from the tube steel beams at an elevation just above the false ceiling. The grating and its supports were designed to resist impact loading from the roof panel and joists after they are failed by the design blast load. The building framing could be strengthened to resist the blast load, but the existing joists and metal panel roof system could not be strengthened by the required amount to resist the design blast load.

9.11 BLAST-RESISTANT SHELL UPGRADES

When options for reinforcing an existing structure are not feasible, an independent shell or "cocoon" structure can be built around the existing building with reinforced concrete or a steel frame and blast-resistant panels. The reinforced concrete option can be either cast-in-place or precast wall and roof panels. Obviously, this is usually the most expensive retrofit option. In some cases, large amounts of buried cables around the building may preclude the placement of a new foundation system outside the building. However, factors that make shells an attractive option include the following:

- Interruption of interior ongoing operations is minimized because the bulk of the work is done outside of the building.
- A shell can meet almost any specific blast-resistance criteria. This is not true for other upgrade options where certain upper limits will apply.
- Constructability of connections can be less of a problem for shell building upgrades. The logistics of adequately connecting new upgrade components to the existing building can be quite difficult. Access to the critical joints in buildings that require reinforcement to upgrade the building blast capacity can sometimes be virtually impossible.

Suggestions on some of the shell upgrade options are shown in Figure 9-19. A number of issues, however, need to be addressed. The following considerations are not all-inclusive:

- Foundations: A gap should be maintained between the new and the existing wall to prevent the existing wall or blocks from being knocked into the building when the outside panel deflects under blast loading. The width of this gap affects the location of the footing for the outer shell. Ideally, the shell will rest on its own new footing. But a thick wall required by high blast load may require a large footing that could encroach on the existing foundation. In this case, techniques will include staggering the horizontal level of the footings or perhaps using piers or piles adjacent to the existing footing.
- Structural: The depth of structural steel columns should fit in the gap between the new and the existing building walls.

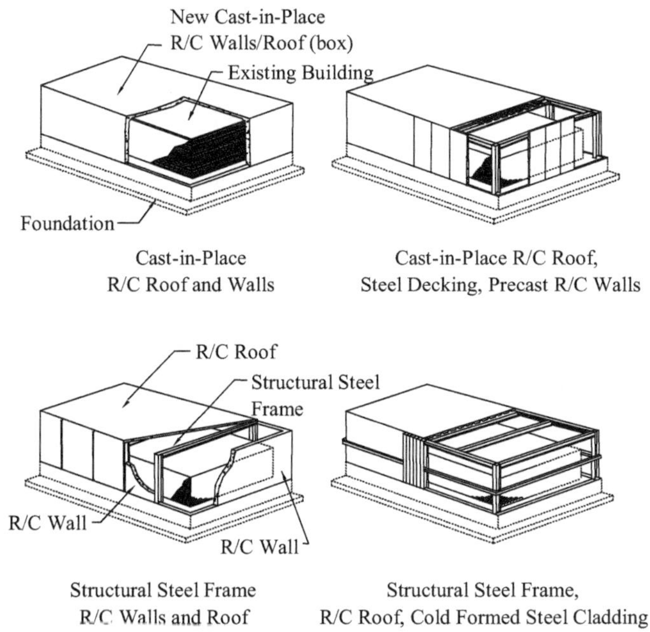

Figure 9-19. Blast-resistant shell around an existing building.

- Ancillaries: Pipe racks and cable trays that are placed adjacent to the existing building will require attention. Penetrations in the new wall are necessary for power cables and instrument lines. These openings should not adversely affect the pressure rating of the building.
- Air gap: A gap should be maintained between the shield wall and the existing building wall. Refer to Section 9.8 for detailed guidance.
- Existing HVAC and air intake stack issues discussed previously must also be addressed with this upgrade approach.

9.12 WALL AND ROOF UPGRADES WITH ENERGY-ABSORBING SYSTEMS

Typically, upgrades consist of adding strength and/or ductility to a structural component. However, these conventional options might not always be feasible for several reasons, including cost-effectiveness, accessibility, and increased foundation load from the upgrade. In these cases, an energy absorption technique may be considered. This technique is commonly used to resist impact loads and it may therefore also be applicable for upgrades to improve the blast resistance of a building. An energy absorption technique that has been used to upgrade building for blast loads is placing a shield system in front of an existing wall or roof component that is attached to the existing structure with crushable, energy-absorbing components. The energy-absorbing components directly transfer a load into the existing structure that is limited to their yield capacity. This transferred load is of much lower magnitude and longer duration compared to the blast load. Information on blast tests and design methods for use of this concept to upgrade buildings to resist blast loads is available in the literature (Whitney 1996, Lavarnway and Pollino 2015, Oswald 2018). Whitney (1996) describes energy-absorbing connectors that were designed, tested in a shock tube, and used for upgrade of a building at a petrochemical plant.

The energy-absorbing components may be crushable struts, sandwich materials, honeycomb material, polypropylene fibers, or other material that is designed to yield in compression at a prescribed load that can be resisted by the existing structure and maintain the yield resistance even when pushed to large strains. The shield wall spans between supports from the energy-absorbing components. The crushable components must be able to undergo a deflection large enough so that their strain energy absorbs the energy of the applied blast load at a deflection that does not exceed their strain capacity or the available space between the shield wall and the existing structure. If their strain capacity is exceeded, the energy-absorbing component will fail or "lock-up" so that it acts as a rigid component that directly transfers the applied blast load into the existing structure. The additional pressure created by the sudden entrapment of any unvented air between the shield wall and the existing structure must be considered. A vent area into the existing building must usually be provided to avoid this additional pressure.

Figure 9-20 shows different shapes of struts fabricated with aluminum or steel to yield at a constant load as they deform to absorb energy. Figure 9-21

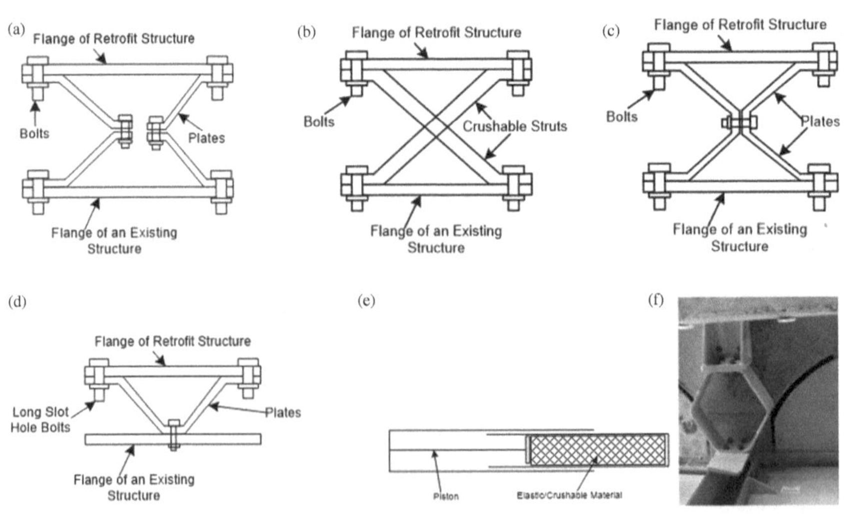

Figure 9-20. Crushable struts: (a) "V" strut, (b) "X" strut, (c) double "U" strut, (d) single "U" strut, (e) piston type, (f) bent plate.

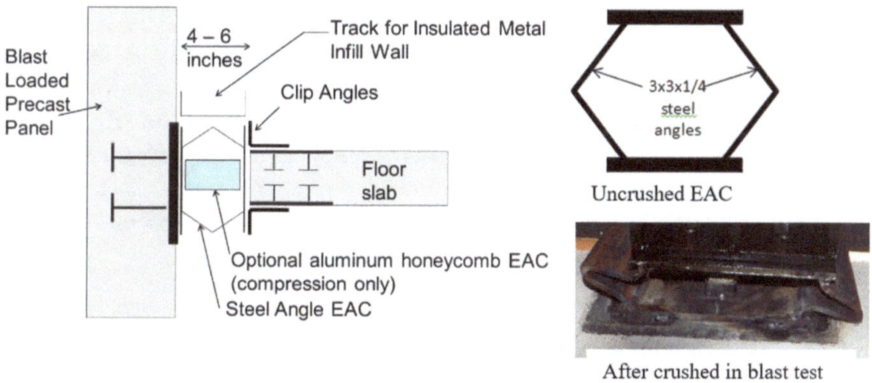

Figure 9-21. Energy-absorbing connectors for new concrete panels.

shows energy-absorbing connectors between a new concrete panel and an existing building with limited lateral load–resisting capacity and limited space available for a high-strength connection to an upgraded wall. These approaches absorb much of the blast energy with crushable elements that limit the peak blast load transmitted into the existing structure. Other similar energy-absorbing components can also be used.

9.13 WINDOW UPGRADES

Windows can be a significant hazard to the occupants of existing buildings. Choosing the most appropriate option to upgrade windows requires a knowledge of the relationship between glass strength, blast loads, and the interior hazard from failed windows. Window failure from blast loads is sometimes acceptable with certain industry clients if the glass is not thrown into the building with sufficient velocity to injure building occupants. Only limited blast test data are available from glazing manufacturers on specific window products that provide protection against glazing hazards. These data include a validation of specific applied blast loads on a specific window size, glazing layup, frame, bite, and frame anchorage.

In general, upgrading of windows may include the following options:

- Elimination of windows. A common requirement is that no window is allowed within 200 ft (61 m) from a potential blast source. A large number of petrochemical control rooms now use closed-circuit TV monitors to watch the process units.
- Replacement of windows with laminated glass, which consists of two or more plies of annealed or heat-strengthened glass bonded by a clear inner layer of polyvinyl butyral. The glass must have a large bite into the frame [i.e., up to 1 in. (25 mm)] or be attached to the frame with high-strength structural

silicone sealants so that the entire window is not thrown out of the frame into the building by a blast load. The frame attachment to the building wall must also resist high-reaction forces.

- Placement of a minimum 4 mi (100 μm) thick safety film on windows. It should be noted that the application of the film does not improve the strength of the glass but only reduces the number of glass fragments. Care must be exercised not to trade off small glass hazard with blunt impact hazard from a whole sheet of glass attached together by the film. The film can be attached to the window frame with a batten system or with high-strength structural silicone so that the whole sheet of filmed glass is not thrown into the building when the glass breaks. Film strength degrades with time and should typically be replaced at the end of the manufacturer's warranty period.
- Reduce the span width of the open glass area by adding support struts or mullions, as shown in Figure 9-22.
- Installation of a "catch system" to block larger glass fragments, or even the entire windowpane and frame. The system must be able to stop the entire window missile within a reasonable distance. Numerous catch systems are available or under development by vendors. These catch systems should be used only when there are test data validating the effectiveness of the catch system and the catch system attachments to the building for a blast load similar to the design blast load, or they should be used only when the system is designed with an approach that is validated with test data. A more generic

Figure 9-22. Window retrofit using mullions and a film.

Figure 9-23. Window film and a catch bar retrofit.

catch bar system, as shown in Figure 9-23, has also been found to be effective when combined with a daylight film application. The bar requires very strong connections to the wall, so that it yields in flexure before the connection fails.

- Narrow [e.g., 12 in. (300 mm) wide] vertical-oriented blast-resistant windows placed in new shield, blocked-in, or cocoon walls allow light into the building without significantly sacrificing the blast-resistant wall design.

The effects of negative pressure and rebound can be very important for glazing if there is concern regarding window fragments outside the building and should therefore be included in the upgrade design. It should be noted that even if a window is upgraded with a higher-strength glass type or a battened film system, the structural integrity of the window frame and the frame attachments to the building structure must be investigated. If the frame attachment to the wall is not able to withstand reaction forces from the blast-resistant window, the entire window frame will become a hazard instead of the small glass fragments. The window frame, mullions, and the attachments of these components to the building should always be designed to have a larger blast capacity than the supported window system.

9.14 DOOR UPGRADES

Conventional hollow metal doors with a 1.5 h fire rating (i.e., Class B fire doors) have a low blast capacity. These doors fail in a number of modes, beginning with failure during rebound in the unseating direction when the door is overloaded by blast to a moderate extent. If these doors have a cylindrical latch, they may withstand a rebound force of 50 psf, or 0.35 psi (2.4 kPa). Doors with a mortised latch may be adequate for a rebound force of 100 psf, or 0.7 psi (4.8 kPa). When conventional doors are severely overloaded by the blast load, the combined deformation of the door panel and the frame allow the door panel to be blown into the building. This is illustrated in Figure 9-24 for a blast load that was sufficient to deform the door through the frame but not large enough to throw it as a missile.

In some cases, existing doors can be strengthened to resist the blast load. Attention must be paid to the light steel door frame, because large blast loads cause the door frame to deform, allowing the door to be pushed through the door frame into the building. In general, test data are necessary to show that an

Figure 9-24. Conventional hollow metal door damage under a low blast load.

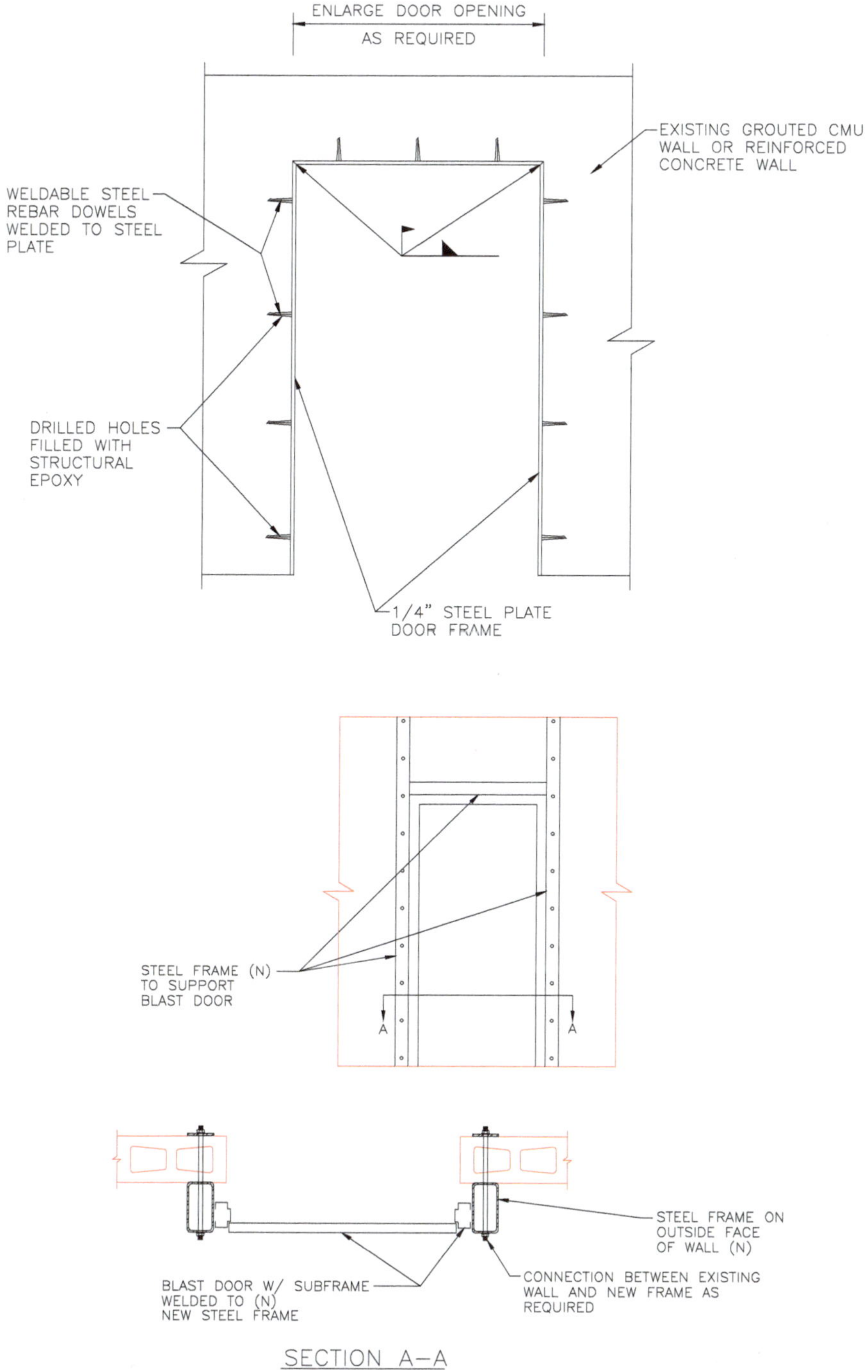

Figure 9-25. Concepts for providing a blast door frame in existing masonry or a reinforced concrete wall.

upgraded conventional door and frame can resist a specified blast load. In low-occupancy buildings, doors may be attached to steel framing with cables that allow the door to fail but catch the door before it is thrown well into the building. The cable attachment to the door must be made with through bolts and large plate washers that engage a large amount of door material, because conventional doors have very thin gauge steel panels. Also, the steel cable must be designed to resist the portion of the blast load applied to the door before it is caught by the cable without yielding, because steel cable is manufactured from nonductile wire.

Typically, the most practical approach is to replace inadequate doors with blast-resistant doors supplied by a specialty door vendor. Unlike conventional doors, blast doors are typically provided as a complete assembly, including the door, frame, hardware, and accessories. This is because all the components are dependent on each other to provide the overall blast resistance. Refer to Chapter 8 for performance requirements and design details for blast-resistant doors. The primary challenge associated with using a new blast-resistant door is often how to achieve an adequate attachment of the new door frame to the surrounding building (i.e., it is challenging to achieve an adequate connection). Figure 9-25 shows several concepts for attaching a new blast-resistant door frame to an existing masonry or reinforced concrete wall. Conservatively, the door frame attachment can be designed to have an ultimate strength that resists twice the peak applied blast pressure on the door. This requirement can be reduced if there is a detailed analysis to determine the maximum dynamic forces transferred by the door panel to the door frame. A separate steel frame supported by the building framing and the foundation can also be constructed inside the building to support a new blast-resistant door.

References

ACI (American Concrete Institute). 2017. *Guide for the design and construction of externally bonded FRP systems for strengthening concrete structures*, ACI 440.2R-17. Farmington Hills, MI: ACI.

ASCE. 2023. *Seismic evaluation and retrofit of existing buildings*, ASCE 41-23. Reston, VA: ASCE.

ASCE/FEMA (ASCE/Federal Emergency Management Agency). 2000. *Prestandard and commentary for the seismic rehabilitation of buildings*, FEMA 356. Washington, DC: FEMA.

ASTM International. 2014. *Standard specification for structural bolts, alloy steel, heat treated, 150 ksi minimum tensile strength*, ASTM A490-12. West Conshohocken, PA: ASTM.

ASTM. 2016. *Standard specification for structural bolts, steel, heat treated, 120/105 ksi minimum tensile strength (withdrawn 2016)*, ASTM A325-14. West Conshohocken, PA: ASTM.

ASTM. 2023. *Standard specification for carbon steel bolts, studs, and threaded rod 60 000 psi tensile strength*, ASTM A307-21. West Conshohocken, PA: ASTM.

Bazan, M., and E. Conrath. 2022. "Implications of blast-resistant retrofits on the blast response of masonry walls." In *Proc., 22nd Int. Symp. on Military Aspects of Blast and Shock (MABS)*. Austrasse, Spiez, Switzerland: MABS.

Bazan, M., E. Conrath, and C. Oswald. 2012. "Implications of blast-resistant retrofits on the blast response of masonry walls." In *Proc., Military Aspects of Blast and Shock (MABS) Conf.* Austrasse, Spiez, Switzerland: MABS.

CCPS (Center for Chemical Process Safety). 2012. *Guidelines for evaluating process plant buildings for external explosions, fires, and toxic releases.* 2nd ed. New York: American Institute of Chemical Engineers.

ICC (International Code Council). 2021. *Plastic.* Chapter 26. Washington, DC: ICC.

Lavarnway, D., and M. Pollino. 2015. "Mitigation of air-blast pressure impulses on building envelopes through blast resistant ductile connectors." *J. Eng. Arch.* 3 (2): 9–24.

NFPA (National Fire Protection Association). 2023. *Standard on explosion protection by deflagration venting,* NFPA 68. Quincy, MA: NFPA.

Oswald, C. 2002. "Test data and design guidance for steel panels that limit blast load transferred into frame." In *Proc., 30th Annual Department of Defense Explosives Safety Board (DDESB) Seminar - Department of Defense Armed Services Explosives Safety Board (ASESB).*

Oswald, C. 2018. "Blast testing on energy absorbing connectors for blast resistant design." In *Proc., 15th Int. Conf. on Structures under Shock and Impact.* Wessex Institute of Technology. https://www.wessex.ac.uk/conferences/2018/susi-2018.

SAC (Structural Engineering Association of California). 1995. *Parametric analytical investigations of ground motion and structural response, Northridge Earthquake of January 17, 1994,* edited by S. Campbell, M. Englehardt, J. Hall, G. Hart, et al. Richmond, CA: SAC Joint Venture.

SidePlate Systems. 1998. *Blast resistant strengthening of pre-engineered steel frame buildings using SidePlate moment connection technology.* Mission Viejo, CA: SlidePlate Systems.

USACE/NAVFAC/AFCESA (US Army Corp of Engineers/Naval Facilities Engineering Command/Air Force Civil Engineer Support Agency). 2014. *Structures to resist the effects of accidental explosions,* UFC 3-340-02, Change 2. Washington, DC: National Institute of Building Sciences.

US DoD (US Department of Defense). 2005. *Retrofit of existing buildings to resist explosive effects,* UFC 4-023-02. Washington, DC: National Institute of Building Sciences.

Whitney, M. G. 1996. *Blast damage mitigation using reinforced concrete panels and energy absorbing connectors.* San Antonio: Wilfred Baker Engineering.

CHAPTER 10

Shear Wall Building Design Example

10.1 INTRODUCTION

The following is a sample blast design for a control building using reinforced concrete walls and a structural steel frame for vertical support. There are two blast load cases, one applied to the long side of the building, and the other applied to the short side. The explosion source and side-on overpressure [6 psi (41 kPa) for 0.05 s] are determined by others. Reflected and side-on blast design parameters for some components are taken from Chapter 3, Appendix 3A.

In this example, blast loads and dynamic properties are computed on a unit force basis in contrast to Chapter 11 calculations. Blast loads are directly applied to a component over an appropriate tributary area.

Structural code provisions, as applicable, are from

- ACI 318-19 (ACI 2022), *Building Code Requirements for Structural Concrete*
- ANSI/AISC 360-16 (ANSI/AISC 2019), *Specification for Structural Steel Buildings*
- ANSI/AISC 341-16 (ANSI/AISC 2016), *Seismic Provisions for Structural Steel Buildings*

Additional references are

- Biggs (1964), *Introduction to Structural Dynamics*
- Blodgett (1966), *Design of Welded Structures*
- Budynas and Sadegh (2020), Roark's *Formulas for Stress and Strain*
- AISC Manual (AISC 2017), 15th Edition, *AISC Manual of Steel Construction*
- AISC Guide (AISC 2024), *AISC Design Guide 1, Column Base Plates*
- UFC 3-340-02 (USACE/NAVFAC/AFCESA 2014), *Structures to Resist the Effects of Accidental Explosions*

For brevity, evaluation of conventional loads is not included in this example. Designs for blast doors and foundation are also not included in this example.

10.2 STRUCTURAL SYSTEM

The structure in this example is of cast-in-place concrete wall construction, as described in Chapter 4, Section 4.3.7. Vertical loads are resisted by a structural steel frame. Lateral loads are resisted by the concrete roof diaphragm and by the side shear walls.

10.2.1 Description of the Structure

- One-story structure with reinforced concrete walls and roof slab supported by steel framing.
- Plan dimensions are 93 ft (28.3 m) by 67 ft (20.4 m).
- Wall height is 12 ft (3.66 m).
- Columns support roof girders, which have an 18 ft (5.49 m) spacing across the long dimension of the building.
- Roof beams span between roof girders with 8 ft (2.44 m) spacing.
- Roof slab consists of a 5 in. (127 mm) full concrete section poured over a 2 in. (51 mm) metal deck.
- Foundation for walls consists of a series of vertical and battered piles around the perimeter; foundation for columns consists of a pier, pile cap, and vertical piles.

10.2.2 Framing Plan

For the framing plan, refer to Figure 10-1.

10.2.3 Components for Blast Design

The design will proceed component by component. The tracking of reactions to supported structural elements is usually not utilized. Each component will be designed as an independent uncoupled structural member.

Lateral load–resisting components include the front wall, back wall, side/shear walls, roof diaphragm, and foundation. Vertical load–resisting components include the roof slab, roof beams, roof girders, columns, and foundation. A section through the reinforced concrete shear wall is shown in Figure 10-2. This section applies specifically to the shorter shear walls, where the longer exterior walls would show a roof girder beyond.

Determine required member sizes for

1. Exterior walls (out-of-plane loads),
2. Roof slab (out-of-plane loads),
3. Roof slab (in-plane loads),
4. Side walls (in-plane loads),

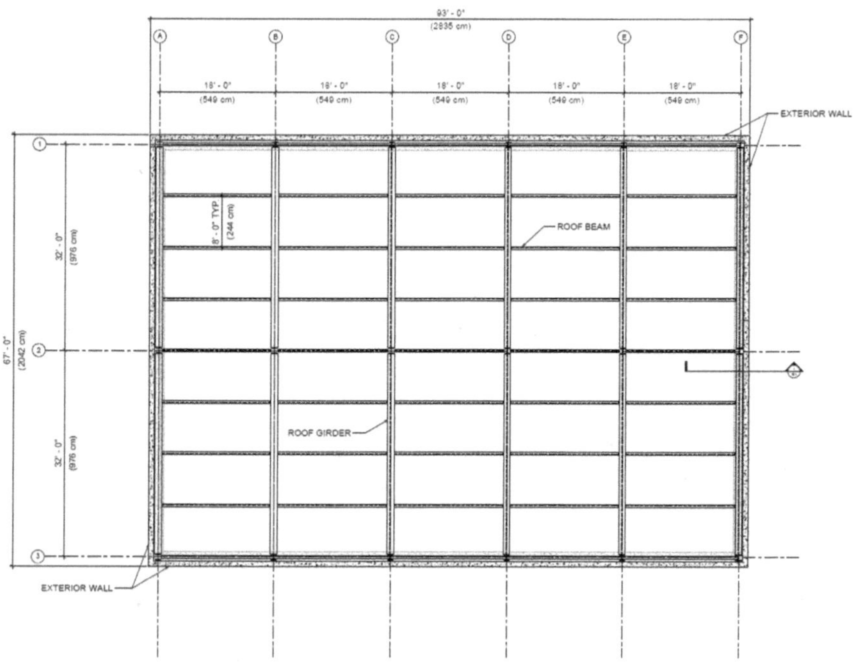

Figure 10-1. Roof framing plan.

5. Roof beams (and their connections),
6. Roof girders (and their connections), and
7. Columns (and their base connections).

10.2.4 General Solution Procedure

1. Calculate design blast loads,
2. Select trial sizes,
3. Compute bending resistance,
4. Compute shear resistance,
5. Compute other SDOF properties,
6. Compute the response and compare it to deformation limits,
7. Revise section as required, and
8. Design connections for steel components.

Notes: This general solution procedure is not rigid. It varies in some regard from the member design process steps presented in Chapter 7, Section 7.3, as well as the steps used in the Chapter 11 calculations.

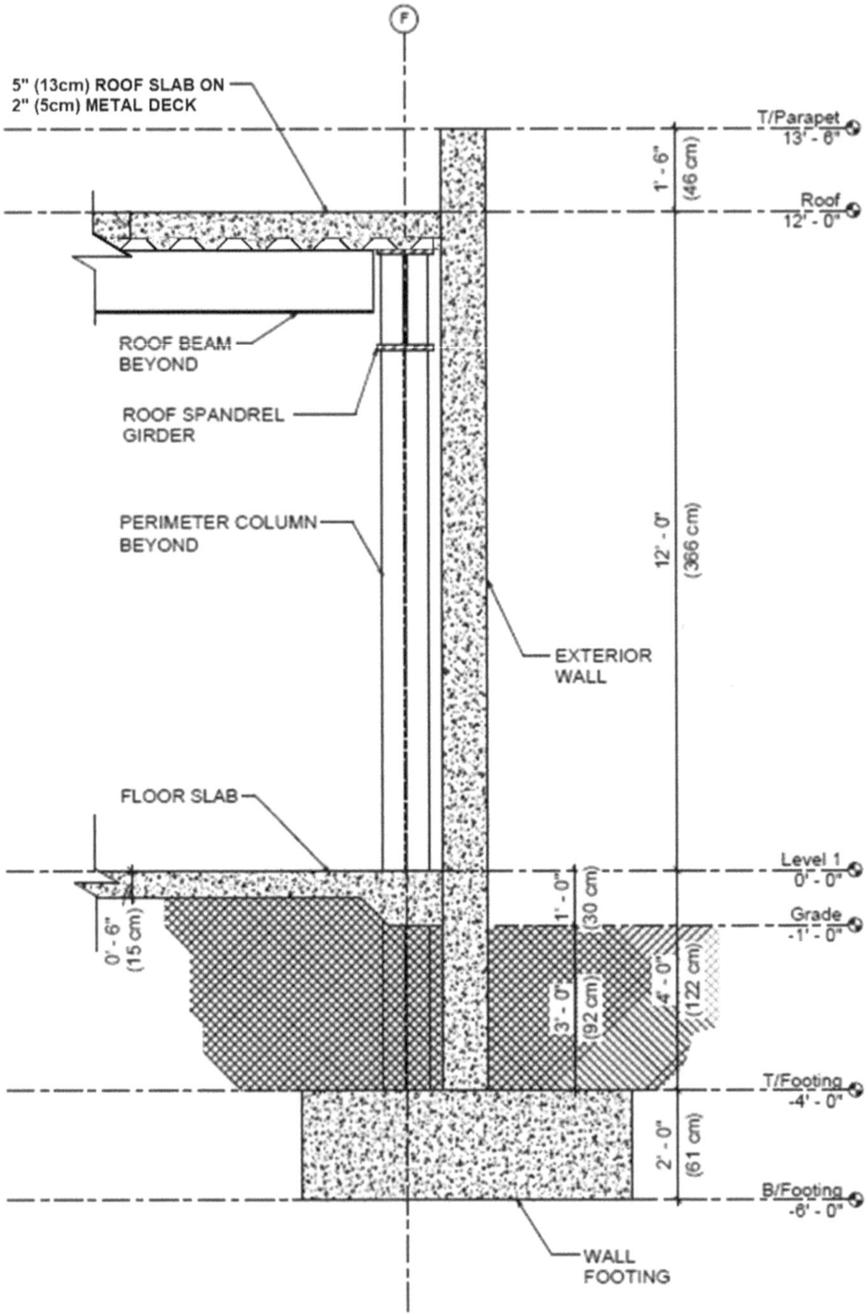

Figure 10-2. Wall section.

10.3 DESIGN DATA

10.3.1 Material Properties

As is typical for blast design in the petrochemical industry, commonly used structural materials will be used.

Concrete: $f_c' = 4{,}000$ psi (27.6 MPa)

Reinforcing steel: ASTM A615/A615M-22 (ASTM 2024), Grade 60, $F_y = 60$ ksi (414 MPa)

Structural steel (W shapes): ASTM A992/A992M-01 (ASTM 2017), $F_y = 50$ ksi (345 MPa)

Structural steel (plates): ASTM A36/A36M-19 (ASTM 2019), $F_y = 36$ ksi (248 MPa), $F_u = 58$ ksi (400 MPa)

Strength increase factors (SIFs) and dynamic increase factors (DIFs) are taken from Chapter 5, Tables 5A-1 through 5A-3. These values are summarized in Table 10-1.

Because of the low permissible dynamic response, $F_{ds} = F_{dy}$ (Chapter 5, Tables 5A-4 and 5A-5).

The ratio of steel modulus of elasticity to that of concrete is given by $n = E_s/E_c = (29{,}000 \text{ ksi})/(3{,}605 \text{ ksi}) = 8.04$. This quantity will be important when determining effective section properties later in this chapter.

10.3.2 Design Loads

Roof dead load: 10 psf (0.5 kPa)

This value is based on supported weight associated with nonstructural elements including roofing materials, insulation, and mechanical systems.

Roof live loads are assumed to be negligible at the time of a potential blast incident.

The design blast loads are usually taken from those calculated in Chapter 3, Appendix 3A. Some additional blast loads are considered for roof beams, roof

Table 10-1. Dynamic Material Properties.

Material	*F_y or f_c'*	*SIF*	*DIF*	*Modulus of elasticity, E_s or E_c*
Concrete	4 ksi (27.6 MPa)	1.0	1.19[a]	3,605 ksi (24,900 MPa)
Reinforcing steel	60 ksi (414 MPa)	1.1	1.22[a]	29,000 ksi (200,000 MPa)
Structural steel (angles, plates)	36 ksi (248 MPa)	1.1	1.29	29,000 ksi (200,000 MPa)
Structural steel (W shapes)	50 ksi (345 MPa)	1.1	1.19[b]	29,000 ksi (200,000 MPa)

[a]For diagonal tension shear capacity checks, DIF = 1.0. For direct shear capacity checks, DIF = 1.1.
[b]For tension/compression of W shapes, DIF = 1.12.

Table 10-2. Response Limits.

Element	Ductility ratio, μ_a	Support rotation, θ_a (degree)
Reinforced concrete wall (with rebar at each face and $RI < 0.05$)	—	2
Reinforced concrete wall (with rebar at each face and $RI \geq 0.05$)	—	1
Reinforced concrete roof slab (with a single layer of rebar)	—	1
Reinforced concrete diaphragm (flexure controlled)	3	—
Reinforced concrete shear wall (flexure controlled)	—	1
Reinforced concrete shear wall/ diaphragm (shear controlled)	1.3	—
Roof beams	3	2
Roof girders	1.5	1

girders, and columns based on unique parameters for those elements related to blast wave sweeping.

Case A, explosion occurs on the long side of the building.

Case B, explosion occurs on the short side of the building.

10.3.3 Building Performance Requirements: Deformation Limits

A low response range is selected to maximize the reuse of the building with minimal cost of repairs. Corresponding response limits for the members to be designed are taken from Chapter 5, Appendix 5B. These values are summarized in Table 10-2. The column limit can conservatively be taken as $\mu_a = 1$ based on the critical buckling load for pure element compression.

10.4 EXTERIOR WALLS (OUT-OF-PLANE LOADS)

The exterior concrete walls are 12 ft (3,658 mm) from the top of the floor slab to the top of the roof slab. The shorter walls are 67 ft (20,420 mm) long, a 5.6 to 1 length to height ratio. Therefore, all walls will be analyzed as one-way beams pinned at the base by a crossed reinforcing configuration at the slab level (similar to Figure 7-12) and pinned at the top because of a thinner roof slab (Figure 7-13). The wall is assumed to be unrestrained (for axial forces) at the top end and will not respond in tension membrane action.

Note: If the wall panel aspect ratio were less than 2 to 1, the panel could be analyzed as a two-way span using resistance equations and transformation factors from the applicable tables in UFC 3-340-02.

Notes: The response of the parapet wall above the roof slab is not evaluated in this example. The blast wave would quickly wrap around this short wall section for negligible net effect. If the parapet has a more substantial height, this effect on the wall section should be checked. One approach is to analyze the parapet as a cantilever with flexural response to the net applied blast load and the resulting base reaction added to the load acting on the roof diaphragm. The relative timing of the peak reaction forces of the parapet and main span may be considered when adding these reactions. The additional reaction from the parapet could potentially have a significant effect on related connection design.

Span, $L = 12$ ft or 144 in. (3,658 mm) from the top of the floor slab to the top of the roof slab

Use a nominal design width, $b = 1$ ft or 12 in. (305 mm).

10.4.1 Calculate Front Wall Design Blast Load (from Load Case A)

Front wall loading results in the maximum wall response.

From the Appendix 3A calculation:

$P_r = 13.8$ psi (95 kPa)

$P_s = 6.8$ psi (47 kPa)

$t_c = 0.034$ s

$t_d = 0.05$ s

$t_e = 0.042$ s

In addition, the intermediate point where the cleared reflection load and stagnation load intersect is 2.2 psi (15 kPa) at $t_c = 0.034$ s. The blast-loaded area tributary to the exterior wall is the wall height of 144 in. (3,658 mm) multiplied by the nominal design width of 12 in. (305 mm).

Peak load:

$$F_o = (13.8\text{ psi})(144\text{ in. span})(12\text{ in. width})/(1{,}000\text{ lb/kip}) = 23.8\text{ kips }(106\text{ kN})$$

Intermediate load point:

$$F_{\text{int}} = (2.2\text{ psi})(144\text{ in. span})(12\text{ in. width})/(1{,}000\text{ lb/kip}) = 3.8\text{ kips }(16.9\text{ kN})$$

The total load can then be represented by either the bilinear-cleared loading profile or an equivalent triangular loading profile shown subsequently.

Note: Using the simplified equivalent triangular load is usually a more conservative approach for design because it has the same peak pressure and it applies the same impulse over a shorter duration.

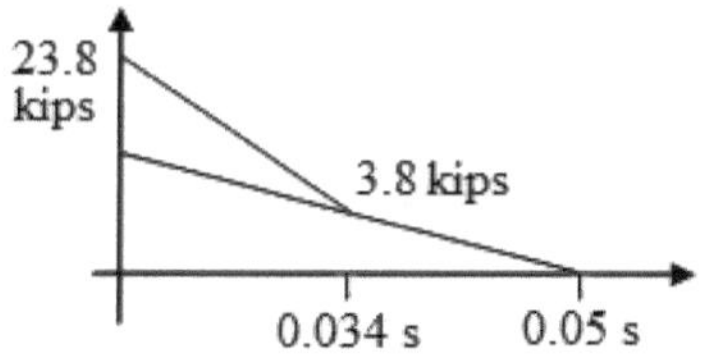

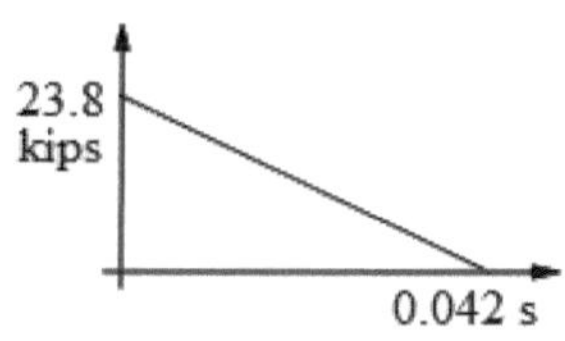

10.4.2 Calculate Side Wall Design Blast Load

Considering a uniform wall construction around the building, side wall loading will not directly result in the maximum wall response. However, some additional thought should be given when the side walls are subjected to both in-plane and out-of-plane loads to ensure adequate overall system response. See related discussion in Section 10.7.2.

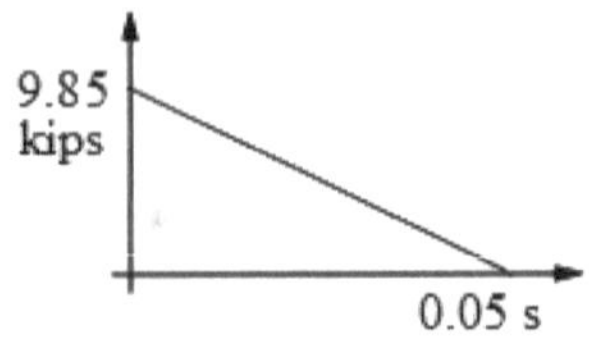

From the Appendix 3A calculation:

$P_a = 5.7$ psi (39 kPa)

$t_r = 0$ s

$t_d = 0.05$ s

Peak load:

$F_o = (5.7 \text{ psi})(144 \text{ in. span})(12 \text{ in. width})/(1{,}000 \text{ lb/kip}) = 9.85 \text{ kips } (43.8 \text{ kN})$

10.4.3 Calculate Rear Wall Design Blast Load

Ignore this case because it will not control the response.

10.4.4 Select Trial Sizes

Notes: The following trial dimensions and material proportions may be obtained from trial calculations, by inspecting similar structures, or from experience. The results of this dynamic calculation will determine the adequacy of the trial size.

Wall thickness, $h = 10$ in. (254 mm)

#5 @ 6 in. (152 mm), each face, vertical

#5, diameter = 0.625 in. (15.9 mm); #5, area = 0.307 in.^{2} (198 mm^2)

#3 @ 6 in. (152 mm), each face, horizontal

#3, diameter = 0.375 in. (9.5 mm); #3, area = 0.110 in.^{2} (71.3 mm^2)

Vertical bars are positioned inside of horizontal bars. Take clear covers as minimum values per ACI 318 (ACI 2022), 0.75 in. (19 mm) for the non-exposed interior face and 1.5 in. (38.1 mm) for the exposed exterior face.

10.4.5 Compute Bending Resistance

For dynamic bending: (Chapter 5, Appendix 5A)

$$F_{ds} = F_{dy} = (\text{SIF})(\text{DIF})\, F_y = (1.1)(1.22)(60 \text{ ksi}) = 80.5 \text{ ksi } (555 \text{ MPa})$$
$$f'_{dc} = (\text{SIF})(\text{DIF})\, f'_c = (1.0)(1.19)(4 \text{ ksi}) = 4.76 \text{ ksi } (32.8 \text{ MPa})$$

For the nominal design width:

$A_s = (0.307 \text{ in.}^2)(12 \text{ in./ft})/(6 \text{ in. bar spacing}) = 0.614 \text{ in.}^2 \ (396 \text{ mm}^2)$

For bending tension on the inside face: (ACI 318, Table 20.5.1.3.1)

$$d = (10 \text{ in. thick}) - (0.75 \text{ in. clear}) - (0.375 \text{ in. horizontal bar}) - (0.625 \text{ in. vertical bar})/2 = 8.56 \text{ in. } (217 \text{ mm})$$

$$A_{s,\,min} = 3\frac{\sqrt{f'_{dc}}}{F_{ds}}bd = 3\frac{\sqrt{4{,}760 \text{ psi}}}{80{,}500 \text{ psi}}(12 \text{ in.})(8.56 \text{ in.}) \text{ (ACI 318, Section 9.6.1.2)}$$
$$= 0.264 \text{ in.}^2 \ (170 \text{ mm}^2) < A_s, \quad \textbf{OK}$$

$$A_{s,\,min} = 200\ bd/F_{ds} = 200\ (12 \text{ in.})(8.56 \text{ in.})/(80{,}500 \text{ psi}) \text{ (ACI 318, Section 9.6.1.2)}$$
$$= 0.255 \text{ in.}^2 \ (165 \text{ mm.}^2) < A_s, \quad \textbf{OK}$$

$$d_{sb} = A_s F_{ds}/0.85\ f'_{dc}\ b$$
$$= (0.614 \text{ in.}^2)(80.5 \text{ ksi})/(0.85)(4.76 \text{ ksi})(12 \text{ in.})$$
$$= 1.02 \text{ in. } (25.8 \text{ mm})$$

Note: In ACI 318, the parameter "d_{sb}" is defined as "a."

$$M_p = A_s F_{ds}\,[d - d_{sb}/2]$$
$$= (0.614 \text{ in.}^2)(80.5 \text{ ksi})\ [(8.56 \text{ in.}) - (1.02 \text{ in.})/2]$$
$$= 398 \text{ kip·in. } (45.0 \text{ kN·m})$$

Positive (inward) bending resistance based on pinned ends:

$R_b = 8\,M_p/L = 8\ (398 \text{ kip·in.})/(144 \text{ in.}) = 22.1 \text{ kips } (98.3 \text{ kN})$ (Chapter 6, Table 6-1)

For bending tension on the outside face: (ACI 318, Table 20.5.1.3.1)

$$d = (10 \text{ in. thick}) - (1.5 \text{ in. clear}) - (0.375 \text{ in. horizontal bar}) - (0.625 \text{ in. vertical bar})/2 = 7.81 \text{ in. } (198 \text{ mm})$$

$$A_{s,\,min} = 3\frac{\sqrt{f'_{dc}}}{F_{ds}}bd = 3\frac{\sqrt{4{,}760 \text{ psi}}}{80{,}500 \text{ psi}}(12 \text{ in.})(7.81 \text{ in.}) \text{ (ACI 318, Section 9.6.1.2)}$$
$$= 0.241 \text{ in.}^2 \ (155 \text{ mm}^2) < A_s, \quad \textbf{OK}$$

$$A_{s,\,min} = 200\ bd/F_{ds} = 200\ (12 \text{ in.})(7.81 \text{ in.})/(80{,}500 \text{ psi}) \text{ (ACI 318, Section 9.6.1.2)}$$
$$= 0.233 \text{ in.}^2 \ (150 \text{ mm}^2) < A_s, \quad \textbf{OK}$$

$$M_p = A_s F_{ds}\,[d - d_{sb}/2]$$
$$= (0.614 \text{ in.}^2)(80.5 \text{ ksi})\ [(7.81 \text{ in.}) - (1.02 \text{ in.})/2]$$
$$= 361 \text{ kip·in. } (40.8 \text{ kN·m})$$

Rebound (outward) bending resistance based on pinned ends:

$R_{b\text{-}} = 8\,M_p/L = 8\ (361 \text{ kip·in.})/(144 \text{ in.}) = 20.0 \text{ kips } (89.2 \text{ kN})$ (Chapter 6, Table 6-1)

Note: Equations from ACI 318 throughout this chapter have been modified to incorporate dynamic material properties. Equations with static properties also need to be satisfied to meet code requirements.

10.4.6 Compute Shear Resistance

For dynamic shear: (Chapter 5, Appendix 5A)

$$f'_{dc} = (\text{SIF})(\text{DIF})\ f'_c = (1.0)(1.0)(4\text{ ksi}) = 4.0\text{ ksi (27.6 MPa)}$$

A DIF value of 1.0 is used for evaluating diagonal tension shear capacity. Because positive or rebound bending can occur, calculate shear resistance on the basis of the smaller of d based on inside tension or outside tension. In this case, use $d = 7.81$ in (198 mm).

$$\begin{aligned} V_n &= 2\sqrt{f'_{dc}}\ bd \text{ (ACI 318, Table 22.5.5.1)} \\ &= 2\sqrt{4{,}000\text{ psi}}\ (12\text{ in.})(7.81\text{ in.})/(1{,}000\text{ lb/kip}) \\ &= 11.9\text{ kips (52.7 kN)} \end{aligned}$$

The critical section for shear is at a distance d from the support.

$$\begin{aligned} R_s &= V_n\,L/(0.5\,L - d) \\ &= (11.9\text{ kips})(144\text{ in.})/[0.5\ (144\text{ in.}) - (7.81\text{ in.})] \\ &= 26.6\text{ kips (118 kN)} \end{aligned}$$

10.4.7 Resistance and Permissible Response

Because $R_b < R_s$, bending controls.

Positive resistance, $R_u = R_b =$ **22.1 kips** (98.3 kN)

Rebound resistance, $R_{u-} = -R_{b-} =$ **−20.0 kips** (−89.2 kN)

Reinforcement index, $RI = A_s/bd \times F_{ds}/f'_{dc}$ [Chapter 5, Equation (5-1)]

$$= (0.614\text{ in.}^2)/(12\text{ in.})(8.56\text{ in.}) \times (80.5\text{ ksi})/(4.76\text{ ksi}) = 0.10$$

Response limit: $\theta_a = 1$ degree (based on no shear reinforcing and $RI > 0.05$)

Note: The inbound d value was used to calculate the reinforcement index, RI, because the inbound response typically controls the design.

10.4.8 Compute Other Single-Degree-of-Freedom Properties

Moment of inertia will be based on positive (inward) bending.

Gross moment of inertia:

$$I_g = bh^3/12 = (12\text{ in.})(10\text{ in.})^3/12 = 1{,}000\text{ in.}^4\ (416.2 \times 10^6\text{ mm}^4)$$

Transformed rebar area:

$$nA_s = (8.04)(0.614\text{ in.}^2) = 4.94\text{ in.}^2\ (3{,}180\text{ mm}^2)$$

Location of transformed neutral axis:

$$d_{na} = \frac{-nA_s + \sqrt{nA_s\,(nA_s + 2bd)}}{b}$$

$$= \frac{-4.94\text{ in.}^2 + \sqrt{4.94\text{ in.}^2\,(4.94\text{ in.}^2 + 2(12\text{ in.})(8.56\text{ in.}))}}{12\text{ in.}}$$

$$= 2.27\text{ in}(57.8\text{ mm})$$

Cracked moment of inertia:

$$I_{cr} = bd_{na}{}^3/3 + nA_s\,(d - d_{na})^2$$
$$= (12\text{ in.})(2.27\text{ in.})^3/3 + (4.94\text{ in.}^2)(8.56\text{ in.} - 2.27\text{ in.})^2$$
$$= 242\text{ in.}^4\,(100.8 \times 10^6\text{ mm}^4)$$

Average moment of inertia:

$$I_a = (I_g + I_{cr})/2 = (1{,}000\text{ in.}^4 + 242\text{ in.}^4)/2 = 621\text{ in.}^4\,(258.5 \times 10^6\text{ mm}^4)$$

Elastic stiffness:

$$K_e = 384\,E_c\,I_a/5\,L^3 \text{ (Chapter 6, Table 6-1)}$$
$$= 384\,(3{,}605\text{ ksi})(621\text{ in.}^4)/5\,(144\text{ in.})^3$$
$$= \mathbf{57.6\ kip/in.}\ (10.1\text{ kN/mm})$$

Elastic yield deflection: $y_e = R_u/K_e = (22.1\text{ kips})/(57.6\text{ kip/in.}) = \mathbf{0.384\ in.}$ (9.75 mm)

Mass:

$$\text{weight} = (0.15\text{ kcf})(0.833\text{ ft thick})(1\text{ ft nominal width})(12\text{ ft span})$$
$$= 1.5\text{ kips }(6.67\text{ kN})$$

$$M = \text{weight/g}$$
$$= 1.5\text{ kips}/(386\text{ in./s}^2)$$
$$= \mathbf{0.00389\ kip{\cdot}s^2/in.}\ (680\text{ kg})$$

Note: The unusual mass unit of kip·s²/in. is used for consistency with the respective resistance and stiffness units of kip and kip/in. In metric units, $M = \text{weight/g} = 6.67\text{ kN}/9.81\text{ m/s}^2 = 0.68\text{ kN·s}^2\text{/m} = 680\text{ kg}$.

Load-mass transformation factors: (Chapter 6, Table 6-1)

$$\text{elastic, } K_{LM} = K_M/K_L = 0.50/0.64 = \mathbf{0.78}$$

$$\text{plastic, } K_{LM} = K_M/K_L = 0.33/0.50 = \mathbf{0.66}$$

Assuming some plastic deformation, use an average value for the load-mass factor when calculating the natural period:

$$K_{LM} = (0.78 + 0.66)/2 = \mathbf{0.72}$$

Equivalent mass:

$$M_e = K_{LM}\,M = 0.72\,(0.00389\text{ kip·s}^2\text{/in.}) = 0.0028\text{ kip·s}^2\text{/in. }(490\text{ kg})$$

Natural period: [Chapter 6, Equation (6-8)]

$$t_n = 2\pi\sqrt{\frac{M_e}{K_e}} = 2\pi\sqrt{\frac{(0.0028\ \text{kip}\cdot\text{s}^2/\text{in.})}{(57.6\ \text{kip/in.})}} = \mathbf{0.0438\ s}$$

10.4.9 Compute Response and Compare to Deformation Limits (Chart Solution for Front Wall)

Note: Both the chart and the numerical integration approaches need not be used. A chart solution is presented for the front wall to illustrate implementation.

To use Chapter 6, Figure 6-6, a linearly decaying, triangular waveform must be assumed. The loading profile to the right, as determined in Section 10.4.1, is used.

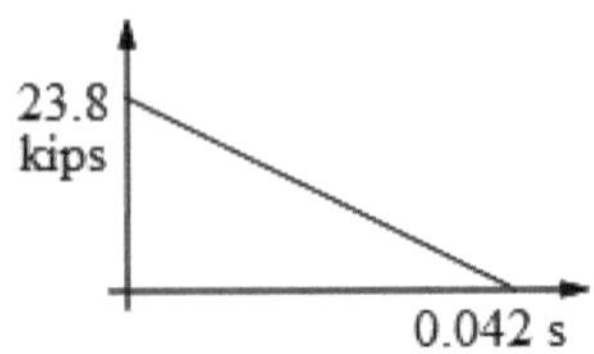

$$t_d/t_n = (0.042\ \text{s})/(0.0438\ \text{s}) = 0.96$$
$$R_u/F_o = (22.1\ \text{kips})/(23.8\ \text{kips}) = 0.93$$

Using these values, $\mu_d \approx 2.2$. (Chapter 6, Figure 6-6)
Maximum deflection: $y_m = \mu_d\, y_e = 2.2\ (0.384\ \text{in.}) = 0.845\ \text{in.}\ (21.5\ \text{mm})$
Support rotation: (Chapter 5, Figure 5-9)

$\theta_d = \tan^{-1}(y_m/0.5\,L) = \tan^{-1}[(0.845\ \text{in.})/(0.5)(144\ \text{in.})] = 0.67$ degrees $< \theta_a$, **OK**

The response is adequate. A revision is not necessary.

10.4.10 Compute the Response and Compare It to Deformation Limits (Numerical Integration Solution for the Front Wall)

The numerical integration procedure can handle the bilinear loading profile determined in Section 10.4.1 in lieu of an equivalent triangular loading.

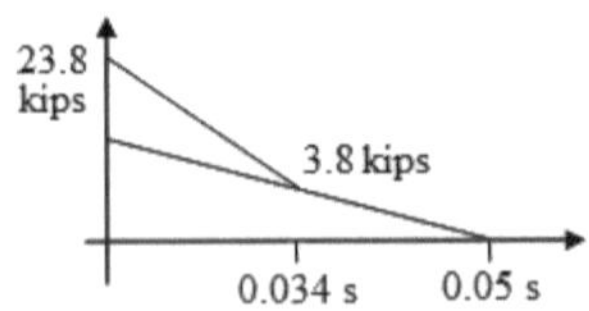

$F_o = 23.8$ kips (106 kN)

$F_{int} = 3.8$ kips (16.9 kN)

$t_c = 0.034$ s

$t_d = 0.05$ s

Time increment:

$$t_n/10 = 0.0438 \text{ s}/10 \approx 0.0044 \text{ s}$$

However, it is recommended to use at most a 0.001 s time step to obtain at least 10 increments of loading in each section of the pressure-time history. Consider a time range of 0.150 s to capture multiple deflection cycles and component behavior following load termination.

Numerical integration follows the approach outlined in Chapter 6, Appendix 6A. For the numerical integration through the time of the initial rebound, refer to Table 10-3. For corresponding output plots extending multiple cycles, see Figure 10-3.

Peak positive deflection: $y_m = +0.731$ in. (18.6 mm) at $t = 0.025$ s

Peak rebound deflection: $y_{m-} = +0.078$ in. (1.97 mm) at $t = 0.050$ s

Note: Without damping being applied, the front wall SDOF model oscillates about a permanent plastic deflection of 0.35 in. (8.8 mm).

Notes: Performing some sensitivity trials with the numerical integration method outlined in Appendix 6A, a time step of 0.001 s is sufficiently small to provide a converged solution and smooth output. In most cases, a time step of 0.0001 s provides adequate output. Larger time steps are considered in this example, however, because of the space limitations of the tabulated output results.

Support rotation: (Chapter 5, Figure 5-9)

$$\theta_d = \tan^{-1}(y_m/0.5\,L) = \tan^{-1}[(0.731 \text{ in.})/(0.5)(144 \text{ in.})] = 0.58 \text{ degrees} < \theta_a, \quad \textbf{OK}$$

The response is adequate. A revision is not necessary.

10.4.11 Compute the Response and Compare It to Deformation Limits (Numerical Integration Solution for the Side Wall)

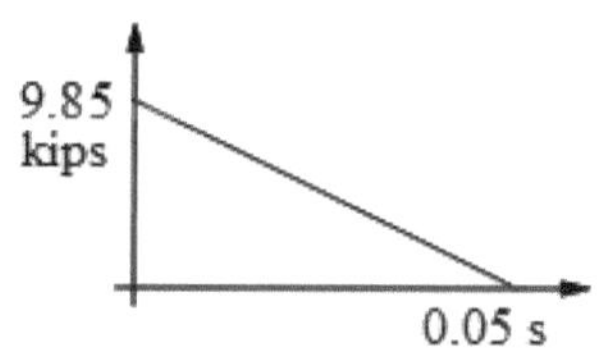

$F_o = 9.85$ kips (44 kN)

$t_r = 0$ s

$t_d = 0.05$ s

Time increment:

$t_n/10 = 0.0438 \text{ s}/10 \approx 0.0044$ s

A smaller time step can be used to potentially obtain a more refined solution with smoother output plots. Consider a time range of 0.150 s to capture multiple deflection cycles and component behavior following load termination.

Table 10-3. Front Wall Numerical Integration.

Example:	Section 10.4 Front Wall								KEY = 0 when elastic
									KEY = +1 when positive plastic
time increment:	0.001	s	**Blast Load**						KEY = -1 when rebound plastic
mass:	0.00389	k-s^2/in	**t0:**	0	s	**P0:**	23.8	kips	Z = 1 when continued elastic
stiffness:	57.6	k/in	**t1:**	0.034	s	**P1:**	3.8	kips	Z = 2 at start of positive plastic deformation
damping:	0	k-s/in	**t2:**	0.05	s	**P2:**	0	kips	Z = 3 when increasing plastic deformation
positive resistance:	22.1	kips	**Dynamic Reaction Factors**			**BC: S-S**			Z = 4 at end of positive plastic deformation
rebound resistance:	-20.0	kips	**Crel:**	0.39		**Cfel:**	0.11		Z = 6 at start of rebound plastic deformation
static load:	0	kips	**Crpl:**	0.38		**Cfpl:**	0.12		Z = 8 at end of rebound plastic deformation

time (s)	load (kips)	y (in)	v (in/s)	a (in/s^2)	ye+ (in)	ye- (in)	R (kips)	KEY -	KLM	V (kips)	eff K (k/in)	incr F (kips)	incr y (in)	incr v (in/s)	Z -
0.000	23.800	0.000	0.00	7844	0.384	-0.347	0.000	0	0.78	2.618	18263	70.8	0.004	7.71	
0.001	23.212	0.004	7.71	7576	0.384	-0.347	0.223	0	0.78	2.640	18263	208.7	0.011	7.37	1
0.002	22.624	0.015	15.08	7166	0.384	-0.347	0.882	0	0.78	2.832	18263	339.2	0.019	6.89	1
0.003	22.035	0.034	21.97	6619	0.384	-0.347	1.951	0	0.78	3.185	18263	459.7	0.025	6.28	1
0.004	21.447	0.059	28.26	5947	0.384	-0.347	3.401	0	0.78	3.686	18263	568.0	0.031	5.56	1
0.005	20.859	0.090	33.81	5163	0.384	-0.347	5.193	0	0.78	4.320	18263	662.0	0.036	4.72	1
0.006	20.271	0.126	38.53	4281	0.384	-0.347	7.281	0	0.78	5.069	18263	739.9	0.041	3.80	1
0.007	19.682	0.167	42.33	3318	0.384	-0.347	9.614	0	0.78	5.915	18263	800.3	0.044	2.81	1
0.008	19.094	0.211	45.14	2292	0.384	-0.347	12.138	0	0.78	6.834	18263	842.1	0.046	1.76	1
0.009	18.506	0.257	46.90	1223	0.384	-0.347	14.794	0	0.78	7.805	18263	864.3	0.047	0.68	1
0.010	17.918	0.304	47.57	131	0.384	-0.347	17.520	0	0.78	8.804	18263	866.7	0.047	-0.42	1
0.011	17.329	0.352	47.16	-964	0.384	-0.347	20.254	0	0.78	9.805	18263	849.2	0.046	-1.50	1
0.012	16.741	0.398	45.66	-2087	0.384	-0.347	22.100	1	0.66	10.407	15404	686.6	0.045	-2.20	2
0.013	16.153	0.443	43.45	-2316	0.384	-0.347	22.100	1	0.66	10.336	15404	651.0	0.042	-2.43	3
0.014	15.565	0.485	41.02	-2545	0.384	-0.347	22.100	1	0.66	10.266	15404	611.7	0.040	-2.66	3
0.015	14.976	0.525	38.36	-2775	0.384	-0.347	22.100	1	0.66	10.195	15404	569.0	0.037	-2.89	3
0.016	14.388	0.562	35.47	-3004	0.384	-0.347	22.100	1	0.66	10.125	15404	522.7	0.034	-3.12	3
0.017	13.800	0.596	32.36	-3233	0.384	-0.347	22.100	1	0.66	10.054	15404	472.9	0.031	-3.35	3
0.018	13.212	0.626	29.01	-3462	0.384	-0.347	22.100	1	0.66	9.983	15404	419.6	0.027	-3.58	3
0.019	12.624	0.653	25.43	-3691	0.384	-0.347	22.100	1	0.66	9.913	15404	362.7	0.024	-3.81	3
0.020	12.035	0.677	21.63	-3920	0.384	-0.347	22.100	1	0.66	9.842	15404	302.4	0.020	-4.03	3
0.021	11.447	0.697	17.59	-4149	0.384	-0.347	22.100	1	0.66	9.772	15404	238.4	0.015	-4.26	3
0.022	10.859	0.712	13.33	-4378	0.384	-0.347	22.100	1	0.66	9.701	15404	171.0	0.011	-4.49	3
0.023	10.271	0.723	8.83	-4608	0.384	-0.347	22.100	1	0.66	9.630	15404	100.0	0.006	-4.72	3
0.024	9.682	0.730	4.11	-4837	0.384	-0.347	22.100	1	0.66	9.560	15404	25.5	0.002	-4.95	3
0.025	9.094	0.731	-0.84	-5066	0.384	-0.347	22.100	1	0.66	9.489	15404	-52.5	-0.003	-5.18	3
0.026	8.506	0.728	-6.02	-4416	0.731	0.000	21.904	0	0.78	9.478	18263	-150.4	-0.008	-4.43	4
0.027	7.918	0.720	-10.45	-4453	0.731	0.000	21.429	0	0.78	9.228	18263	-231.4	-0.013	-4.43	1
0.028	7.329	0.707	-14.88	-4406	0.731	0.000	20.699	0	0.78	8.879	18263	-311.6	-0.017	-4.34	1
0.029	6.741	0.690	-19.22	-4276	0.731	0.000	19.717	0	0.78	8.431	18263	-389.5	-0.021	-4.17	1
0.030	6.153	0.669	-23.40	-4065	0.731	0.000	18.488	0	0.78	7.887	18263	-463.5	-0.025	-3.92	1
0.031	5.565	0.643	-27.32	-3777	0.731	0.000	17.026	0	0.78	7.252	18263	-532.3	-0.029	-3.60	1
0.032	4.976	0.614	-30.91	-3418	0.731	0.000	15.347	0	0.78	6.533	18263	-594.5	-0.033	-3.21	1
0.033	4.388	0.582	-34.12	-2994	0.731	0.000	13.472	0	0.78	5.737	18263	-649.0	-0.036	-2.75	1
0.034	3.800	0.546	-36.87	-2513	0.731	0.000	11.425	0	0.78	4.874	18263	-694.4	-0.038	-2.19	1
0.035	3.562	0.508	-39.07	-1870	0.731	0.000	9.235	0	0.78	3.994	18263	-728.4	-0.040	-1.53	1
0.036	3.325	0.468	-40.60	-1191	0.731	0.000	6.938	0	0.78	3.071	18263	-750.1	-0.041	-0.84	1
0.037	3.087	0.427	-41.44	-489	0.731	0.000	4.572	0	0.78	2.123	18263	-759.0	-0.042	-0.13	1
0.038	2.850	0.386	-41.57	221	0.731	0.000	2.178	0	0.78	1.163	18263	-755.0	-0.041	0.57	1
0.039	2.612	0.344	-40.99	928	0.731	0.000	-0.203	0	0.78	0.208	18263	-738.1	-0.040	1.27	1
0.040	2.375	0.304	-39.72	1617	0.731	0.000	-2.531	0	0.78	-0.726	18263	-708.7	-0.039	1.95	1
0.041	2.137	0.265	-37.78	2275	0.731	0.000	-4.766	0	0.78	-1.624	18263	-667.2	-0.037	2.58	1
0.042	1.900	0.228	-35.19	2891	0.731	0.000	-6.871	0	0.78	-2.471	18263	-614.6	-0.034	3.17	1
0.043	1.662	0.195	-32.02	3451	0.731	0.000	-8.809	0	0.78	-3.253	18263	-551.8	-0.030	3.70	1
0.044	1.425	0.165	-28.32	3947	0.731	0.000	-10.550	0	0.78	-3.958	18263	-479.9	-0.026	4.16	1
0.045	1.187	0.138	-24.17	4367	0.731	0.000	-12.063	0	0.78	-4.574	18263	-400.4	-0.022	4.54	1
0.046	0.950	0.116	-19.63	4705	0.731	0.000	-13.326	0	0.78	-5.093	18263	-314.8	-0.017	4.83	1
0.047	0.712	0.099	-14.80	4954	0.731	0.000	-14.319	0	0.78	-5.506	18263	-224.6	-0.012	5.03	1
0.048	0.475	0.087	-9.77	5109	0.731	0.000	-15.027	0	0.78	-5.808	18263	-131.6	-0.007	5.14	1
0.049	0.237	0.080	-4.63	5168	0.731	0.000	-15.442	0	0.78	-5.996	18263	-37.5	-0.002	5.15	1
0.050	0.000	0.078	0.52	5128	0.731	0.000	-15.561	0	0.78	-6.069	18263	56.1	0.003	5.10	1
0.051	0.000	0.081	5.62	5070	0.731	0.000	-15.384	0	0.78	-6.000	18263	148	0.008	4.99	1

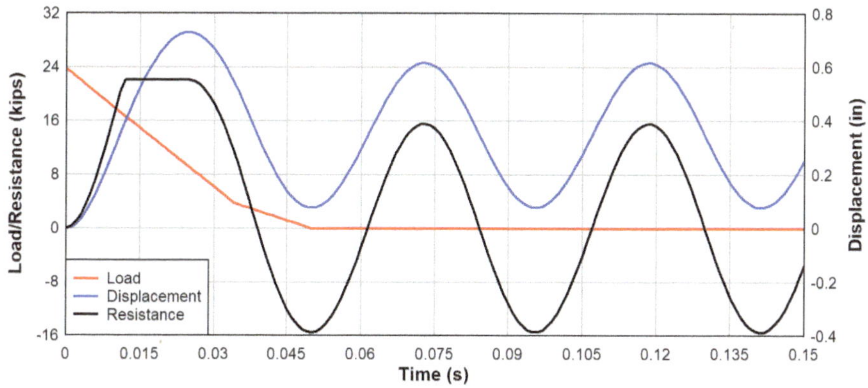

Figure 10-3. Output plots from front wall numerical integration.

Performing some sensitivity trials, a time step of 0.001 s appears to be required to provide a converged solution and smooth output. For corresponding output plots extending multiple cycles, see Figure 10-4.

Peak positive deflection: $y_m = +0.271$ in. (6.88 mm) at $t = 0.021$ s

Peak rebound deflection: $y_{m-} = -0.157$ in. (−3.99 mm) at $t = 0.091$ s

Support rotation: (Chapter 5, Figure 5-9)

$\theta_d = \tan^{-1} (y_m/0.5\ L) = \tan^{-1}$ [(0.271 in.)/(0.5)(144 in.)] $= 0.22$ degrees $< \theta_a$, **OK**

The response is adequate. A revision is not necessary. Refer to the discussion in Section 10.7.2 for the behavior of walls subjected to both in-plane and out-of-plane loads.

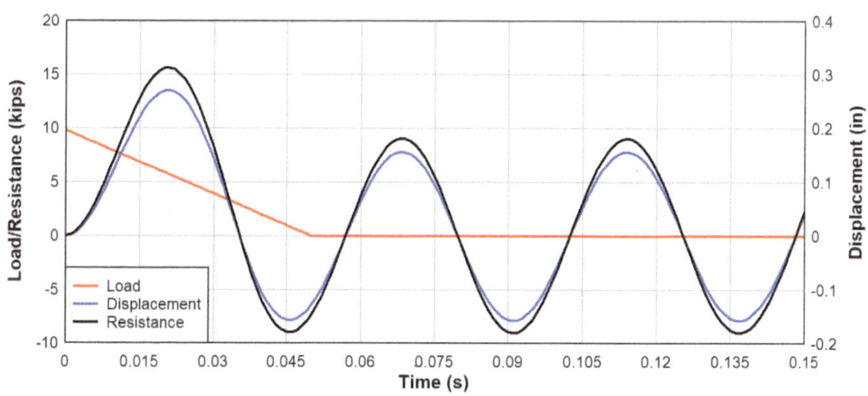

Figure 10-4. Output plots from side wall numerical integration.

10.5 ROOF SLAB (OUT-OF-PLANE LOADS)

The roof panels are 18 ft (5,486 mm) by 8 ft (2,438 mm), a 2.3 to 1 ratio. Therefore, the roof will be analyzed as a one-way beam fixed at both ends by adjoining roof slab spans or by thicker side walls. Typically, the roof slab would not be constructed monolithically with the walls, and the roof slab would be analyzed with fixed-simple supports to determine that it meets the deformation limits. The fixed-fixed span where the roof panel is continuous over two beams would also be checked, as shown subsequently, to verify that the roof slab reactions do not exceed the shear strength of the roof slab at the critical shear locations.

Note: If the roof panel aspect ratio were less than 2 to 1, the panel could be analyzed as a two-way span using resistance equations and transformation factors from the applicable tables in UFC 3-340-02.

A noncomposite deck will be used as a form only. According to the manufacturer's literature, composite metal decking is not intended for dynamic loads. A 2 in. (51 mm) deep metal deck is selected.

Span, $L = 8$ ft or 96 in. (2,438 mm) from center to center of supporting beams

Use a nominal design width, $b = 1$ ft or 12 in. (305 mm).

10.5.1 Calculate Design Blast Load Case A (Explosion on Long Side of the Building)

From the Appendix 3 calculation:

$P_a = 5.4$ psi (37 kPa)

$t_r = 0.006$ s

$t_d = 0.05$ s

The blast-loaded area tributary to the roof slab is the span of 96 in. (2,438 mm) multiplied by the nominal design width of 12 in. (305 mm).

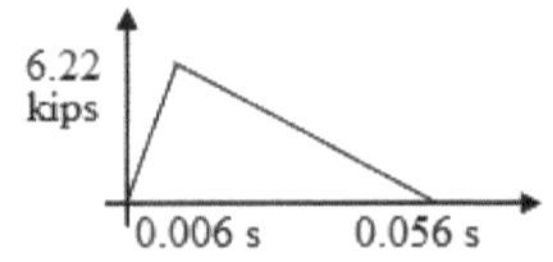

Peak load:

$$F_o = (5.4 \text{ psi})(96 \text{ in. span})(12 \text{ in. width})/(1{,}000 \text{ lb/kip})$$
$$= 6.22 \text{ kips } (27.7 \text{ kN})$$

10.5.2 Calculate Design Blast Load Case B (Explosion on Short Side of the Building)

From the Appendix 3 calculation:

$P_{so} = 6$ psi (41 kPa)

$t_d = 0.05$ s

$L_w = 66$ ft or 792 in. (20,120 mm)

$L_1 = 1$ ft (305 mm), which leads to $L_w/L_1 = 66$ and $C_e = 1.0$ per Chapter 3, Figure 3-10, and, therefore, $t_r = 0.0$ s.

Peak overpressure: [Chapter 3, Equation (3-11)]

$$P_a = C_e\,(P_{so}) + C_d\,(q_o) = (1.0)\,(6\text{ psi}) + (-0.4)\,(0.8\text{ psi}) = 5.7\text{ psi (39 kPa)}$$

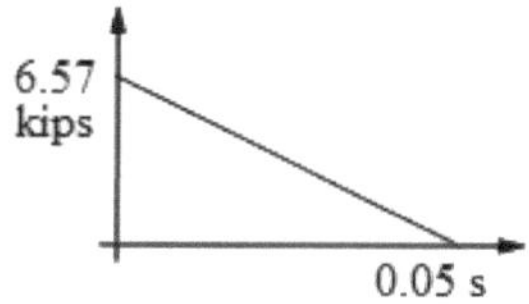

Peak load:

$$F_o = (5.7\text{ psi})(96\text{ in. span})(12\text{ in. width})/(1{,}000\text{ lb/kip})$$
$$= 6.57\text{ kips (29.2 kN)}$$

10.5.3 Select Trial Sizes

Note: The following trial dimensions and material proportions may be obtained from trial calculations, by inspecting similar structures, or from experience. The results of this dynamic calculation will determine the adequacy of the trial size.

5 in. (127 mm) concrete slab plus metal deck

#3 @ 8 in. (203 mm), each way, at the center of the main slab section thickness

#3, diameter = 0.375 in. (9.5 mm); #3, area = 0.110 in.^{2} (71.3 mm^2)

10.5.4 Compute Bending Resistance

For dynamic bending: (Chapter 5, Appendix 5A)

$$F_{ds} = F_{dy} = (\text{SIF})(\text{DIF})\,F_y = (1.1)(1.22)(60\text{ ksi}) = 80.5\text{ ksi (555 MPa)}$$
$$f'_{dc} = (\text{SIF})(\text{DIF})\,f'_c = (1.0)(1.19)(4\text{ ksi}) = 4.76\text{ ksi (32.8 MPa)}$$

$$A_s = (0.110\text{ in.}^2)(12\text{ in.})/(8\text{ in.}) = 0.166\text{ in.}^2\ (107\text{ mm}^2)$$

For bending tension on the bottom face:

$$d \approx (5\text{ in. slab})/2 = 2.5\text{ in. (63.5 mm)}$$

$$A_{s,\,\min} = 3\,\frac{\sqrt{f'_{dc}}}{F_{ds}}\,bd = 3\,\frac{\sqrt{4{,}760\text{ psi}}}{80{,}500\text{ psi}}\,(12\text{ in.})(2.5\text{ in.})\ \text{(ACI 318, Section 9.6.1.2)}$$
$$= 0.077\text{ in.}^2\ (49.8\text{ mm}^2) < A_s,\quad \textbf{OK}$$

$$A_{s,\,\min} = 200\,bd/F_{ds} = 200(12\text{ in.})(2.5\text{ in.})/(80{,}500\text{ psi})\ \text{(ACI 318, Section 9.6.1.2)}$$
$$= 0.075\text{ in.}^2\ (48.1\text{ mm}^2) < A_s,\quad \textbf{OK}$$

$$d_{sb} = A_s\,F_{ds}/0.85\,f'_{dc}\,b$$
$$= (0.166\text{ in.}^2)(80.5\text{ ksi})/(0.85)(4.76\text{ ksi})(12\text{ in.})$$
$$= 0.275\text{ in. (7.0 mm)}$$

$$M_{p,\,\mathrm{bott}} = A_s F_{ds}\,[d - d_{sb}/2]$$
$$= (0.166\ \mathrm{in.}^2)(80.5\ \mathrm{ksi})\ [(2.5\ \mathrm{in.}) - (0.275\ \mathrm{in.})/2]$$
$$= 31.5\ \mathrm{kip{\cdot}in.}\ (3.6\ \mathrm{kN{\cdot}m})$$

For bending tension on the top face:

$$d \approx (5\ \mathrm{in.\ slab})/2 + (2\ \mathrm{in.\ deck})/2 = 3.5\ \mathrm{in.}\ (88.9\ \mathrm{mm})$$

$$A_{s,\,\min} = 3\,\frac{\sqrt{f'_{dc}}}{F_{ds}}\,bd = 3\,\frac{\sqrt{4{,}760\ \mathrm{psi}}}{80{,}500\ \mathrm{psi}}\ (12\ \mathrm{in.})(3.5\ \mathrm{in.})\ \text{(ACI 318, Section 9.6.1.2)}$$
$$= 0.108\ \mathrm{in.}^2\ (69.7\ \mathrm{mm}^2) < A_s,\quad \textbf{OK}$$

$$A_{s,\,\min} = 200\ bd/F_{ds} = 200\ (12\ \mathrm{in.})(3.5\ \mathrm{in.})/(80{,}500\ \mathrm{psi})\ \text{(ACI 318, Section 9.6.1.2)}$$
$$= 0.104\ \mathrm{in.}^2\ (67.3\ \mathrm{mm}^2) < A_s,\quad \textbf{OK}$$

$$M_{p,\,\mathrm{top}} = A_s f_{ds}\,[d - d_{sb}/2]$$
$$= (0.166\ \mathrm{in.}^2)(80.5\ \mathrm{ksi})\ [(3.5\ \mathrm{in.}) - (0.27\ \mathrm{in.})/2]$$
$$= 44.9\ \mathrm{kip{\cdot}in.}\ (5.1\ \mathrm{kN{\cdot}m})$$

Positive (downward) and negative (upward) bending resistance based on fixed ends:

$$R_b = R_{b\text{-}} = 8\,(M_{p,\,\mathrm{bott}} + M_{p,\,\mathrm{top}})/L\ \text{(Chapter 6, Table 6-2)}$$
$$= 8\,[(31.5\ \mathrm{kip{\cdot}in.}) + (44.9\ \mathrm{kip{\cdot}in.})]/96\ \mathrm{in.}$$
$$= 6.36\ \mathrm{kips}\ (28.3\ \mathrm{kN})$$

10.5.5 Compute Shear Resistance

For dynamic shear: (Chapter 5, Appendix 5A)

$$f'_{dc} = (\mathrm{SIF})(\mathrm{DIF})\ f'_c = (1.0)(1.0)(4\ \mathrm{ksi}) = 4.0\ \mathrm{ksi}\ (27.6\ \mathrm{MPa})$$

A DIF value of 1.0 is used for evaluating diagonal tension shear capacity. Because positive or rebound bending can occur, calculate shear resistance on the basis of the smaller of top or bottom tension. In this case, use $d = 2.5$ in. (63.5 mm).

$$V_n = 2\sqrt{f'_{dc}}\ bd\ \text{(ACI 318, Table 22.5.5.1)}$$
$$= 2\sqrt{4{,}000\ \mathrm{psi}}\ (12\ \mathrm{in.})(2.5\ \mathrm{in.})/(1{,}000\ \mathrm{lb/kip})$$
$$= 3.79\ \mathrm{kips}\ (16.9\ \mathrm{kN})$$

The critical section for shear is at a distance d from the support.

$$R_s = V_n\,L/[0.5\,L - d]$$
$$= (3.79\ \mathrm{kips})(96\ \mathrm{in.})/[0.5\,(96\ \mathrm{in.}) - 2.5\ \mathrm{in.}]$$
$$= 8.01\ \mathrm{kips}\ (35.6\ \mathrm{kN})$$

10.5.6 Resistance and Permissible Response

Because $R_b < R_s$, bending controls.

Positive and rebound resistance, $R_u = -R_{u\text{-}} = R_b =$ **6.36 kips** (28.3 kN)

$$RI = A_s/bd \times F_{ds}/f'_{dc} \qquad \text{[Equation (5-1)]}$$
$$= (0.166\ \mathrm{in.}^2)/(12\ \mathrm{in.})(2.5\ \mathrm{in.}) \times (80.5\ \mathrm{ksi})/(4.76\ \mathrm{ksi}) = 0.093$$

Response limit: $\theta_a = 1°$ (based on no shear reinforcing and $RI > 0.05$)
Note: The inbound d value was used to calculate the reinforcement index, RI.

10.5.7 Compute Other Single-Degree-of-Freedom Properties

Moment of inertia will be based on positive (downward) bending of 5 in. (127 mm) concrete slab.

Gross moment of inertia:

$$I_g = bh^3/12 = (12\text{ in.})(5\text{ in.})^3/12 = 125\text{ in.}^4\ (52{,}030{,}000\text{ mm}^4)$$

Transformed rebar area:

$$nA_s = (8.04)(0.166\text{ in.}^2) = 1.33\text{ in.}^2\ (860\text{ mm}^2)$$

Location of transformed neutral axis:

$$\begin{aligned} d_{na} &= \frac{-nA_s + \sqrt{nA_s(nA_s + 2bd)}}{b} \\ &= \frac{-1.37\text{ in.}^2 + \sqrt{1.37\text{ in.}^2(1.37\text{ in.}^2 + 2(12\text{ in.})(2.5\text{ in.}))}}{12\text{ in.}} \\ &= 0.642\text{ in.}(16.3\text{ mm}) \end{aligned}$$

Cracked moment of inertia:

$$\begin{aligned} I_{cr} &= bd_{na}{}^3/3 + n\,A_s\,(d - d_{na})^2 \\ &= (12\text{ in.})(0.642\text{ in.})^3/3 + (1.33\text{ in.}^2)(2.5\text{ in.} - 0.642\text{ in.})^2 \\ &= 5.66\text{ in.}^4\ (2{,}356{,}000\text{ mm}^4) \end{aligned}$$

Average moment of inertia:

$$I_a = (I_g + I_{cr})/2 = (125\text{ in.}^4 + 5.66\text{ in.}^4)/2 = 65.3\text{ in.}^4\ (27{,}190{,}000\text{ mm}^4)$$

Effective "bilinear" stiffness: (Chapter 6, Table 6-2)

$$\begin{aligned} K_e &= 307\,E_c I_a/L^3 \\ &= 307\,(3{,}605\text{ ksi})(65.3\text{ in.}^4)/(96\text{ in.})^3 \\ &= \mathbf{81.7\ kip/in.}\ (14.3\text{ kN/mm}) \end{aligned}$$

Effective elastic yield deflection:

$$y_e = R_u/K_e = (6.36\text{ kips})/(81.7\text{ psi/in.}) = \mathbf{0.0778\ in.}\ (1.98\text{ mm})$$

Slab weight is based on a 5 in. (127 mm) slab plus a 1 in. (25 mm) average concrete thickness in the deck, for an effective thickness of 6 in. (152 mm).

Mass:

$$\text{weight} = (0.15\text{ kcf})(0.5\text{ ft thick})(1\text{ ft width})(8\text{ ft span}) = 0.6\text{ kip}\ (2.67\text{ kN})$$

$$\begin{aligned} \text{static load} &= (\text{slab weight}) + (\text{tributary weight}) = 0.6\text{ kip} + (8\text{ ft})(1\text{ ft}) \\ &\quad (0.010\text{ ksf dead load}) = 0.68\text{ kip}\ (3.0\text{ kN}) \end{aligned}$$

M = static load/g
= (0.68 kip)/(386 in./s^2)
= **0.00176 kip·s^2/in.** (308 kg)

Static (dead load) deflection:

y_{st} = static load/K_e = (0.68 kip)/(81.7 kip/in.) = **0.0083 in.** (0.21 mm)

Load-mass transformation factors: (Chapter 6, Table 6-2)

elastic, $K_{LM} = K_M/K_L = 0.41/0.53 = \mathbf{0.77}$

plastic, $K_{LM} = K_M/K_L = 0.33/0.50 = \mathbf{0.66}$

Assuming some plastic deformation, use an average value for the load-mass factor when calculating the natural period:

$K_{LM} = (0.77 + 0.66)/2 = \mathbf{0.715}$

Equivalent mass:

$M_e = K_{LM}\, M = 0.715$ (0.00176 kip·s^2/in.) = 0.00126 kip·s^2/in. (221 kg)

Natural period: [Equation (6-8)]

$$t_n = 2\pi\sqrt{\frac{M_e}{K_e}} = 2\pi\sqrt{\frac{(0.00126\ \text{kip-s}^2/\text{in.})}{(81.7\ \text{kip/in.})}} = \mathbf{0.0247\ s}$$

10.5.8 Compute Response and Compare to Deformation Limits (Numerical Integration Solution for Load Case A)

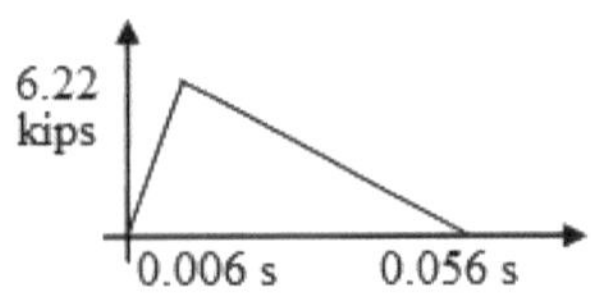

F_o = 6.22 kips (27.7 kN)

t_r = 0.006 s

t_d = 0.05 s

Time increment:

$t_n/10$ = 0.0247 s/10 ≈ 0.0025 s

However, it is recommended to use at most a 0.0005 s time step to obtain at least 10 increments of loading in the smallest, first section of the pressure-time history. Consider a time range of 0.100 s to capture multiple deflection cycles and component behavior following load termination.

For corresponding output plots extending multiple cycles, see Figure 10-5.

Peak positive deflection: y_m = +0.291 in. (7.39 mm) at t = 0.028 s

Peak rebound deflection: y_{m-} = +0.191 in. (4.86 mm) at t = 0.0665 s

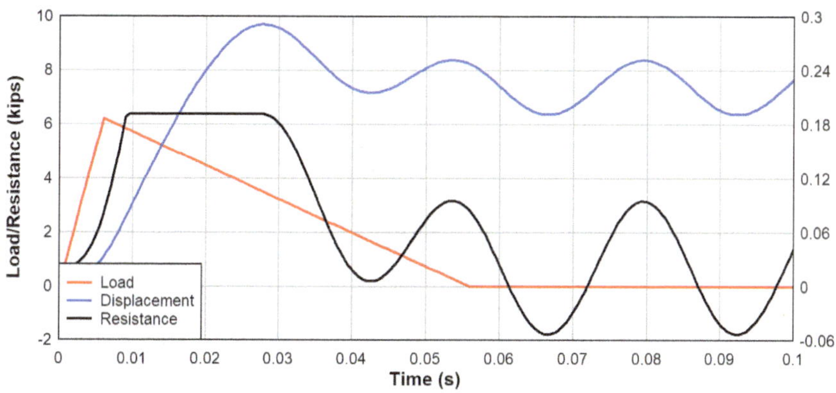

Figure 10-5. Output plots from roof slab numerical integration (Load Case A).

Note: The selected rebound peak is the first to occur after the dissipation of the applied load.

Support rotation: (Figure 5-9)

$$\theta_d = \tan^{-1}(y_m/0.5\,L) = \tan^{-1}[(0.291\text{ in.})/(0.5)(96\text{ in.})] = 0.35° < \theta_a,\ \textbf{OK}$$

Refer to the discussion in Section 10.6.9 for the behavior of roof slabs subjected to both in-plane and out-of-plane loads.

10.5.9 Compute Response and Compare to Deformation Limits (Numerical Integration Solution for Load Case B)

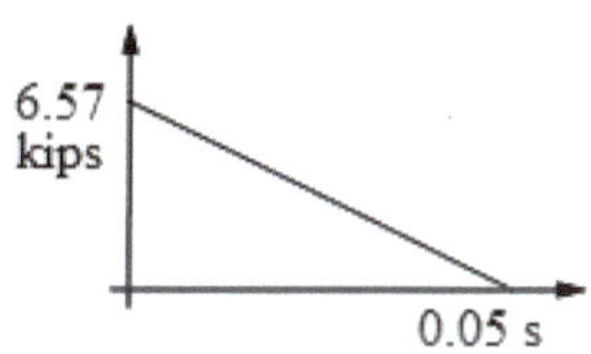

$P_o = 6.57$ kips (29.2 kN)

$t_r = 0$ s

$t_d = 0.05$ s

Time increment:

$t_n/10 = 0.0247\text{ s}/10 \approx 0.0025$ s

A smaller time step can be used to potentially obtain a more refined solution with smoother output plots. Consider a time range of 0.100 s to capture multiple deflection cycles and component behavior following load termination.

Performing some sensitivity trials, a time step of 0.0005 s appears to be required to provide a converged solution and smooth output. For the numerical integration through the time of peak deflection, refer to Table 10-4. For corresponding output plots extending multiple cycles, see Figure 10-6.

Table 10-4. Roof Slab Numerical Integration (Load Case B).

Example:	Section 10.5 Roof Slab							KEY = 0 when elastic
								KEY = +1 when positive plastic
time increment:	0.0005 s	**Blast Load**						KEY = -1 when rebound plastic
mass:	0.00176 k-s²/in	**t0:**	0 s	**P0:**	6.57 kips			Z = 1 when continued elastic
stiffness:	81.7 k/in	**t1:**	0.05 s	**P1:**	0 kips			Z = 2 at start of positive plastic deformation
damping:	0 k-s/in	**t2:**	0 s	**P2:**	0 kips			Z = 3 when increasing plastic deformation
positive resistance:	6.36 kips	**Dynamic Reaction Factors**		**BC: S-S**				Z = 4 at end of positive plastic deformation
rebound resistance:	-6.36 kips	**Crel:**	0.39	**Cfel:**	0.11			Z = 6 at start of rebound plastic deformation
static load:	0.68 kips	**Crpl:**	0.38	**Cfpl:**	0.12			Z = 8 at end of rebound plastic deformation

time (s)	load (kips)	y (in)	v (in/s)	a (in/s²)	ye+ (in)	ye- (in)	R (kips)	KEY -	KLM	V (kips)	eff K (k/in)	Incr F (kips)	incr y (in)	incr v (in/s)	Z -
0.0000	6.570	0.0083	0.00	4786	0.078	-0.078	0.000	0	0.78	0.723	33029	19.6	0.001	2.37	
0.0005	6.504	0.0089	2.37	4703	0.078	-0.078	0.729	0	0.78	1.000	33029	58.4	0.002	2.31	1
0.0010	6.439	0.0107	4.69	4550	0.078	-0.078	0.873	0	0.78	1.049	33029	95.9	0.003	2.22	1
0.0015	6.373	0.0136	6.90	4329	0.078	-0.078	1.110	0	0.78	1.134	33029	131.5	0.004	2.09	1
0.0020	6.307	0.0176	9.00	4044	0.078	-0.078	1.435	0	0.78	1.254	33029	164.8	0.005	1.94	1
0.0025	6.242	0.0226	10.93	3699	0.078	-0.078	1.843	0	0.78	1.405	33029	195.3	0.006	1.75	1
0.0030	6.176	0.0285	12.68	3300	0.078	-0.078	2.326	0	0.78	1.587	33029	222.5	0.007	1.54	1
0.0035	6.110	0.0352	14.22	2851	0.078	-0.078	2.876	0	0.78	1.794	33029	245.9	0.007	1.30	1
0.0040	6.044	0.0427	15.52	2360	0.078	-0.078	3.485	0	0.78	2.024	33029	265.4	0.008	1.05	1
0.0045	5.979	0.0507	16.57	1834	0.078	-0.078	4.141	0	0.78	2.273	33029	280.5	0.008	0.78	1
0.0050	5.913	0.0592	17.35	1281	0.078	-0.078	4.835	0	0.78	2.536	33029	291.0	0.009	0.50	1
0.0055	5.847	0.0680	17.85	708	0.078	-0.078	5.555	0	0.78	2.810	33029	296.9	0.009	0.21	1
0.0060	5.782	0.0770	18.06	125	0.078	-0.078	6.289	0	0.78	3.089	33029	297.9	0.009	-0.08	1
0.0065	5.716	0.0860	17.97	31	0.078	-0.078	6.360	1	0.66	3.103	27878	250.6	0.009	0.00	2
0.0070	5.650	0.0950	17.97	-26	0.078	-0.078	6.360	1	0.66	3.095	27878	250.4	0.009	-0.03	3
0.0075	5.585	0.1040	17.95	-82	0.078	-0.078	6.360	1	0.66	3.087	27878	249.8	0.009	-0.06	3
0.0080	5.519	0.1129	17.89	-139	0.078	-0.078	6.360	1	0.66	3.079	27878	248.9	0.009	-0.08	3
0.0085	5.453	0.1219	17.81	-195	0.078	-0.078	6.360	1	0.66	3.071	27878	247.5	0.009	-0.11	3
0.0090	5.387	0.1307	17.70	-252	0.078	-0.078	6.360	1	0.66	3.063	27878	245.7	0.009	-0.14	3
0.0095	5.322	0.1395	17.56	-308	0.078	-0.078	6.360	1	0.66	3.055	27878	243.6	0.009	-0.17	3
0.0100	5.256	0.1483	17.39	-365	0.078	-0.078	6.360	1	0.66	3.048	27878	241.0	0.009	-0.20	3
0.0105	5.190	0.1569	17.19	-422	0.078	-0.078	6.360	1	0.66	3.040	27878	238.1	0.009	-0.22	3
0.0110	5.125	0.1655	16.97	-478	0.078	-0.078	6.360	1	0.66	3.032	27878	234.8	0.008	-0.25	3
0.0115	5.059	0.1739	16.71	-535	0.078	-0.078	6.360	1	0.66	3.024	27878	231.0	0.008	-0.28	3
0.0120	4.993	0.1822	16.43	-591	0.078	-0.078	6.360	1	0.66	3.016	27878	226.9	0.008	-0.31	3
0.0125	4.928	0.1903	16.12	-648	0.078	-0.078	6.360	1	0.66	3.008	27878	222.4	0.008	-0.34	3
0.0130	4.862	0.1983	15.78	-704	0.078	-0.078	6.360	1	0.66	3.000	27878	217.5	0.008	-0.37	3
0.0135	4.796	0.2061	15.42	-761	0.078	-0.078	6.360	1	0.66	2.992	27878	212.2	0.008	-0.39	3
0.0140	4.730	0.2137	15.02	-817	0.078	-0.078	6.360	1	0.66	2.984	27878	206.5	0.007	-0.42	3
0.0145	4.665	0.2211	14.60	-874	0.078	-0.078	6.360	1	0.66	2.977	27878	200.4	0.007	-0.45	3
0.0150	4.599	0.2283	14.15	-931	0.078	-0.078	6.360	1	0.66	2.969	27878	193.9	0.007	-0.48	3
0.0155	4.533	0.2353	13.67	-987	0.078	-0.078	6.360	1	0.66	2.961	27878	187.0	0.007	-0.51	3
0.0160	4.468	0.2420	13.16	-1044	0.078	-0.078	6.360	1	0.66	2.953	27878	179.8	0.006	-0.54	3
0.0165	4.402	0.2484	12.63	-1100	0.078	-0.078	6.360	1	0.66	2.945	27878	172.1	0.006	-0.56	3
0.0170	4.336	0.2546	12.06	-1157	0.078	-0.078	6.360	1	0.66	2.937	27878	164.0	0.006	-0.59	3
0.0175	4.271	0.2605	11.47	-1213	0.078	-0.078	6.360	1	0.66	2.929	27878	155.6	0.006	-0.62	3
0.0180	4.205	0.2661	10.85	-1270	0.078	-0.078	6.360	1	0.66	2.921	27878	146.7	0.005	-0.65	3
0.0185	4.139	0.2713	10.20	-1327	0.078	-0.078	6.360	1	0.66	2.913	27878	137.5	0.005	-0.68	3
0.0190	4.073	0.2763	9.52	-1383	0.078	-0.078	6.360	1	0.66	2.906	27878	127.8	0.005	-0.71	3
0.0195	4.008	0.2808	8.82	-1440	0.078	-0.078	6.360	1	0.66	2.898	27878	117.8	0.004	-0.73	3
0.0200	3.942	0.2851	8.08	-1496	0.078	-0.078	6.360	1	0.66	2.890	27878	107.4	0.004	-0.76	3
0.0205	3.876	0.2889	7.32	-1553	0.078	-0.078	6.360	1	0.66	2.882	27878	96.6	0.003	-0.79	3
0.0210	3.811	0.2924	6.53	-1609	0.078	-0.078	6.360	1	0.66	2.874	27878	85.3	0.003	-0.82	3
0.0215	3.745	0.2954	5.71	-1666	0.078	-0.078	6.360	1	0.66	2.866	27878	73.7	0.003	-0.85	3
0.0220	3.679	0.2981	4.86	-1722	0.078	-0.078	6.360	1	0.66	2.858	27878	61.7	0.002	-0.88	3
0.0225	3.614	0.3003	3.99	-1779	0.078	-0.078	6.360	1	0.66	2.850	27878	49.3	0.002	-0.90	3
0.0230	3.548	0.3021	3.08	-1836	0.078	-0.078	6.360	1	0.66	2.843	27878	36.5	0.001	-0.93	3
0.0235	3.482	0.3034	2.15	-1892	0.078	-0.078	6.360	1	0.66	2.835	27878	23.3	0.001	-0.96	3
0.0240	3.416	0.3042	1.19	-1949	0.078	-0.078	6.360	1	0.66	2.827	27878	9.8	0.000	-0.99	3
0.0245	3.351	0.3046	0.20	-2005	0.078	-0.078	6.360	1	0.66	2.819	27878	-4.2	0.000	-1.02	3
0.0250	3.285	0.3044	-0.81	-1736	0.305	0.149	6.348	0	0.78	2.837	33029	-20.6	-0.001	-0.87	4

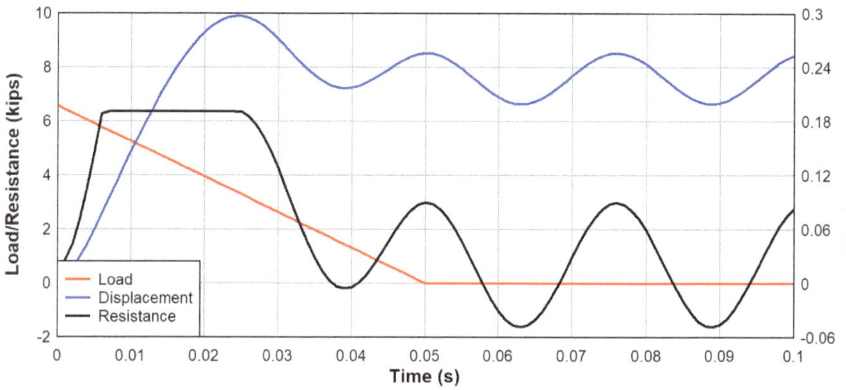

Figure 10-6. Output plots from roof slab numerical integration (Load Case B).

Peak positive deflection: $y_m = +0.305$ in. (7.74 mm) at $t = 0.0245$ s

Peak rebound deflection: $y_{m-} = +0.207$ in. (5.26 mm) at $t = 0.063$ s

Note: The selected rebound peak is the first to occur after the dissipation of the applied load. The minimum deflection achieved is the initial static deflection of 0.0083 in. (0.21 mm).

Support rotation: (Figure 5-9)

$$\theta_d = \tan^{-1}(y_m/0.5\,L) = \tan^{-1}[(0.305 \text{ in.})/(0.5)(96 \text{ in.})] = 0.36° < \theta_a, \quad \textbf{OK}$$

Refer to the discussion in Section 10.6.9 for the behavior of roof slabs subjected to both in-plane and out-of-plane loads.

Note: Reduction in roof slab/diaphragm proportions to achieve a higher support rotation, and therefore a more optimal design, would be restricted by the need for workable dimensions for all reinforcement types involved, compliance with minimum flexural steel code requirements, and the ability to transfer in-plane loads.

10.6 ROOF SLAB (IN-PLANE LOADS)

The roof diaphragm is designed to transfer wall loads to the side shear walls. The diaphragm is fixed at both ends by continuous attachment to the walls. The center of mass coincides with the center of rigidity, indicating no incidental torsion.

Note: Boundary conditions may more resemble simple supports depending on the detailing between the diaphragm and the walls. In such a case, the factors used within the following calculations should be modified accordingly.

Span, L = (93 ft outer perimeter-to-outer perimeter) or 1,116 in. (28,350 mm)

Depth = (67 ft outer perimeter-to-outer perimeter) or 804 in. (20,420 mm)

Determine the width of the composite (wall) flange: (ACI 318, Table 6.3.2.1)

a. (10 in. wall) × 6 = 60 in. (1,524 mm) [CONTROLS]

b. (144 in. wall span)/2 = 72 in. (1,829 mm)

c. (1,116 in. span)/12 = 93 in. (2,360 mm)

Use an effective width, $b = 5$ ft or 60 in. (1,524 mm).

Initially, a static check of the roof diaphragm will be performed on the basis of ultimate resistance of the walls. If this check confirms an adequate roof diaphragm, no further action is required. In the event this check indicates an inadequate roof diaphragm, a dynamic run can be performed with the roof slab acting as a deep beam subject to the blast load on the long (controlling) side of the building. If a dynamic analysis approach is used, it is beneficial to minimize the chord reinforcement if possible so that the roof diaphragm is controlled by flexural response and can, therefore, have a higher allowable maximum response limit.

10.6.1 Static Check of Roof Diaphragm

The load that needs to be carried by the roof diaphragm is based on $R_b = 22.1$ kips (98.3 kN). On a pressure basis, with $L = 12$ ft (3,658 mm) and $b = 1$ ft (305 mm), $R_b = 12.8$ psi (88.2 kPa). Considering half of the wall load transfers into the roof diaphragm, the load demand totals 12.8 psi × 1,116 in. × 144 in./2/(1,000 lb/kip) = 1,028 kips (4,570 kN) or 514 kips (2,285 kN) into each side of the diaphragm. Initially considering the dynamic shear capacity of the diaphragm concrete alone and using only half of the roof thickness owing to cracking from out-of-plane loading, $V_n = 2\lambda\,(f'_{dc})^{1/2}\,A_{cv} = 2 \times 1.0 \times (4{,}400\ \text{psi})^{1/2} \times 2.5\ \text{in.} \times 804\ \text{in.}$/(1,000 lb/kip) = 267 kips (1,190 kN). As this value is less than the demand of 514 kips (2,285 kN), it is judged that the roof diaphragm is inadequate based on a static analysis, and it should be checked dynamically.

Notes: The diaphragm in-plane shear strength checks made previously are conservative. The contribution of the typical diaphragm steel can be considered using Equation 12.5.3.3 in ACI 318 (ACI 2022). This would add $\rho_t\,F_{ds}\,A_{cv}$ = (0.110 in.²/8 in./2.5 in.) × 72.6 ksi × 2.5 in. × 804 in. = 806 kips (3,585 kN) to V_n, resulting in an updated $V_n = 1{,}073$ kips (4,770 kN). This value well exceeds the demand of 514 kips (2,285 kN). Moreover, using only half of the roof thickness to account for out-of-plane cracking behavior is likely conservative because cracking would occur in different locations within the roof slab from these loading conditions. Nevertheless, dynamic checks of the roof diaphragm are made in the following sections for illustrative purposes.

10.6.2 Dynamic Check of Roof Diaphragm: Calculate Design Blast Load Case A (Applied to Long Side of Building)

From the Appendix 3 calculation:

$P_r = 13.8$ psi (95 kPa)

$P_s = 6.8$ psi (47 kPa)

$t_c = 0.034$ s

$t_d = 0.05$ s

In addition, the intermediate point where the cleared reflection load and stagnation load intersect is 2.2 psi (15 kPa) at $t_c = 0.034$ s. The blast-loaded area tributary to the roof diaphragm is half the exterior wall height of 144 in. (3,658 mm) multiplied by the building width of 1,116 in. (28,350 mm).

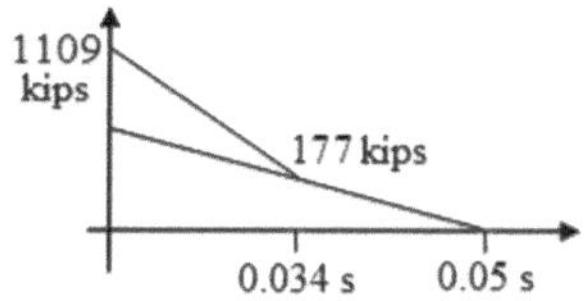

Peak load:

$$F_o = (13.8 \text{ psi})(144 \text{ in. height} \times 0.5)(1{,}116 \text{ in. width})/(1{,}000 \text{ lb/kip})$$
$$= 1{,}109 \text{ kips } (4{,}930 \text{ kN})$$

Intermediate load point:

$$F_{\text{int}} = (2.2 \text{ psi})(144 \text{ in. height} \times 0.5)(1{,}116 \text{ in. width})/(1{,}000 \text{ lb/kip})$$
$$= 177 \text{ kips } (786 \text{ kN})$$

The total load can then be represented by the bilinear cleared loading profile.

Note: The rear wall load, with the applicable arrival delay, could be applied to minimize the roof diaphragm response.

10.6.3 Dynamic Check of Roof Diaphragm: Calculate Design Blast Load Case B (Applied to Short Side of Building)

Ignore this case because it will not control the response.

10.6.4 Select Trial Sizes

Concrete roof slab:

5 in. (127 mm) full thickness, plus concrete within 2 in. (51 mm) steel decking for an average of 6 in. (152 mm)

Roof reinforcing (used to resist shear):

#3 @ 8 in. (203 mm), at the center of the main slab section thickness

#3, diameter = 0.375 in. (9.5 mm); #3, area = 0.110 in.^{2} (71.3 mm^2)

Chord reinforcing (used to resist bending):

Assume that 10 in. (254 mm) concrete walls are attached to roof slab with chord reinforcing consisting of 4 #8 bars bending into the center of the wall.

#8, diameter = 1 in. (25.4 mm); #8, area = 0.785 in.^{2} (507 mm^2)

$A_s = (4)(0.785 \text{ in.}^2) = 3.14 \text{ in.}^2$ (2,030 mm^2)

10.6.5 Compute Bending Resistance

For dynamic bending: (Chapter 5, Appendix 5A)

$$F_{ds} = F_{dy} = \text{(SIF)(DIF)}\ F_y = (1.1)(1.22)(60\text{ ksi}) = 80.5\text{ ksi (555 MPa)}$$
$$f'_{dc} = \text{(SIF)(DIF)}\ f'_c = (1.0)(1.19)(4\text{ ksi}) = 4.76\text{ ksi (32.8 MPa)}$$

$$d = (804\text{ in depth}) - (10\text{ in. wall})/2 = 799\text{ in. (20,290 mm)}$$

$$\begin{aligned} d_{sb} &= A_s\, F_{ds}/0.85\ f'_{dc}\ b \\ &= (3.14\text{ in.}^2)(80.5\text{ ksi})/(0.85)(4.76\text{ ksi})(60\text{ in.}) \\ &= 1.04\text{ in. (26.5 mm) [within the thickness of the wall]} \end{aligned}$$

$$\begin{aligned} M_p &= A_s\, F_{ds}\, [d - d_{sb}/2] \\ &= (3.14\text{ in.}^2)(80.5\text{ ksi})\ [(799\text{ in.}) - (1.02\text{ in.})/2] \\ &= 202{,}000\text{ kip·in. (22,820 kN·m)} \end{aligned}$$

$$\begin{aligned} R_b &= 8\ (M_{ps} + M_{pc})/L = 16\ M_p/L\ \text{(Chapter 6, Table 6-2)} \\ &= 16\ (202{,}000\text{ kip·in.})/(1{,}116\text{ in.}) \\ &= 2{,}896\text{ kips (12,880 kN)} \end{aligned}$$

10.6.6 Compute Shear Resistance

For dynamic shear: (Chapter 5, Appendix 5A)

$$F_{ds} = F_{dy} = \text{(SIF)(DIF)}\ F_y = (1.1)(1.1)(60\text{ ksi}) = 72.6\text{ ksi (501 MPa)}$$
$$f'_{dc} = \text{(SIF)(DIF)}\ f'_c = (1.0)(1.1)(4\text{ ksi}) = 4.4\text{ ksi (30.3 MPa)}$$

A DIF value of 1.1 is used for evaluating the direct (as opposed to diagonal tension) shear capacity of the reinforcing steel and concrete.

$$\begin{aligned} V_n &= [2\lambda\,(f'_{dc})^{1/2} + \rho_t\, F_{ds}]\ A_{cv}\ \text{(ACI 318, Equation 12.5.3.3)} \\ &= [(2.0)(1.0)(4{,}400\text{ psi})^{1/2}/(1{,}000\text{ lb/kip}) + (0.11\text{ in.}^2)/(8\text{ in.})/(2.5\text{ in.}) \times (72.6\text{ ksi})] \\ &\quad \times (2.5\text{ in.} \times 804\text{ in.}) \\ &= 1{,}073\text{ kips (4,770 kN)} \end{aligned}$$

$$R_s = 2\ V_n = 2\ (1{,}073\text{ kips}) = 2{,}145\text{ kips (9,540 kN)}$$

10.6.7 Resistance and Permissible Response

Note: Roof diaphragm resistance could be controlled by conventional deep beam action or upper bounds placed on diaphragm shear capacity. Applicable code criteria provided in ACI 318, Sections 9.9 and 12.5, should be checked.

Because $R_s < R_b$, shear controls.

Positive and rebound resistance, $R_u = -R_{u-} = R_s =$ **2,145 kips** (9,540 kN)

Response Limit: $\mu_a = 1.3$

Notes: If acceptable per the conventional load design, the amount of chord reinforcement could potentially be decreased to make the diaphragm flexure-controlled. This would permit a higher ductility limit to be used.

10.6.8 Compute Other Single-Degree-of–Freedom Properties

Because the roof diaphragm is a deep and relatively short beam, the stiffness must include shear deformations. Compute the total midspan deflection for an arbitrary load of 1,000 lb/in. (175 N/mm).

For moment of inertia calculations, to approximate the effect of roof cracking caused by the out-of-plane loads, use half the roof slab thickness.

Chord width:

Take $b = 6$ (wall thickness) $= 6$ (10 in.) $= 60$ in. (1,524 mm). The effective section is shown in Figure 10-7.

Effective chord reinforcing:

$$nA_s = (8.04)(3.07\ \text{in.}^2) = 24.7\ \text{in.}^2\ (15{,}920\ \text{mm}^2)$$

Moment of inertia:

$$\begin{aligned} I &= \Sigma\, bh^3/12 + \Sigma\,(\text{area})(\text{distance})^2 \\ &= 2\,(60\ \text{in.})(10\ \text{in.})^3/12 + (2.5\ \text{in.})(784\ \text{in.})^3/12 \\ &\quad + 2\,[(60\ \text{in.})(10\ \text{in.}) + 24.7\ \text{in.}^2](397\ \text{in.})^2 \\ &= 297.3 \times 10^6\ \text{in.}^4\ (123.8 \times 10^{12}\ \text{mm}^4) \end{aligned}$$

Flexural deflection: (AISC 2017 Manual, Beam Tables)

$$\begin{aligned} \Delta_f &= (\text{load})\, L^4/384\, E_c\, I \\ &= (1{,}000\ \text{lb/in.})(1{,}116\ \text{in.})^4/384\,(3{,}605{,}000\ \text{psi})(297.3 \times 10^6\ \text{in.}^4) \\ &= 0.0038\ \text{in.}\ (0.096\ \text{mm}) \end{aligned}$$

Shear modulus:

$$G = E_c/2\,(1 + \nu) = (3{,}605{,}000\ \text{psi})/2\,(1 + 0.2) = 1{,}502{,}000\ \text{psi}\ (10{,}360\ \text{MPa})$$

Shear deflection: (Budynas and Sadegh 2020, Roark's Formula, p. 166)

$$\begin{aligned} \Delta_s &= 0.125\,(6/5)(\text{load})\, L^2/(\text{area})\, G \\ &= 0.125\,(6/5)(1{,}000\ \text{lb/in.})(1{,}116\ \text{in.})^2/(5\ \text{in.})(799\ \text{in.})(1{,}502{,}000\ \text{psi}) \\ &= 0.0311\ \text{in.}\ (0.791\ \text{mm}) \end{aligned}$$

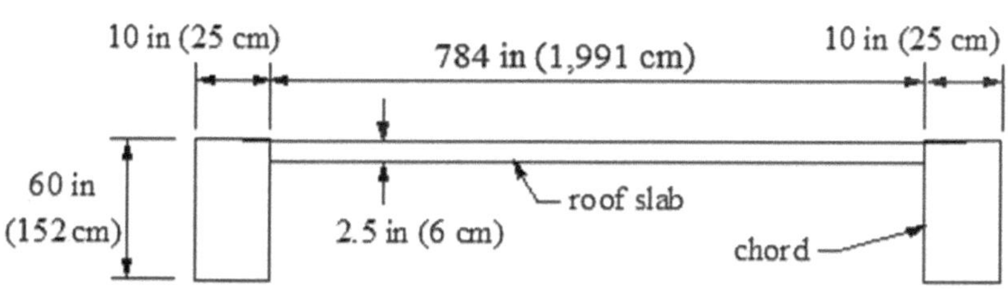

Figure 10-7. Effective roof slab section for determining single-degree-of-freedom properties.

Effective stiffness:

$$\begin{aligned} K_e &= (\text{load})\ L/(\Delta_f + \Delta_s) \\ &= (1{,}000\ \text{lb/in.})(0.001\ \text{kip/lb})(1{,}116\ \text{in.})/(0.0038\ \text{in.} + 0.0311\ \text{in.}) \\ &= \mathbf{31{,}980\ kip/in.}\ (5{,}600\ \text{kN/mm}) \end{aligned}$$

Effective elastic yield deflection:

$$y_e = R_u/K_e = (2{,}145\ \text{kips})/(31{,}980\ \text{kip/in.}) = 0.067\ \text{in.}\ (1.70\ \text{mm})$$

Note: As the shear deflection is much larger than the flexural deflection, the shear deflection can be used to directly calculate the effective stiffness without significant error.

Mass:

$$\begin{aligned} \text{weight} &= (\text{diaphragm weight}) + (\text{tributary wall weight}) \\ &= (0.15\ \text{kcf})(1{,}116\ \text{in.})[(6\ \text{in.})(784\ \text{in.}) + (10\ \text{in.}) \\ &\quad (144\ \text{in.} \times 0.5)(2\ \text{walls})]/(1{,}728\ \text{in.}^3/\text{ft}^3) \\ &= 595\ \text{kips}\ (2{,}650\ \text{kN}) \end{aligned}$$

$$\begin{aligned} M &= \text{weight/g} \\ &= 595\ \text{kips}/(386\ \text{in./s}^2) \\ &= \mathbf{1.54\ kip{\cdot}s^2/in.}\ (270{,}000\ \text{kg}) \end{aligned}$$

Load-mass transformation factors: (Chapter 6, Table 6-2)

elastic, $K_{LM} = K_M/K_L = 0.41/0.53 = \mathbf{0.77}$

Equivalent mass:

$$M_e = K_{LM}\, M = 0.77\ (1.54\ \text{kip}{\cdot}\text{s}^2/\text{in.}) = 1.19\ \text{kip}{\cdot}\text{s}^2/\text{in.}\ (207{,}900\ \text{kg})$$

Natural period: [Equation (6-8)]

$$t_n = 2\pi\sqrt{\frac{M_e}{K_e}} = 2\pi\sqrt{\frac{(1.19\ \text{kip}\cdot\text{s}^2/\text{in.})}{(31{,}980\ \text{kip/in.})}} = \mathbf{0.0383\ s}$$

10.6.9 Compute Response and Compare to Deformation Limits (Numerical integration Solution for Load Case A)

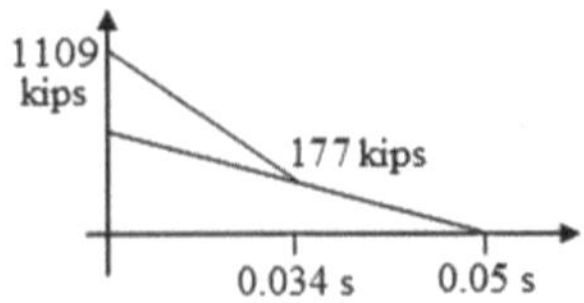

$F_o = 1{,}109$ kips (4,930 kN)

$F_{\text{int}} = 177$ kips (786 kN)

$t_c = 0.034$ s

$t_d = 0.05$ s

Time increment:

$t_n/10 = 0.0383\ \text{s}/10 \approx 0.0038\ \text{s}$

However, it is recommended to use at most a 0.001 s time step to obtain at least 10 increments of loading in each section of the pressure-time history. Consider a time range of 0.150 s to capture multiple deflection cycles and component behavior following load termination.

For corresponding output plots extending multiple cycles, see Figure 10-8.

Peak positive deflection: $y_m = +0.054$ in. (1.38 mm) at $t = 0.017$ s

Peak rebound deflection: $y_{m\text{-}} = -0.035$ in. (−0.89 mm) at $t = 0.076$ s

Note: The selected rebound peak is the first to occur after the dissipation of the applied load.

Ductility:

$\mu_d = y_m/y_e = (0.054\ \text{in.})/(0.067\ \text{in.}) = 0.81 < \mu_a$, **OK**

$\mu_{d\text{-}} = y_{m\text{-}}/y_e = (0.035\ \text{in.})/(0.067\ \text{in.}) = 0.52 < \mu_a$, **OK**

The Response is adequate. A revision is not necessary.

Note: An interaction check can be performed using Equation (7-2) to confirm that the response resulting from both in-plane and out-of-plane loads is acceptable. In the event the interaction check yields a result greater than 1.0, the roof slab response could still be deemed adequate after considering aspects such as the phasing of the in-plane and out-of-plane responses or conservative assumptions used in each separate analysis. This check is often not performed because of engineering judgment related to these and other aspects of the building's global response behavior.

10.7 SIDE WALLS (IN-PLANE LOADS)

Each side shear wall is a cantilever, which transfers roof diaphragm reactions to the floor slab and foundation. Using a 16 ft (4,877 mm) height is a bit conservative because some amount of the lateral force is removed at the floor slab level.

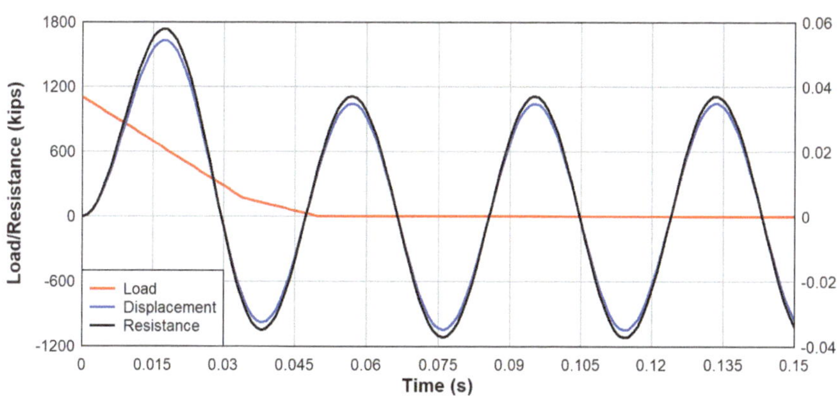

Figure 10-8. Output plots from roof slab numerical integration (in-plane loads).

Height = 16 ft or 192 in. (4,877 mm)

Length = 67 ft or 804 in. (20,420 mm)

A static check of the shear walls will be performed initially based on the ultimate resistance of the walls subject to a reflected load. If this check confirms adequate shear walls, no further action is required. In the event this check indicates inadequate shear walls, a dynamic run can be performed with the shear walls acting as cantilevered deep beams subjected to an equivalent load acting along their top edges, as distributed through the roof diaphragm.

10.7.1 Static Check of Shear Walls

The load that needs to be carried by the shear walls is based on $R_b = 22.1$ kips (98.3 kN). On a pressure basis, with $L = 12$ ft (3,658 mm) and $b = 1$ ft (305 mm), $R_b = 12.8$ psi (88.2 kPa). Considering half of the wall load transfers into the roof diaphragm, the load demand totals 12.8 psi × 1,116 in. × 144 in./2/(1,000 lb/kip) = 1,028 kips (4,570 kN), or 514 kips (2,285 kN) into each shear wall. Initially considering the dynamic shear capacity of the shear wall concrete alone and using only half of the wall thickness due to cracking from out-of-plane loading, $V_n = 2\lambda (f'_{dc})^{1/2} A_{cv} = 2 \times 1.0 \times (4{,}400 \text{ psi})^{1/2} \times 5 \text{ in.} \times 804 \text{ in.}/(1{,}000 \text{ lb/kip}) = 533$ kips (2,370 kN). As this value exceeds the demand of 514 kips (2,285 kN), it is judged that the shear walls are adequate on the basis of a static analysis, and they do not need to be checked dynamically.

Note: This example assumes that the entire length of each side wall uniformly carries the lateral load from the roof diaphragm. A more detailed analysis would consider the effect of openings in the side walls (e.g., doors and windows). Wall sections between openings can be analyzed as individual piers carrying a larger proportion of the lateral loads based on their relative stiffness.

10.7.2 Interaction Check for In-Plane and Out-of-Plane Response

Note: An interaction check can be performed using Equation (7-2) to confirm that the response resulting from both in-plane and out-of-plane loads is acceptable. In the event the interaction check yields a result greater than 1.0, the side wall response could still be deemed adequate after considering aspects such as the phasing of the in-plane and out-of-plane responses or conservative assumptions used in each separate analysis. This check is often not performed because of engineering judgment related to these and other aspects of the building's global response behavior.

10.8 ROOF BEAMS

Each interior roof beam supports a roof slab width of 8 ft (2,438 mm).

The roof beam is connected to the roof slab to prevent separation during rebound. In this case, the connection is to be designed to prevent composite action between the roof slab and the roof beam.

Note: Because composite action greatly increases the bending capacity, while not increasing the beam's shear capacity, ignoring this effect could be highly unconservative. The designer should be cautious not to create a condition where a stronger component cannot adequately transfer potentially developed loads through its cross section or at its end supports.

Span, $L = 18$ ft or 216 in. (5,486 mm) with pinned connections at each end

Beam spacing, $L_s = 8$ ft or 96 in. (2,438 mm)

10.8.1 Calculate Design Blast Load Case A (Perpendicular to Span of Beam)

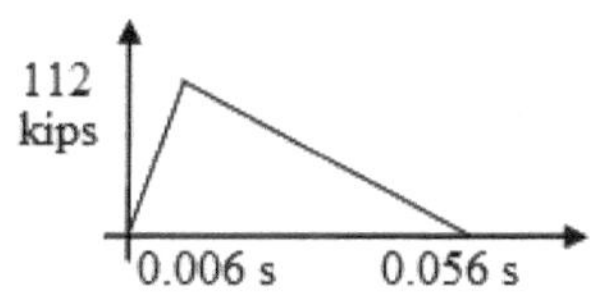

From the Appendix 3 calculation:

$P_a = 5.4$ psi (37 kPa)

$t_r = 0.006$ s

$t_d = 0.05$ s

Peak load:

$F_o = (5.4 \text{ psi})(216 \text{ in. span})(96 \text{ in. width})/(1{,}000 \text{ lb/kip}) = 112$ kips (498 kN)

10.8.2 Calculate Design Blast Load Case B (Parallel to Span of Beam)

From the Appendix 3 calculation:

$P_{so} = 6$ psi (41 kPa)

$t_d = 0.05$ s

$L_w = 66$ ft or 792 in. (20,120 mm)

$L_1 = L = 216$ in. (5,486 mm)

Equivalent load coefficient: (Chapter 3, Figure 3-10)

$L_w/L_1 = (792 \text{ in.})/(216 \text{ in.}) = 3.7$; therefore, $C_e = 0.87$.

Reduced dynamic wind pressure: [Chapter 3, Equation (3-4)]

$q_o = 0.022\,(C_e\,P_{so})^2 = 0.022\,(0.87 \times 6 \text{ psi})^2 = 0.6$ psi (4 kPa)

Equivalent peak overpressure: [Chapter 3, Equation (3-11)]

$P_a = C_e\,P_{so} + C_d\,q_o = (0.87)(6 \text{ psi}) + (-0.4)(0.6 \text{ psi}) = 5.0$ psi (34 kPa)

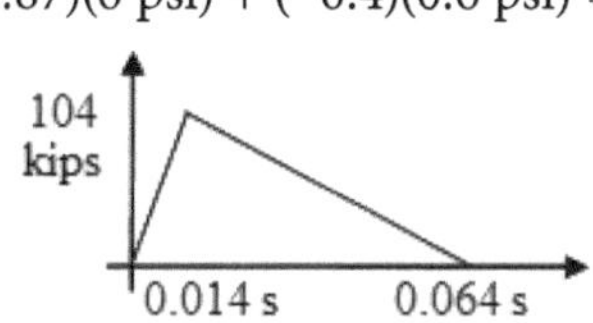

Peak load:

F_o = (5.0 psi)(216 in. span)(96 in. width)/(1,000 lb/kip) = 104 kips (461 kN)

Rise time:

$t_r = L_1/U$ = (18 ft)/(1,312 ft/s) = 0.014 s

10.8.3 Select Trial Size

Try W14 × 26: (AISC 2017 Manual)

Beam depth: d = 13.9 in. (353 mm)

Flange width/thickness: $b_f/2t_f = \lambda_f = 6.0$

Web thickness: t_w = 0.255 in. (6.47 mm)

Web depth/thickness: $h/t_w = \lambda_w = 48.1$

Radius of gyration: r_y = 1.08 in. (27.4 mm)

Moment of inertia: I_x = 245 in.^{4} (102 × 10^6 mm^4)

Plastic modulus: Z_x = 40.2 in.^{3} (658,800 mm^3)

10.8.4 Compute Bending Resistance

For dynamic bending: (Chapter 5, Appendix 5A)

$F_{ds} = F_{dy}$ = (SIF)(DIF) F_y = (1.1)(1.19)(50 ksi) = 65.5 ksi (451 MPa)

Check the flange and web of the section for compactness limits. Because roof beams are considered secondary members, width-thickness ratios do not need to meet AISC seismic requirements.

Check the flange: (AISC 360, Table B4.1)

$$\lambda_p = 0.38\sqrt{\frac{E_s}{F_{ds}}} = 0.38\sqrt{\frac{(29{,}000\text{ ksi})}{(65.5\text{ ksi})}} = 8.0 > \lambda_f, \quad \textbf{OK}$$

Check web: (AISC 360, Table B4.1)

$$\lambda_p = 3.76\sqrt{\frac{E_s}{F_{ds}}} = 3.76\sqrt{\frac{(29{,}000\text{ ksi})}{(65.5\text{ ksi})}} = 79.1 > \lambda_w, \quad \textbf{OK}$$

The roof beams are continuously braced by the concrete roof slab; therefore, the unbraced length is OK.

Note: If an 8 ft unbraced length is determined to be too long, additional lateral bracing can be provided to brace the bottom flange of the roof beams; otherwise, a reduced rebound resistance should be considered in the following calculations:

Moment capacity: (AISC 360, Equation F2-1)

$M_p = Z_x F_{ds}$ = (40.2 in.^{3})(65.5 ksi) = 2,630 kip·in. (297 kN·m)

$R_b = 8\, M_p/L$ = 8 (2,630 kip·in.)/(216 in.) = 97.4 kips (433 kN)

10.8.5 Compute Shear Resistance

For dynamic shear: (Chapter 5, Appendix 5A)

$F_{ds} = F_{dy} = \text{(SIF)(DIF)}\ F_y = (1.1)(1.19)(50\ \text{ksi}) = 65.5\ \text{ksi}\ (451\ \text{MPa})$

Because $h/t_w < 260$, $k_v = 5$. (AISC 360, Section G2.1)
Check web slenderness criteria: (AISC 360, Equation G2-3)

$$\lambda_v = 1.1\sqrt{\frac{k_v E_s}{F_{ds}}} = 1.1\sqrt{\frac{(5)(29{,}000\ \text{ksi})}{(65.5\ \text{ksi})}} = 51.8 > \lambda_w;$$

therefore, $C_v = 1.0$.

$V_n = 0.6\ F_{ds}\ dt_w\ C_v$ (AISC 360, Equation G2-1)
$= 0.6\ (65.5\ \text{ksi})(13.9\ \text{in.})(0.255\ \text{in.})(1.0)$
$= 139$ kips (619 kN)

$R_s = 2\ V_n = 2\ (139\ \text{kips}) = 278$ kips (1,240 kN)

10.8.6 Resistance and Permissible Response

Because $R_b < R_s$, bending controls.
Positive and rebound resistance, $R_u = -R_{u\text{-}} = R_b =$ **97.4 kips** (433 kN)
Response limits: $\mu_a = 3$, $\theta_a = 2$ degrees

10.8.7 Compute Other Single-Degree-of-Freedom Properties

Effective stiffness: (Chapter 6, Table 6-1)

$K_e = 384\ E_s\ I_x/5\ L^3$
$= 384\ (29{,}000\ \text{ksi})(245\ \text{in.}^4)/(5)(216\ \text{in.})^3$
$=$ **54.1 kip/in.** (9.48 kN/mm)

Elastic yield deflection: $y_e = R_u/K_e = (97.4\ \text{kips})/(54.1\ \text{kip/in.}) =$ **1.80 in.** (45.7 mm)

Static load = (beam weight) + (tributary slab weight)
= (0.026 klf)(18 ft) + (8 ft)(18 ft)[(0.075 ksf slab) + (0.010 ksf dead load)]
= 0.5 kip + 12.2 kips = 12.7 kips (56.5 kN)

Mass:

M = static load/g
= (12.7 kips)/(386 in./s^2)
= **0.0329 kip·s^2/in.** (5,760 kg)

Static (dead load) deflection:

y_{st} = static load/K_e = (12.7 kips)/(54.1 kip/in.) = **0.235 in.** (5.96 mm)

Load-mass transformation factors: (Chapter 6, Table 6-1)

elastic, $K_{LM} = K_M/K_L = 0.50/0.64 =$ **0.78**
plastic, $K_{LM} = K_M/K_L = 0.33/0.50 =$ **0.66**

Assuming some plastic deformation, use an average value for the load-mass factor when calculating the natural period:

$K_{LM} = (0.78 + 0.66)/2 = \mathbf{0.72}$

Equivalent mass:

$M_e = K_{LM}\, M = (0.72)(0.0329 \text{ kip·s}^2\text{/in.}) = 0.0237 \text{ kip·s}^2\text{/in.}$ (4,150 kg)

Natural period: [Equation (6-8)]

$$t_n = 2\pi\sqrt{\frac{M_e}{K_e}} = 2\pi\sqrt{\frac{(0.0237 \text{ kip}\cdot\text{s}^2/\text{in.})}{(54.1 \text{ kip/in.})}} = \mathbf{0.131\ s}$$

10.8.8 Compute Response and Compare to Deformation Limits (Numerical Integration Solution for Load Case A)

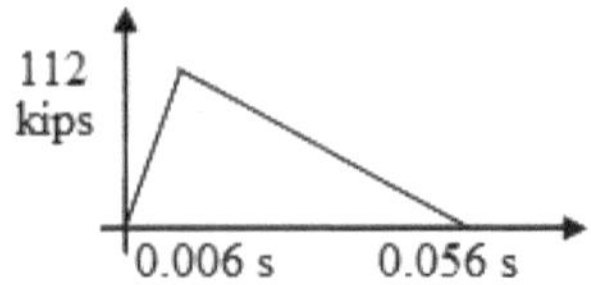

$F_o = 112$ kips (498 kN)

$t_r = 0.006$ s

$t_d = 0.05$ s

Time increment:

$t_n/10 = 0.131 \text{ s}/10 \approx 0.013$ s

However, it is recommended to use at most a 0.0006 s time step to obtain at least 10 increments of loading in the smallest, first section of the pressure–time history. A smaller time step can be used to potentially obtain a more refined solution with smoother output plots. Consider a time range of 0.400 s to capture multiple deflection cycles and component behavior following load termination.

For output plots of key parameters, see Figure 10-9.

Peak positive deflection: $y_m = +2.55$ in. (64.8 mm) at $t = 0.057$ s

Peak rebound deflection: $y_{m-} = -0.58$ in. (−14.7 mm) at $t = 0.125$ s

Note: The selected rebound peak is the first to occur after the dissipation of the applied load.

Ductility:

$\mu_d = y_m/y_e = (2.55 \text{ in.})/(1.80 \text{ in.}) = 1.42 < \mu_a$, **OK**

Support rotation: (Figure 5-9)

$\theta_d = \tan^{-1}(y_m)/(0.5\, L) = \tan^{-1}[(2.55 \text{ in.})/(0.5)(216 \text{ in.})] = 1.35° < \theta_a$, **OK**

The response is adequate. A revision is not necessary.

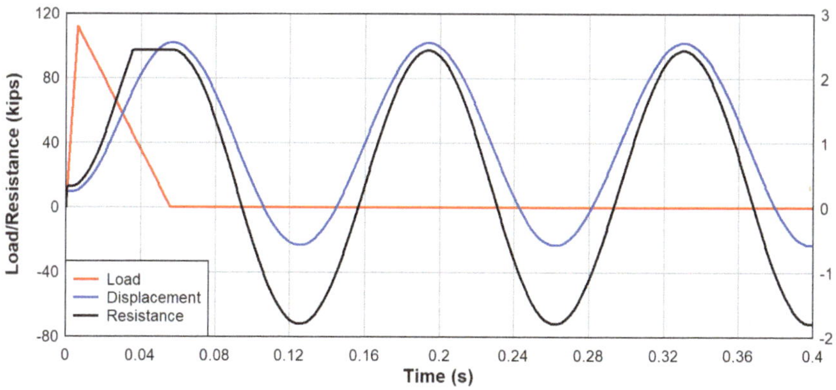

Figure 10-9. Output plots from roof beam numerical integration (Load Case A).

10.8.9 Compute Response and Compare to Deformation Limits (Numerical Integration Solution for Load Case B)

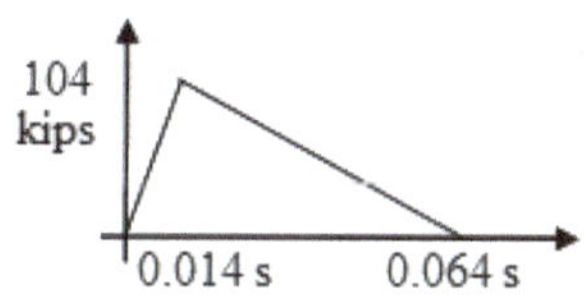

$F_o = 104$ kips (461 kN)

$t_r = 0.014$ s

$t_d = 0.05$ s

Time increment:

$t_n/10 = 0.131 \text{ s}/10 \approx 0.013 \text{ s}$

However, it is recommended to use at most a 0.0014 s time step to obtain at least 10 increments of loading in the smallest, first section of the pressure-time history. A smaller time step can be used to potentially obtain a more refined solution with smoother output plots. Consider a time range of 0.400 s to capture multiple deflection cycles and component behavior following load termination.

For the numerical integration through the time of peak positive deflection, refer to Table 10-5. For output plots of key parameters, see Figure 10-10.

Peak positive deflection: $y_m = +2.65$ in. (67.3 mm) at $t = 0.063$ s

Peak rebound deflection: $y_{m-} = -0.48$ in. (−12.3 mm) at $t = 0.132$ s

Note: The selected rebound peak is the first to occur after the dissipation of the applied load.

Ductility:

$\mu_d = y_m/y_e = (2.65 \text{ in.})/(1.80 \text{ in.}) = 1.47 < \mu_a$, **OK**

Table 10-5. Roof Beam Numerical Integration (Load Case B).

Example:	Section 10.8 Roof Beam							KEY = 0 when elastic
								KEY = +1 when positive plastic
time increment:	0.0014 s	**Blast Load**						KEY = -1 when rebound plastic
mass:	0.0329 k-s²/in	**t0:**	0 s	**P0:**	0 kips			Z = 1 when continued elastic
stiffness:	54.1 k/in	**t1:**	0.014 s	**P1:**	104 kips			Z = 2 at start of positive plastic deformation
damping:	0 k-s/in	**t2:**	0.064 s	**P2:**	0 kips			Z = 3 when increasing plastic deformation
positive resistance:	97.4 kips	**Dynamic Reaction Factors**		**BC: S-S**				Z = 4 at end of positive plastic deformation
rebound resistance:	-97.4 kips	**Crel:**	0.39	**Cfel:**	0.11			Z = 6 at start of rebound plastic deformation
static load:	12.7 kips	**Crpl:**	0.38	**Cfpl:**	0.12			Z = 8 at end of rebound plastic deformation

time (s)	load (kips)	y (in)	v (in/s)	a (in/s²)	ye+ (in)	ye- (in)	R (kips)	KEY -	KLM	V (kips)	eff K (k/in)	incr F (kips)	incr y (in)	incr v (in/s)	Z -
0.0000	0.000	0.2348	0.00	0	1.800	-1.800	0.000	0	0.78	0.000	78611	10.4	0.000	0.28	
0.0014	10.400	0.2349	0.28	405	1.800	-1.800	12.707	0	0.78	6.100	78611	72.8	0.001	0.85	1
0.0028	20.800	0.2358	1.13	808	1.800	-1.800	12.757	0	0.78	7.263	78611	197.2	0.003	1.41	1
0.0042	31.200	0.2383	2.54	1208	1.800	-1.800	12.893	0	0.78	8.460	78611	383.3	0.005	1.97	1
0.0056	41.600	0.2432	4.51	1603	1.800	-1.800	13.157	0	0.78	9.707	78611	630.1	0.008	2.52	1
0.0070	52.000	0.2512	7.03	1992	1.800	-1.800	13.590	0	0.78	11.020	78611	936.8	0.012	3.05	1
0.0084	62.400	0.2631	10.08	2372	1.800	-1.800	14.235	0	0.78	12.416	78611	1302.0	0.017	3.58	1
0.0098	72.800	0.2797	13.66	2742	1.800	-1.800	15.131	0	0.78	13.909	78611	1724.2	0.022	4.09	1
0.0112	83.200	0.3016	17.75	3101	1.800	-1.800	16.318	0	0.78	15.516	78611	2201.7	0.028	4.58	1
0.0126	93.600	0.3296	22.34	3447	1.800	-1.800	17.833	0	0.78	17.251	78611	2732.5	0.035	5.06	1
0.0140	104.000	0.3644	27.40	3779	1.800	-1.800	19.713	0	0.78	19.128	78611	3301.1	0.042	5.15	1
0.0154	101.088	0.4064	32.55	3577	1.800	-1.800	21.985	0	0.78	19.694	78611	3851.9	0.049	4.86	1
0.0168	98.176	0.4554	37.40	3361	1.800	-1.800	24.636	0	0.78	20.407	78611	4369.3	0.056	4.54	1
0.0182	95.264	0.5110	41.95	3130	1.800	-1.800	27.643	0	0.78	21.260	78611	4851.3	0.062	4.21	1
0.0196	92.352	0.5727	46.16	2886	1.800	-1.800	30.982	0	0.78	22.242	78611	5295.7	0.067	3.86	1
0.0210	89.440	0.6400	50.02	2631	1.800	-1.800	34.626	0	0.78	23.343	78611	5700.8	0.073	3.50	1
0.0224	86.528	0.7126	53.52	2365	1.800	-1.800	38.549	0	0.78	24.552	78611	6064.8	0.077	3.12	1
0.0238	83.616	0.7897	56.63	2088	1.800	-1.800	42.723	0	0.78	25.860	78611	6386.4	0.081	2.72	1
0.0252	80.704	0.8709	59.36	1804	1.800	-1.800	47.118	0	0.78	27.254	78611	6664.1	0.085	2.32	1
0.0266	77.792	0.9557	61.68	1511	1.800	-1.800	51.704	0	0.78	28.722	78611	6896.8	0.088	1.91	1
0.0280	74.880	1.0435	63.59	1213	1.800	-1.800	56.451	0	0.78	30.253	78611	7083.6	0.090	1.49	1
0.0294	71.968	1.1336	65.07	910	1.800	-1.800	61.326	0	0.78	31.834	78611	7223.7	0.092	1.06	1
0.0308	69.056	1.2255	66.13	602	1.800	-1.800	66.297	0	0.78	33.452	78611	7316.4	0.093	0.63	1
0.0322	66.144	1.3185	66.76	293	1.800	-1.800	71.332	0	0.78	35.095	78611	7361.5	0.094	0.19	1
0.0336	63.232	1.4122	66.95	-18	1.800	-1.800	76.398	0	0.78	36.751	78611	7358.7	0.094	-0.24	1
0.0350	60.320	1.5058	66.71	-329	1.800	-1.800	81.463	0	0.78	38.406	78611	7308.0	0.093	-0.68	1
0.0364	57.408	1.5987	66.03	-638	1.800	-1.800	86.492	0	0.78	40.047	78611	7209.7	0.092	-1.11	1
0.0378	54.496	1.6905	64.92	-945	1.800	-1.800	91.454	0	0.78	41.661	78611	7064.2	0.090	-1.54	1
0.0392	51.584	1.7803	63.38	-1248	1.800	-1.800	96.315	0	0.78	43.237	78611	6872.0	0.087	-1.96	1
0.0406	48.672	1.8677	61.43	-1659	1.800	-1.800	97.400	1	0.66	42.853	66471	5605.5	0.084	-2.42	2
0.0420	45.760	1.9521	59.01	-1793	1.800	-1.800	97.400	1	0.66	42.503	66471	5371.9	0.081	-2.60	3
0.0434	42.848	2.0329	56.41	-1927	1.800	-1.800	97.400	1	0.66	42.154	66471	5120.8	0.077	-2.79	3
0.0448	39.936	2.1099	53.61	-2062	1.800	-1.800	97.400	1	0.66	41.804	66471	4852.2	0.073	-2.98	3
0.0462	37.024	2.1829	50.63	-2196	1.800	-1.800	97.400	1	0.66	41.455	66471	4566.1	0.069	-3.17	3
0.0476	34.112	2.2516	47.47	-2330	1.800	-1.800	97.400	1	0.66	41.105	66471	4262.6	0.064	-3.36	3
0.0490	31.200	2.3157	44.11	-2464	1.800	-1.800	97.400	1	0.66	40.756	66471	3941.6	0.059	-3.54	3
0.0504	28.288	2.3750	40.57	-2598	1.800	-1.800	97.400	1	0.66	40.407	66471	3603.1	0.054	-3.73	3
0.0518	25.376	2.4292	36.84	-2732	1.800	-1.800	97.400	1	0.66	40.057	66471	3247.2	0.049	-3.92	3
0.0532	22.464	2.4781	32.92	-2866	1.800	-1.800	97.400	1	0.66	39.708	66471	2873.8	0.043	-4.11	3
0.0546	19.552	2.5213	28.81	-3000	1.800	-1.800	97.400	1	0.66	39.358	66471	2482.9	0.037	-4.29	3
0.0560	16.640	2.5587	24.52	-3134	1.800	-1.800	97.400	1	0.66	39.009	66471	2074.5	0.031	-4.48	3
0.0574	13.728	2.5899	20.04	-3268	1.800	-1.800	97.400	1	0.66	38.659	66471	1648.7	0.025	-4.67	3
0.0588	10.816	2.6147	15.37	-3403	1.800	-1.800	97.400	1	0.66	38.310	66471	1205.4	0.018	-4.86	3
0.0602	7.904	2.6328	10.51	-3537	1.800	-1.800	97.400	1	0.66	37.960	66471	744.6	0.011	-5.05	3
0.0616	4.992	2.6440	5.46	-3671	1.800	-1.800	97.400	1	0.66	37.611	66471	266.4	0.004	-5.23	3
0.0630	2.080	2.6480	0.23	-3805	1.800	-1.800	97.400	1	0.66	37.262	66471	-228.5	-0.003	-5.39	3
0.0644	0.000	2.6446	-5.16	-3293	2.648	-0.953	97.214	0	0.78	37.913	78611	-821.4	-0.010	-4.60	4

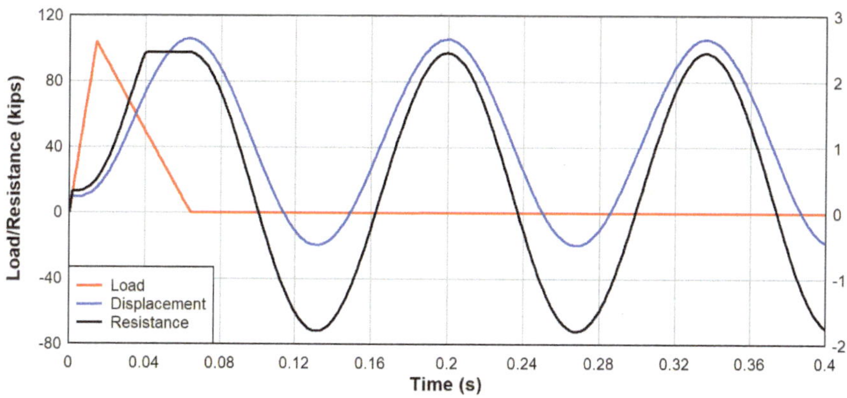

Figure 10-10. Output plots from roof beam numerical integration (Load Case B).

Support rotation: (Figure 5-9)

$\theta_d = \tan^{-1}(y_m)/(0.5\ L) = \tan^{-1}[(2.65\text{ in.})/(0.5)(216\text{ in.})] = 1.40° < \theta_a$, **OK**

The response is adequate. A revision is not necessary.

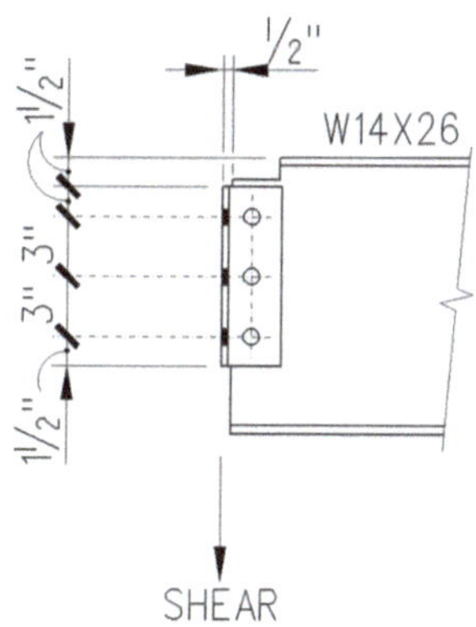

10.9 ROOF BEAM CONNECTIONS

The roof beams connect to the girder web or column flange.

Roof beam resistance: (Section 10.8.6)

$R_u = 97.4$ kips (433 kN)

Connection design load:

$V_u = R_u/(2\text{ beam ends}) = (97.4\text{ kips})/2 = 48.7$ kips (217 kN)

Try:

AISC Standard All-Bolted Shear Connection

ASTM A325 bearing bolts, three rows

threads in shear plane

diameter, $d_b = 0.75$ in. (19 mm)

Nominal shear strength: $F_{nv} = 48$ ksi (330 MPa) (AISC 360, Table J3.2)

Nominal unthreaded area:

$$A_b = \pi\, d_b^2/4 = \pi\,(0.75 \text{ in.})^2/4 = 0.442 \text{ in.}^2\ (285 \text{ mm}^2)$$

10.9.1 Bolt Shear Strength (Beam to Clip Angle)

SIF (A325 High-Strength Bolt) = 1.0 (conservative value)

DIF (A325 High-Strength Bolt) = 1.05 (Section 5.5.4)

Bolt shear strength:

$$\begin{aligned} V_n &= (\text{SIF})(\text{DIF})\, F_{nv}\, A_b\, N \\ &= (1.0)(1.05)(48 \text{ ksi})(0.442 \text{ in.}^2)(2 \text{ shear planes})(3 \text{ bolts}) \\ &= 134 \text{ kips } (594 \text{ kN}) > V_u, \quad \textbf{OK} \end{aligned}$$

Note: The clip angle to girder connection uses twice the number of bolts in single shear and will not control shear strength.

10.9.2 Clip Angle Shear Strength

The critical section is a vertical plane at the bolt centerline.

Try: L3 × 3 × 3/8

length = 1.5 in. + 2 (3 in.) + 1.5 in. = 9 in. (229 mm)

ASTM A36 steel

Gross section:

$$A_g = (2 \text{ angles})(3/8 \text{ in., thickness})(9 \text{ in., length}) = 6.75 \text{ in.}^2\ (4{,}355 \text{ mm}^2)$$

Bolt hole diameter: (AISC 360, Table J3.3, Section D3)

$$d_h = (13/16 \text{ in., nominal hole diameter}) + (1/16 \text{ in.}) = 0.875 \text{ in. } (22.2 \text{ mm})$$

Net section at bolt holes:

$$\begin{aligned} A_{nv} &= (6.75 \text{ in.}^2) - (2 \text{ angles})(3 \text{ holes})(0.875 \text{ in., diameter})(3/8 \text{ in., thickness}) \\ &= 4.78 \text{ in.}^2\ (3{,}080 \text{ mm}^2) \end{aligned}$$

For dynamic shear: (Chapter 5, Appendix 5A)

$$F_{dy} = (\text{SIF})(\text{DIF})\, F_y = (1.1)(1.29)(36 \text{ ksi}) = 51.1 \text{ ksi } (352 \text{ MPa})$$

$$F_{du} = (\text{SIF})(\text{DIF})\, F_u = (1.1)(1.1)(58 \text{ ksi}) = 70.2 \text{ ksi } (484 \text{ MPa})$$

Shear yielding strength: (AISC 360, Equation J4-3)

$$V_n = 0.6\, F_{dy}\, A_g = 0.6\ (51.1\ \text{ksi})(6.75\ \text{in.}^2) = 207\ \text{kips}\ (920\ \text{kN}) > V_u, \quad \textbf{OK}$$

Shear rupture strength: (AISC 360, Equation J4-4)

$$V_n = 0.6\, F_{du}\, A_{nv} = 0.6\ (70.2\ \text{ksi})(4.78\ \text{in.}^2) = 201\ \text{kips}\ (896\ \text{kN}) > V_u, \quad \textbf{OK}$$

10.9.3 Clip Angle Bearing Strength

Clear distance between holes:

$$L_c = (1\ \text{bolt})[(1.5\ \text{in., edge}) - 0.5\ (0.875\ \text{in., hole diameter})] + (2\ \text{each})[(3\ \text{in., spacing}) - (0.875\ \text{in., hole diameter})] = 5.31\ \text{in.}\ (135\ \text{mm})$$

Bearing strength: (AISC 360, Equation J3-6a)

$$V_n = \min\ [1.2\ L_c\, t_p\, F_{du}\ \text{or}\ 2.4\ d_b\, t_p\, F_{du}]$$

$$1.2\ (5.31\ \text{in.})(3/8\ \text{in.} \times 2\ \text{each})(70.2\ \text{ksi}) = 336\ \text{kips}\ (1{,}490\ \text{kN})$$

$$2.4\ (0.75\ \text{in.} \times 3\ \text{each})(3/8\ \text{in.} \times 2\ \text{each})(70.2\ \text{ksi}) = \textbf{284 kips}\ (1{,}260\ \text{kN})$$

[CONTROLS]

Therefore, $V_n = 284$ kips (1,260 kN) $> V_u$, **OK**

10.9.4 Beam Shear Strength

The critical section is a vertical plane at the bolt centerline.

W14 × 26 properties: (AISC 2017 Manual)

depth, $d = 13.9$ in. (353 mm)

web thickness, $t_w = 0.255$ in. (6.48 mm)

Gross section, assuming that the beam flange is coped 1.25 in. (31.8 mm):

$$A_g = (0.255\ \text{in.})[(13.9\ \text{in.}) - (1.25\ \text{in.})] = 3.23\ \text{in.}^2\ (2{,}080\ \text{mm}^2)$$

Net section at bolt holes:

$$A_{nv} = (3.23\ \text{in.}^2) - (3\ \text{holes})(0.875\ \text{in., diameter})(0.255\ \text{in., thickness}) = 2.56\ \text{in.}^2\ (1{,}650\ \text{mm}^2)$$

For dynamic shear: (Chapter 5, Appendix 5A)

$$F_{dy} = (\text{SIF})(\text{DIF})\ F_y = (1.1)(1.19)(50\ \text{ksi}) = 65.5\ \text{ksi}\ (451\ \text{MPa})$$
$$F_{du} = (\text{SIF})(\text{DIF})\ F_u = (1.1)(1.05)(65\ \text{ksi}) = 75.1\ \text{ksi}\ (518\ \text{MPa})$$

Web shear yielding strength: (AISC 360, Equation J4-3)

$$V_n = 0.6\, F_{dy}\, A_g = 0.6\ (65.5\ \text{ksi})(3.23\ \text{in.}^2) = 127\ \text{kips}\ (563\ \text{kN}) > V_u, \quad \textbf{OK}$$

Web shear rupture strength: (AISC 360, Equation J4-4)

$$V_n = 0.6\, F_{du}\, A_{nv} = 0.6\ (75.1\ \text{ksi})(2.56\ \text{in.}^2) = 115\ \text{kips}\ (512\ \text{kN}) > V_u, \quad \textbf{OK}$$

10.9.5 Beam Bearing Strength

Clear distance between holes:

$$L_c = (1\text{ bolt})[(3\text{ in., edge}) - (1.25\text{ in., cope}) - 0.5\,(0.875\text{ in., hole diameter})] + (2\text{ bolts})[(3\text{ in., spacing}) - (0.875\text{ in., hole diameter})] = 5.56\text{ in. } (141\text{ mm})$$

Web bearing strength: (AISC 360, Equation J3-6a)

$$V_n = \min\,[1.2\,L_c\,t_w\,F_{du}\text{ or }2.4\,d_b\,t_w\,F_{du}]$$
$$1.2\,(5.56\text{ in.})(0.255\text{ in.})(75.1\text{ ksi}) = 128\text{ kips } (568\text{ kN})$$
$$2.4\,(0.75\text{ in.} \times 3\text{ each})(0.255\text{ in.})(75.1\text{ ksi}) = \mathbf{103\ kips}\ (460\text{ kN})\ \textbf{[CONTROLS]}$$

Therefore, $V_n = 103$ kips (460 kN) $> V_u$, **OK**

Use L3 × 3 × 3/8 clip angles with bolted connections as described.

Note: The roof girder web thickness is greater than that of the roof beam and will not control the bearing strength.

Notes: For brevity, calculations for other failure modes such as block shear failure are omitted. These should be checked as well prior to finalizing the connection design.

10.10 ROOF GIRDERS

The roof girders are simply supported at both ends with loads from the roof beams applied at the quarter points.

The roof girder is connected to the roof slab to prevent separation during rebound. In this case, the connection is to be designed to prevent composite action between the roof slab and the roof girder.

Notes: Because composite action greatly increases the bending capacity, while not increasing the girder's shear capacity, ignoring this effect could be very unconservative. The designer should be cautious not to create a condition where a stronger component cannot adequately transfer potentially developed loads through its cross section or at its end supports.

Span, L = 32 ft or 384 in. (9,754 mm) with pinned connections at each end

Girder spacing, L_s = 18 ft or 216 in. (5,486 mm)

10.10.1 Calculate Design Blast Load Case A (Parallel to Span of Girder)

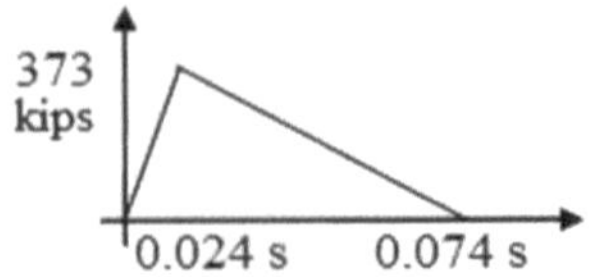

From the Appendix 3 calculation:

$P_{so} = 6$ psi (41 kPa)

$t_d = 0.05$ s

$L_w = 66$ ft or 792 in. (20,120 mm)

The blast-loaded area tributary to the roof girder is the girder span of 384 in. (9,754 mm) multiplied by the girder spacing of 216 in. (5,486 mm).

$L_1 = L = 32$ ft or 384 in. (9,754 mm)

Equivalent load coefficient: (Chapter 3, Figure 3-10)

$L_w/L_1 = (792 \text{ in.})/(384 \text{ in.}) = 2.1$; therefore, $C_e = 0.79$.

Reduced dynamic wind pressure: [Chapter 3, Equation (3-4)]

$q_o = 0.022\,(C_e\,P_{so})^2 = 0.022\,(0.79 \times 6 \text{ psi})^2 = 0.49$ psi (3 kPa)

Equivalent peak overpressure: [Chapter 3, Equation (3-11)]

$P_a = C_e\,P_{so} + C_d\,q_o = (0.79)(6 \text{ psi}) + (-0.4)(0.49 \text{ psi}) = 4.5$ psi (31 kPa)

Peak load:

$F_o = (4.5 \text{ psi})(384 \text{ in.})(216 \text{ in.})/(1{,}000 \text{ lb/kip}) = 373$ kips (1,660 kN)

Rise time:

$t_r = L_1/U = (32 \text{ ft})/(1{,}312 \text{ ft/s}) = 0.024$ s

10.10.2 Load Case B (Perpendicular to Span of Girder)

$L_1 = L_s = 18$ ft or 216 in. (5,486 mm)

Equivalent load coefficient: (Chapter 3, Figure 3-10)

$L_w/L_1 = (792 \text{ in.})/(216 \text{ in.}) = 3.7$; therefore, $C_e = 0.87$.

Reduced dynamic wind pressure: [Chapter 3, Equation (3-4)]

$q_o = 0.022\,(C_e\,P_{so})^2 = 0.022\,(0.87 \times 6 \text{ psi})^2 = 0.6$ psi (4 kPa)

Equivalent peak overpressure: [Chapter 3, Equation (3-11)]

$P_a = C_e\,P_{so} + C_d\,q_o = (0.87)(6 \text{ psi}) + (-0.4)(0.6 \text{ psi}) = 5.0$ psi (34 kPa)

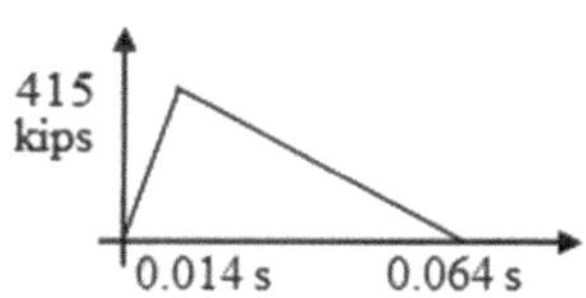

Peak load:

$F_o = (5.0 \text{ psi})(384 \text{ in.})(216 \text{ in.})/(1{,}000 \text{ lb/kip})$
$= 415$ kips (1,845 kN)

Rise time:

$t_r = L_1/U = (18\text{ ft})/(1{,}312\text{ ft/s}) = 0.014\text{ s}$

10.10.3 Select Trial Size

Try W21 × 147: (AISC 2017 Manual)

Beam depth: $d = 22.1$ in. (561 mm)

Flange width/thickness: $b_f/2t_f = \lambda_f = 5.44$

Web thickness: $t_w = 0.72$ in. (18 mm)

Web depth/thickness: $h/t_w = \lambda_w = 26.1$

Radius of gyration: $r_y = 2.95$ in. (75 mm)

Moment of inertia: $I_x = 3{,}630\text{ in.}^4$ $(1{,}511 \times 10^6\text{ mm}^4)$

Plastic modulus: $Z_x = 373\text{ in.}^3$ $(6{,}112{,}000\text{ mm}^3)$

10.10.4 Compute Bending Resistance

For dynamic bending: (Chapter 5, Appendix 5A)

$F_{ds} = F_{dy} = (\text{SIF})(\text{DIF})\, F_y = (1.1)(1.19)(50\text{ ksi}) = 65.5\text{ ksi (451 MPa)}$

Because girders are considered primary members, width-thickness ratios need to meet AISC seismic requirements.

Check the flange: (AISC 341, Table D1.1)

$$\lambda_{ps} = 0.32\sqrt{\frac{E_s}{F_{ds}}} = 0.32\sqrt{\frac{(29{,}000\text{ ksi})}{(65.5\text{ ksi})}} = 6.7 > \lambda_f, \quad \textbf{OK}$$

Check the web: (AISC 341, Table D1.1)

$$\lambda_{ps} = 2.57\sqrt{\frac{E_s}{F_{ds}}} = 2.57\sqrt{\frac{(29{,}000\text{ ksi})}{(65.5\text{ ksi})}} = 54.1 > \lambda_w, \quad \textbf{OK}$$

Unbraced length for plastic design: (AISC 360, Equation F2-5)

The roof girders are braced by the roof beams, which have a spacing of $L_{s,\text{beam}} = 8$ ft or 96 in. (2,438 mm).

$$\begin{aligned} L_p &= 1.76\, r_y\sqrt{\frac{E_s}{F_{ds}}} \\ &= 1.76\,(2.95\text{ in.})\sqrt{\frac{(29{,}000\text{ ksi})}{(65.5\text{ ksi})}} \\ &= 109\text{ in. (2,776 mm)} > L_{s,\text{beam}}, \quad \textbf{OK} \end{aligned}$$

Note: Additional lateral bracing can be provided at girder quarter points to brace the bottom flange of the roof girders; otherwise, a reduced rebound resistance should be considered in the following calculations:

Moment capacity: (AISC 360, Equation F2-1)

$M_p = Z_x F_{ds} = (373 \text{ in.}^3)(65.5 \text{ ksi}) = 24{,}410 \text{ kip·in.} (2{,}758 \text{ kN·m})$

$R_b = 8 M_p/L = 8 (24{,}431 \text{ kip·in.})/(384 \text{ in.}) = 509 \text{ kips} (2{,}260 \text{ kN})$

10.10.5 Compute Shear Resistance

For dynamic shear: (Chapter 5, Appendix 5A)

$F_{ds} = F_{dy} = (\text{SIF})(\text{DIF}) F_y = (1.1)(1.19)(50 \text{ ksi}) = 65.5 \text{ ksi} (451 \text{ MPa})$

$V_n = 0.6 F_{ds} d t_w C_v$ (AISC 360, Equation G2-1)

$= 0.6 (65.5 \text{ ksi})(22.1 \text{ in.})(0.72 \text{ in.})(1.0)$

$= 625 \text{ kips} (2{,}780 \text{ kN})$

$R_s = 2 V_n = 2 (625 \text{ kips}) = 1{,}250 \text{ kips} (5{,}560 \text{ kN})$

10.10.6 Resistance and Permissible Response

Because $R_b < R_s$, bending controls.

Positive and rebound resistance, $R_u = -R_{u\text{-}} = R_b =$ **509 kips** (2,260 kN)

Response limits: $\mu_a = 1.5$, $\theta_a = 1$ degrees

10.10.7 Compute Other Single-Degree-of-Freedom Properties

Chapter 6, Table 6-1, does not include a case for three-point loads. In lieu of a derivation of the needed values, the stiffness and transformation factors for uniform loading will be used as an approximation.

Effective stiffness: (Chapter 6, Table 6-1)

$K_e = 384 E_s I_x/5 L^3$

$= 384 (29{,}000 \text{ ksi})(3{,}630 \text{ in.}^4)/(5)(384 \text{ in.})^3$

$=$ **143 kip/in.** (25.0 kN/mm)

Elastic yield deflection: $y_e = R_u/K_e = (509 \text{ kips})/(143 \text{ kip/in.}) =$ **3.56 in.** (90.4 mm)

Mass:

static load = (girder weight) + (tributary beam weight) + (tributary slab load)

= (0.147 klf)(32 ft) + 3(0.026 klf)(18 ft)

+ (32 ft)(18 ft)[(0.075 ksf slab) + (0.010 ksf dead load)]

= 4.7 kips + 1.4 kips + 49.0 kips = 55.1 kips (245 kN)

$M = \text{static load}/g$

$= (55.1 \text{ kips})/(386 \text{ in./s}^2)$

$=$ **0.143 kip·s²/in.** (25,000 kg)

Static (dead load) deflection:

$y_{st} = \text{static load}/K_e = (55.1 \text{ kips})/(143 \text{ kip/in.}) =$ **0.385 in.** (9.79 mm)

Load-mass transformation factors: (Chapter 6, Table 6-1)

elastic, $K_{LM} = K_M/K_L = 0.50/0.64 = \mathbf{0.78}$
plastic, $K_{LM} = K_M/K_L = 0.33/0.50 = \mathbf{0.66}$

Assuming some plastic deformation, use an average value for the load-mass factor when calculating the natural period:

$$K_{LM} = (0.78 + 0.66)/2 = \mathbf{0.72}$$

Equivalent mass:

$$M_e = K_{LM}\,M = (0.72)(0.143 \text{ kip·s}^2/\text{in.}) = 0.103 \text{ kip·s}^2/\text{in.} \ (18{,}000 \text{ kg})$$

Natural period: [Equation (6-8)]

$$t_n = 2\pi\sqrt{\frac{M_e}{K_e}} = 2\pi\sqrt{\frac{(0.103 \text{ kip-s}^2/\text{in.})}{(143 \text{ kip/in.})}} = \mathbf{0.169 \text{ s}}$$

10.10.8 Compute Response and Compare lto Deformation Limits (Numerical Integration Solution for Load Case A)

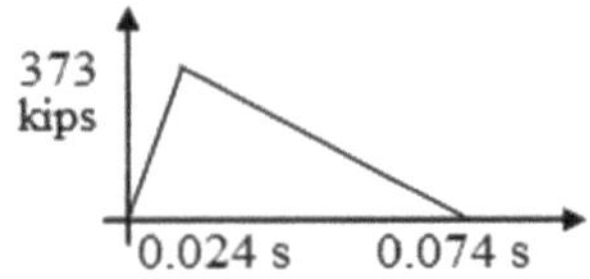

$F_o = 373$ kips (1,660 kN)

$t_r = 0.024$ s

$t_d = 0.05$ s

Time increment:

$$t_n/10 = 0.169 \text{ s}/10 \approx 0.017 \text{ s}$$

However, it is recommended to use at most a 0.0024 s time step to obtain at least 10 increments of loading in the smallest, first section of the pressure-time history. A smaller time step can be used to potentially obtain a more refined solution with smoother output plots. Consider a time range of 0.500 s to capture multiple deflection cycles and component behavior following load termination.

For output plots of key parameters, see Figure 10-11.

Peak positive deflection: $y_m = +3.34$ in. (85.0 mm) at $t = 0.077$ s

Peak rebound deflection: $y_{m-} = -2.57$ in. (−65.4 mm) at $t = 0.163$ s

Note: The selected rebound peak is the first to occur after the dissipation of the applied load.

Ductility:

$$\mu_d = y_m/y_e = (3.34 \text{ in.})/(3.56 \text{ in.}) = 0.94 < \mu_a, \quad \textbf{OK}$$

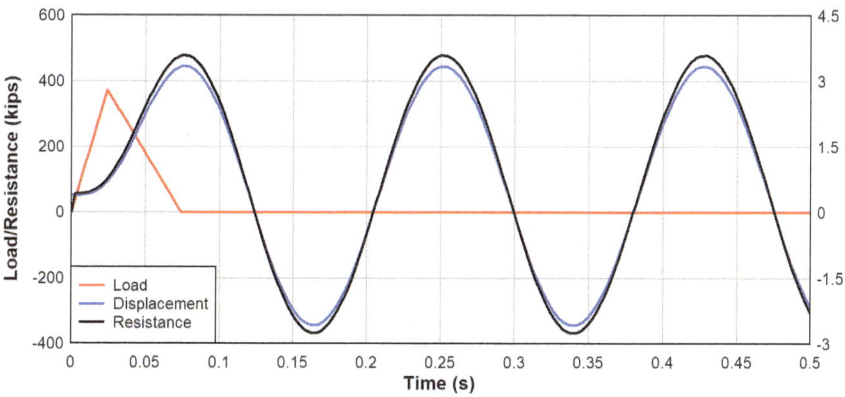

Figure 10-11. Output plots from roof girder numerical integration (Load Case A).

Support rotation: (Figure 5-9)

$\theta_d = \tan^{-1}(y_m)/(0.5\,L) = \tan^{-1}[(3.34\text{ in.})/(0.5 \times 384\text{ in.})] = 1.00\text{ degrees} \approx \theta_a$, **OK**

The response is adequate. A revision is not necessary.

10.10.9 Compute Response and Compare to Deformation Limits (Numerical Integration Solution for Load Case B)

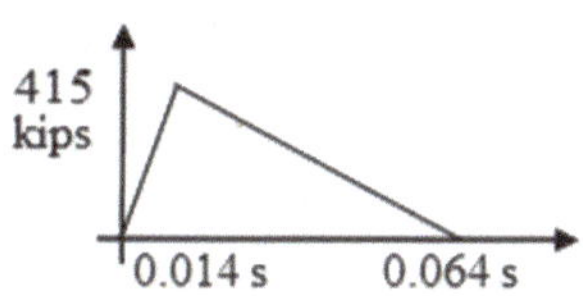

$F_o = 415$ kips (1,845 kN)

$t_r = 0.014$ s

$t_d = 0.05$ s

Time increment:

$t_n/10 = 0.169\text{ s}/10 \approx 0.017\text{ s}$

However, it is recommended to use at most a 0.0014 s time step to obtain at least 10 increments of loading in the smallest, first section of the pressure-time history. A smaller time step can be used to potentially obtain a more refined solution with smoother output plots. Consider a time range of 0.500 s to capture multiple deflection cycles and component behavior following load termination.

For output plots of key parameters, see Figure 10-12.

Peak positive deflection: $y_m = +3.33$ in. (84.5 mm) at $t = 0.070$ s

Peak rebound deflection: $y_{m-} = -2.56$ in. (−64.9 mm) at $t = 0.157$ s

Note: The selected rebound peak is the first to occur after the dissipation of the applied load.

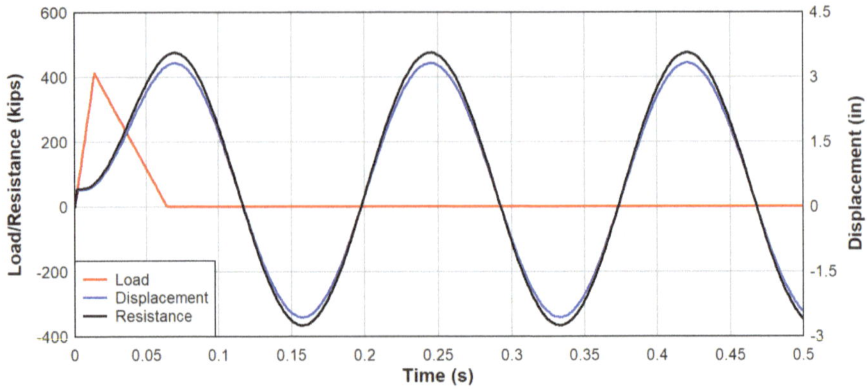

Figure 10-12. Output plots from roof girder numerical integration (Load Case B).

Ductility:

$\mu_d = y_m/y_e = (3.33 \text{ in.})/(3.56 \text{ in.}) = 0.93 < \mu_a$, **OK**

Support rotation: (Chapter 5, Figure 5-9)

$\theta_d = \tan^{-1}(y_m)/(0.5\,L) = \tan^{-1}[(3.33 \text{ in.})/(0.5 \times 384 \text{ in.})] = 0.99 \text{ degrees} < \theta_a$, **OK**

The response is adequate. A revision is not necessary.

Note: As an alternative to the tributary load method, loading can be taken from the dynamic reaction (V) calculated for the roof beam at each time step in Table 10-5. Not performed within this example, see Section 11.7.6 for a similar approach.

10.11 ROOF GIRDER CONNECTIONS

The roof girders connect to the column flange.

Roof beam resistance: (Section 10.10.6)

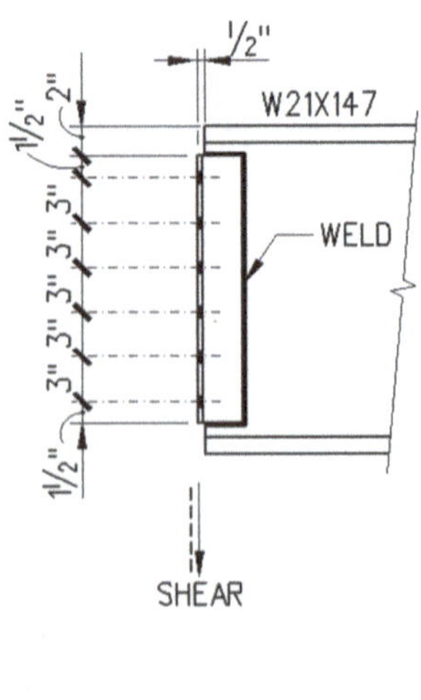

$R_u = 509$ kips (2,260 kN)

Connection design load:

$V_u = R_u/(2 \text{ beam ends})$
$= (509 \text{ kips})/2$
$= 254$ kips (1,130 kN)

Try:

AISC Standard Welded/Bolted Shear Connection

ASTM A325 bearing bolts, six rows

threads in the shear plane

diameter, $d_b = 0.875$ in. (22.2 mm)

Bolt nominal shear strength, $F_{nv} = 48$ ksi (330 MPa) (AISC 360, Table J3.2)

Nominal unthreaded area:

$A_b = \pi d_b^2/4 = \pi\,(0.875 \text{ in.})^2/4 = 0.601 \text{ in.}^2\ (388 \text{ mm}^2)$

10.11.1 Bolt Shear Strength (Clip Angle to Column)

SIF (A325 high-strength bolt) = 1.0 (conservative value)

DIF (A325 high-strength bolt) = 1.05 (Chapter 5, Section 5.5.4)

Bolt shear strength:

$V_n = (\text{SIF})(\text{DIF})\, F_{nv}\, A_b\, N$
$= (1.0)(1.05)(48 \text{ ksi})(0.601 \text{ in.}^2)(1 \text{ shear plane})(12 \text{ bolts})$
$= 364$ kips (1,620 kN) $> V_u$, **OK**

10.11.2 Weld Strength (Girder to Clip Angle) (Blodgett 1966, Section 5.4.3)

Try L4 × 3.5 × 1/2:

Use a 3/8 in. (9.5 mm) fillet weld.

E70XX electrode, $F_{\text{EXX}} = 70$ ksi (483 MPa)

A 4 in. angle leg to attach to the roof girder, $L = 4$ in. (102 mm)

Horizontal weld leg length:

$L_h = (4 \text{ in., angle}) - (0.5 \text{ in., set back}) = 3.5 \text{ in. } (89 \text{ mm})$

Vertical weld leg length:

$L_v = 1.5 \text{ in. } (2 \text{ ends}) + (3 \text{ in.})(5 \text{ spaces}) = 18 \text{ in. } (457 \text{ mm})$

Horizontal weld center:

$n_w = L_h^2/(2b + L_v) = (3.5 \text{ in.})^2/[2\,(3.5 \text{ in.}) + 18 \text{ in.}] = 0.49 \text{ in. } (12 \text{ mm})$

Horizontal weld eccentricity:

$$c_h = L_h - n_w = (3.5\text{ in.}) - (0.49\text{ in.}) = 3.01\text{ in. } (76.5\text{ mm})$$

Vertical weld eccentricity:

$$c_v = L_v/2 = (18\text{ in.})/2 = 9\text{ in. } (229\text{ mm})$$

Polar moment of inertia:

$$\begin{aligned} J_w &= [2L_h + L_v]^3/12 - L_h^2\,[L_h + L_v]^2/[2L_h + L_v] \\ &= [2\,(3.5\text{ in.}) + 18\text{ in.}]^3/12 - (3.5\text{ in.})^2\,[3.5\text{ in.} + 18\text{ in.}]^2/[2\,(3.5\text{ in.}) + 18\text{ in.}] \\ &= 1{,}076\text{ in.}^3\ (17{,}630{,}000\text{ mm}^3) \end{aligned}$$

Torsional horizontal shear:

$$\begin{aligned} f_{vh} &= V_u\,(L - n_w)\,c_v/2\,J_w \\ &= (255\text{ kips})(4\text{ in.} - 0.49\text{ in.})(9\text{ in.})/2\,(1{,}076\text{ in.}^3) \\ &= 3.73\text{ kip/in. } (0.65\text{ kN/mm}) \end{aligned}$$

Torsion vertical shear:

$$\begin{aligned} f_{v1} &= V_u\,(L - n_w)\,c_h/2\,J_w \\ &= (255\text{ kips})(4\text{ in.} - 0.49\text{ in.})(3.01\text{ in.})/2\,(1{,}076\text{ in.}^3) \\ &= 1.25\text{ kip/in. } (0.22\text{ kN/mm}) \end{aligned}$$

Direct shear:

$$\begin{aligned} f_{v2} &= V_u/2\,[2L_h + L_v] = (254\text{ kips})/2\,[2\,(3.5\text{ in.}) + 18\text{ in.}] \\ &= 5.09\text{ kip/in. } (0.89\text{ kN/mm}) \end{aligned}$$

Resultant:

$$\begin{aligned} R_u &= \sqrt{f_{vh}^2 + (f_{v1} + f_{v2})^2} = \sqrt{(3.73\text{ kip/in.})^2 + (1.25\text{ kip/in.} + 5.09\text{ kip/in.})^2} \\ &= 7.35\text{ kip/in. } (1.29\text{ kN/mm}) \end{aligned}$$

SIF (weld) = 1.0 (conservative)

DIF (weld) = 1.05 (Section 5.5.4)

Weld capacity:

$$\begin{aligned} R_s &= 0.6\,(\text{SIF})(\text{DIF})(F_{\text{EXX}})(A_w) \\ &= 0.6\,(1.0)(1.05)(70\text{ ksi})(0.375\text{ in.})(0.707) \\ &= 11.7\text{ kip/in. } (2.05\text{ kN/mm}) > R_u, \quad \textbf{OK} \end{aligned}$$

10.11.3 Clip Angle Shear Strength

The critical section is a vertical plane at the bolt centerline.

L4 × 3.5 × 0.5

length = 1.5 in. + 2 (3 in.) + 1.5 in. = 18 in. (457 mm)

ASTM A36 steel

Gross section:

A_g = (2 angles)(0.5 in., thickness)(18 in., length) = 18.0 in.^{2} (11,610 mm^2)

Bolt hole diameter: (AISC 360, Table J3.3, Section D3)

d_h = (15/16 in., nominal hole diameter) + (1/16 in.) = 1.0 in. (25 mm)

Net section:

A_{nv} = (18.0 in.^{2}) − (2 angles)(6 holes)(1.0 in., diameter)(0.5 in. thickness)
= 12.0 in.^{2} (7,740 mm^2)

For dynamic shear:

F_{dy} = (SIF)(DIF) F_y = (1.1)(1.29)(36 ksi) = 51.1 ksi (352 MPa)
(Chapter 5, Appendix 5A)
F_{du} = (SIF)(DIF) F_u = (1.1)(1.1)(58 ksi) = 70.2 ksi (484 MPa)

Shear yielding strength: (AISC 360, Equation J4-3)

$V_n = 0.6\ F_{dy}\ A_g$ = 0.6 (51.1 ksi)(18.0 in.^{2}) = 552 kips (2,450 kN) > V_u, **OK**

Shear rupture strength: (AISC 360, Equation J4-4)

$V_n = 0.6\ F_{du}\ A_{nv}$ = 0.6 (70.2 ksi)(12.0 in.^{2}) = 505 kips (2,250 kN) > V_u, **OK**

10.11.4 Clip Angle Bearing Strength

Clear distance between holes:

L_c = (1 bolt)[(1.5 in., edge) − 0.5 (1.0 in. hole diameter)]
+ (5 bolts)[(3 in., spacing) − (1.0 in., hole diameter)] = 11.0 in. (279 mm)

Bearing strength: (AISC 360, Equation J3-6a)

V_n = min [$1.2\ L_c\ t_p\ F_{du}$ or $2.4\ d_b\ t_p\ F_{du}$]
1.2 (11.0 in.)(0.5 in. × 2 each)(70.2 ksi) = 926 kips (4,120 kN)
2.4 (0.875 in. × 6 each)(0.5 in. × 2 each)(70.2 ksi) = **884 kips** (3,930 kN)
[CONTROLS]

Therefore, V_n = 884 kips (3,930 kN) > V_u, **OK**

Use L4 × 3.5 × 1/2 clip angles with bolted and welded connections as described.

Note: The column flange thickness is greater than that of the roof girder web and will not control bearing strength.

Notes: For brevity, calculations for other failure modes such as block shear failure are omitted. These should be checked as well prior to finalizing the connection design.

10.12 COLUMNS

The column is pinned at both ends.

Length, L = 12 ft or 144 in. (3,658 mm)

Note: A static design is typically used to determine column adequacy on the basis of the reaction of the members feeding into it. In this case, we take two girders at their ultimate resistance, creating a total demand (V_u) of 509 kips (2,260 kN) upward and downward.

10.12.1 Select Trial Size

Try W10 × 77: (AISC 2017 Manual)

Beam depth: $d = 10.6$ in. (269 mm)

Flange width/thickness: $b_f/2t_f = \lambda_f = 5.86$

Web thickness: $t_w = 0.53$ in. (13 mm)

Web depth/thickness: $h/t_w = \lambda_w = 14.8$

Radius of gyration: $r_y = 2.60$ in. (66 mm)

Moment of inertia: $I_x = 455\text{ in.}^4$ ($189.4 \times 10^6\text{ mm}^4$)

Plastic modulus: $Z_x = 97.6\text{ in.}^3$ ($1{,}599{,}000\text{ mm}^3$)

Area: $A_s = 22.6\text{ in.}^2$ ($14{,}580\text{ mm}^2$)

10.12.2 Compute Compression Resistance

For dynamic compression: (Chapter 5, Appendix 5A)

$$F_{ds} = F_{dy} = (\text{SIF})(\text{DIF})\,F_y = (1.1)(1.12)(50\text{ ksi}) = 61.6\text{ ksi (425 MPa)}$$

Because columns are considered primary members, width-thickness ratios need to meet AISC seismic requirements.

Check the flange: (AISC 341, Table D1.1)

$$\lambda_{ps} = 0.32\sqrt{\frac{E_s}{F_{ds}}} = 0.32\sqrt{\frac{(29{,}000\text{ ksi})}{(61.6\text{ ksi})}} = 6.9 > \lambda_f, \quad \textbf{OK}$$

Check web: (AISC 341, Table D1.1)

$$\lambda_{ps} = 1.57\sqrt{\frac{E_s}{F_{ds}}} = 1.57\sqrt{\frac{(29{,}000\text{ ksi})}{(61.6\text{ ksi})}} = 34.1 > \lambda_w, \quad \textbf{OK}$$

Slenderness ratio: (AISC 360, Section E2)

$$K_y\,L/r_y = (1.0)(144\text{ in.})/(2.60\text{ in.}) = 55.4 < 200, \quad \textbf{OK}$$

$$\lambda_c = 4.71\sqrt{\frac{E_s}{F_{ds}}} = 4.71\sqrt{\frac{(29{,}000\text{ ksi})}{(61.6\text{ ksi})}} = 102\text{ (AISC 360, Section E3)}$$

$$F_e' = \pi^2 E_s/(K_y\,L/r_y)^2 = \pi^2\,(29{,}000\text{ ksi})/55.4^2 = 93.3\text{ ksi (643 MPa)}$$

Compression stress (based on $K_y\,L/r_y < \lambda_c$): (AISC 360, Equation E3-2)

$$f_{cr} = (0.658)^{(F_{ds}/F_e')}\,(F_{ds}) = (0.658)^{(61.6/93.3)}\,(61.6\text{ ksi}) = 46.7\text{ ksi (322 MPa)}$$

Compression capacity: (AISC 360, Equation E3-1)

$P_{nc} = A_s f_{cr} = (22.6 \text{ in.}^2)(46.7 \text{ ksi}) = 1{,}056 \text{ kips } (4{,}700 \text{ kN})$

10.12.3 Compute Tension Resistance

For dynamic tension: (Chapter 5, Appendix 5A)

$F_{ds} = F_{dy} = (\text{SIF})(\text{DIF})\ F_y = (1.1)(1.12)(50 \text{ ksi}) = 61.6 \text{ ksi } (425 \text{ MPa})$

Tensile capacity: (ANSI/AISC 2019, Equation D2-1)

$P_{nt} = A_s\ (F_{ds}) = (22.6 \text{ in.}^2)(61.6 \text{ ksi}) = 1{,}392 \text{ kips } (6{,}190 \text{ kN})$

10.12.4 Compute Demand/Capacity Ratio and Compare to Permissible Limits

Positive resistance: $R_u = P_{nc} =$ **1,056 kips** (4,700 kN)

Rebound resistance: $R_{u\text{-}} = -P_{nt} =$ **−1,392 kips** (−6,190 kN)

Ensure that load demand is significantly lower than these calculated capacity values for both directions.

Demand/capacity ratio (DCR):

$\text{DCR}^+ = V_u/R_u = (509 \text{ kips})/(1{,}056 \text{ kips}) = 0.48 < 1.0,$ **OK**
$\text{DCR}^- = V_{u\text{-}}/R_{u\text{-}} = (-509 \text{ kips})/(-1{,}392 \text{ kips}) = 0.37 < 1.0,$ **OK**

The response is adequate. A revision is not necessary.

Notes: A static design approach using the peak reaction or ultimate resistance from the girders feeding into the columns is typically followed to check the adequacy of interior columns for blast loads. In this example, the tension force acting on the columns is probably significantly lower than 509 kips (2,260 kN) as the girders in Sections 10.10.8 and 10.10.9 do not approach their elastic deflection limit in rebound. This can also be demonstrated by performing a dynamic analysis of the columns, wherein the design blast loads would be identical with those calculated for the girder load cases in Sections 10.10.1 and 10.10.2, the stiffness K can be taken as $A_s\,E/L$, and an appropriate mass M is determined on the basis of the components that the column supports without the need for transformation factors. The resulting column ductility values would be substantially less than the critical value of $\mu_a = 1.0$ in both compression and tension.

10.13 COLUMN BASEPLATE AND ANCHOR BOLT DESIGN

A bolted baseplate connects the column to the foundation.

Design column compression load:

$V_u = 509 \text{ kips } (2{,}260 \text{ kN})$

Design column tension load:

$V_{u-} = -509$ kips (−2,260 kN)

10.13.1 Anchor Bolt Design

Try:

4 each, 1.5 in. (38 mm) diameter bolts

ASTM A325, $F_{nt} = 90$ ksi (620 ksi) (AISC 360, Table J3.2)

SIF (A325 anchor bolt) = 1.0 (conservative)

DIF (A325 anchor bolt) = 1.05 (Chapter 5, Section 5.5.4)

Nominal bolt area:

$A_b = \pi d_b^2/4 = \pi\,(1.5 \text{ in.})^2/4 = 1.77 \text{ in.}^2\ (1{,}140 \text{ mm}^2)$

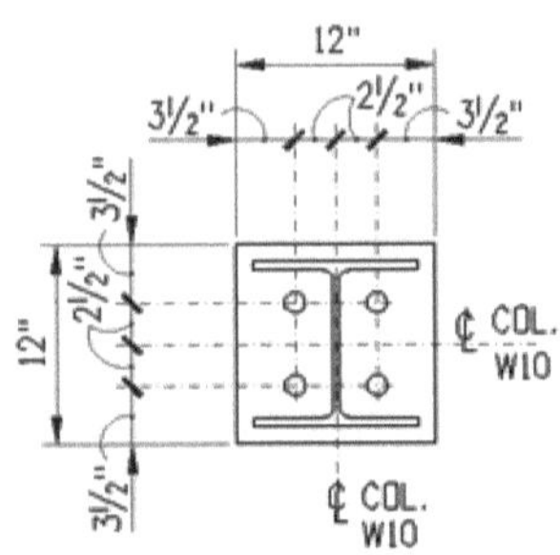

Bolt tensile strength: (AISC 360, Equation J3-1)

$R_n = (\text{SIF})(\text{DIF})\ F_{nt}\ A_b\ N$
$= (1.0)(1.05)(90 \text{ ksi})(1.77 \text{ in.}^2)(4 \text{ bolts})$
$= 668$ kips (2,970 kN) $> V_{u-}$, **OK**

10.13.2 Baseplate Design

Try:

a 12 in. (305 mm) square plate

thickness, $t_p = 1.25$ in. (32 mm)

ASTM A36 steel

For dynamic bending: (Chapter 5, Appendix 5A)

$F_{ds} = F_{dy} = (\text{SIF})(\text{DIF})\ F_y = (1.1)(1.29)(36 \text{ ksi}) = 51.1$ ksi (352 kPa)

W10 × 77 properties: (AISC 2017 Manual)

depth, $d = 10.6$ in. (269 mm)

flange width, $b_f = 10.2$ in. (259 mm)

web thickness, $t_w = 0.53$ in. (13.5 mm)

Condition 1: Column Tension on Baseplate (AISC 2024 Guide)

Select the appropriate equation from $\sqrt{2}\ b_f = \sqrt{2}\ (10.2 \text{ in.}) = 14.4$ in. (366 mm) $> d$:

$$t_{p\ \text{reqd}} = \sqrt{\frac{V_{u-}g_b d}{F_{ds}(d^2 + 2b_f^2)}}$$

$$= \sqrt{\frac{(509 \text{ kips})(5 \text{ in.})(10.6 \text{ in.})}{(51.1 \text{ ksi})[(10.6 \text{ in.})^2 + 2(10.2 \text{ in.})^2]}}$$

$$= 1.28 \text{ in. } (32.6 \text{ mm}) \approx t_p, \quad \textbf{OK}$$

Note: Plate thickness is deemed adequate because the net girder rebound reaction is probably much less than 509 kips, as discussed in Section 10.12.4.

Condition 2: Column Compression on Baseplate (AISC 2017 Manual, Part 14)

Baseplate dimensions:

$$N_p = B_p = 12 \text{ in. } (305 \text{ mm})$$
$$m_1 = [N_p - 0.95\,d]/2 = [(12 \text{ in.}) - 0.95\,(10.6 \text{ in.})]/2 = 0.97 \text{ in. } (25 \text{ mm})$$
$$n_1 = [B_p - 0.8\,b_f]/2 = [(12 \text{ in.}) - 0.8\,(10.2 \text{ in.})]/2 = 1.92 \text{ in. } (49 \text{ mm})$$
$$n' = \sqrt{db_f}\,/4 = \sqrt{(10.6 \text{ in.})(10.2 \text{ in.})}\,/4 = 2.6 \text{ in. } (66 \text{ mm})$$

Conservatively using $P_u/P_p = 1.0$,

$$X = [4\,db_f/(d + b_f)^2]\,(P_u/P_p)$$
$$= [4\,(10.6 \text{ in.})(10.2 \text{ in.})/(10.6 \text{ in.} + 10.2 \text{ in.})^2]\,(1.0)$$
$$= 1.0$$

$$\lambda_{\text{bp}} = \min\left[2\sqrt{X}/(1+\sqrt{1-X}),\ 1.0\right]$$

As $2\sqrt{(1.0)}/(1+\sqrt{1-1.0})$ 1.96, $\lambda_{\text{bp}} = 1.0$.

$$\lambda_{\text{bp}}\,n' = (1.0)(2.6 \text{ in.}) = 2.6 \text{ in. } (66 \text{ mm})$$
$$L = \max\,[m_1, n_1, \lambda_{\text{bp}}\,n'] = 2.6 \text{ in. } (66 \text{ mm})$$

Substituting V_u for P_u,

$$t_{p,\min} = L\sqrt{\frac{2V_u}{\phi F_{ds} B_p N_p}} = (2.6 \text{ in.})\sqrt{\frac{2\,(509 \text{ kips})}{(1.0)(51.1 \text{ ksi})(12 \text{ in.})(12 \text{ in.})}}$$

$$= 0.97 \text{ in. } (24.6 \text{ mm}) < t_p, \quad \textbf{OK}$$

Condition 3: Column Compression on Foundation Pier (ACI 318, Section 22.8.3.2)

Assume that a 30 in. (762 mm) square pier size will be used for interior columns.

For concrete compression: (Chapter 5, Appendix 5A)

$$f'_{ds} = (\text{SIF})(\text{DIF})\ f'_c = (1.0)(1.12)(4{,}000 \text{ psi}) = 4{,}480 \text{ psi } (30.9 \text{ MPa})$$
$$A_1 = B_p\,N_p = (12 \text{ in.})(12 \text{ in.}) = 144 \text{ in.}^2\ (92{,}900 \text{ mm}^2)$$
$$A_2 = s_{\text{pier}}{}^2 = (30 \text{ in.})^2 = 900 \text{ in.}^2\ (580{,}600 \text{ mm}^2)$$

Bearing modification factor: (ACI 318, Table 22.8.3.2)

$$\sqrt{\frac{A_2}{A_1}} = \sqrt{\frac{(900\ \text{in.}^2)}{(144\ \text{in.}^2)}} = 2.5\text{; therefore, use a maximum value of 2.0.}$$

Concrete bearing strength: (ACI 318, Table 22.8.3.2)

$$\begin{aligned} R_n &= 0.85\ f'_{ds}\ A_1 \sqrt{\frac{A_2}{A_1}} \\ &= 0.85\ (4.48\ \text{ksi})(144\ \text{in.}^2)(2.0) \\ &= 1{,}097\ \text{kips}\ (4{,}880\ \text{kN}) > V_u, \quad \textbf{OK} \end{aligned}$$

10.13.3 Column and Base Plate Weld Design

The minimum weld size for a baseplate of more than 3/4 in. (19 mm) thickness is 5/16 in. (7.9 mm) per AISC 360, Table J2.4.

Try:

a 5/16 in. (7.9 mm) fillet weld

E70 electrode, $F_{EXX} = 70$ ksi (484 kPa)

SIF (weld) = 1.0 (conservative)

DIF (weld) = 1.05 (Chapter 5, Section 5.5.4)

Nominal weld strength: (AISC 360, Table J2.5)

$$F_w = 0.6\ (\text{SIF})(\text{DIF})\ F_{EXX} = 0.6\ (1.0)(1.05)(70\ \text{ksi}) = 44.1\ \text{ksi}\ (304\ \text{MPa})$$

Effective weld area:

$$\begin{aligned} A_w &= [(b_f)(4\ \text{sides}) - (t_w)(2\ \text{sides}) + (d)(2\ \text{sides})]\ (\text{weld size})(0.707) \\ &= [(10.2\ \text{in.})(4\ \text{sides}) - (0.53\ \text{in.})(2\ \text{sides}) + (10.6\ \text{in.})(2\ \text{sides})]\ (5/16\ \text{in.})(0.707) \\ &= 13.5\ \text{in.}^2\ (8{,}690\ \text{mm}^2) \end{aligned}$$

Fillet weld effective strength: (AISC 360, Equation J2-3)

$$R_n = F_w A_w = (44.1\ \text{ksi})(13.5\ \text{in.}^2) = 594\ \text{kips}\ (2{,}640\ \text{kN}) > V_u, \quad \textbf{OK}$$

Use PL12 × 12 × 1.25 baseplates with bolted and welded connections as described.

10.14 FOUNDATION

As noted in Chapter 7, foundation design based on conventional loading will usually result in adequate performance during a blast event.

References

ACI (American Concrete Institute). 2022. *Building code requirements for structural concrete and commentary,* ACI 318-19(22) (reapproved 2022). Farmington Hills, MI: ACI.

AISC (American Institute of Steel Construction). 2017. *Steel construction manual.* 15th ed. Chicago: AISC.

AISC. 2024. *Design guide 1: Base connection design for steel structures.* 3rd ed. Chicago: AISC.

ANSI/AISC (American National Standards Institute/American Institute of Steel Construction). 2016. *Seismic provisions for structural steel buildings,* ANSI/AISC 341-16. Chicago: AISC.

ANSI/AISC. 2019. *Specification for structural steel buildings,* ANSI/AISC 360-16. Revised 2019. Chicago: AISC.

ASTM International. 2017. *Standard specification for steel for structural shapes for use in building framing,* ASTM A992/A992M-01. West Conshohocken, PA: ASTM.

ASTM. 2019. *Standard specification for carbon structural steel,* ASTM A36/A36M-19. West Conshohocken, PA: ASTM.

ASTM. 2024. *Standard specification for deformed and plan carbon-steel bars for concrete reinforcement,* ASTM A615/A615M-22. West Conshohocken, PA: ASTM.

Biggs, J. M. 1964. *Introduction to structural dynamics.* New York: McGraw-Hill.

Blodgett, O. W. 1966. *Design of welded structures.* Euclid, OH: James F. Lincoln Arc Welding Foundation.

Budynas, R. G., and A. M. Sadegh. 2020. *Roark's formulas for stress and strain.* 9th ed. New York: McGraw-Hill.

USACE/NAVFAC/AFCESA (US Army Corp of Engineers/Naval Facilities Engineering Command/Air Force Civil Engineer Support Agency). 2014. *Structures to resist the effects of accidental explosions,* UFC 3-340-02, Change 2. Washington, DC: NIBS (National Institute of Building Sciences).

CHAPTER 11

Metal Building Design Example

11.1 INTRODUCTION

The following is a sample blast design for a building using metal cladding and a structural steel frame. In this example, blast loads and dynamic properties are computed on a unit area (i.e., pressure) basis in contrast to Chapter 10 calculations. Blast loads are typically directly applied to a component over an appropriate tributary area, although an alternate method of using the dynamic reaction history from the supported member is considered in Section 11.7.

Structural code provisions, as applicable, are taken from AISC 360-16 (ANSI/AISC 2019), *Specification for Structural Steel Buildings.*

Additional references are Biggs (1964), *Introduction to Structural Dynamics*, and AISC (2017) Manual 15th Edition, *AISC Manual of Steel Construction.*

For brevity, evaluation of conventional loads is not included in this example.

Designs for blast doors and foundation are not included in this example.

11.2 STRUCTURAL SYSTEM

The structure in this example is of Metal Clad Construction as described in Chapter 4, Section 4.3.5.

11.2.1 Description of Structure

- One-story metal frame/metal cladding.
- Plan dimensions are 50 ft (15.2 m) × 100 ft (30.5 m).
- Eave height is 16 ft (4.9 m).
- Rigid frames across short dimension with 20 ft (6.1 m) spacing.
- Braced frames are provided at the roof level and on exterior walls in long dimension.

- Metal deck roof spans over structural steel purlins at 5% slope.
- Purlins span over roof girders and are spliced for continuity.
- Metal siding spans over structural steel girts.
- Girts span between columns.
- Foundation consists of shallow spread footings.

11.2.2 Framing Plan

For the general framing plan and elevation, refer to Figure 11-1.

11.2.3 Components for Blast Design

The structural members and cladding to be examined are shown in Figure 11-2. The metal building cladding will fail in flexure at a low overpressure unless girt spacings are low. Tension membrane response is possible; however, attention must be paid to detailing to ensure that membrane response can be achieved. Tension membrane response can also be exhibited by girts and purlins. For this example problem, all elements will be designed for flexure. Failure of lightweight cladding will not be permissible here, although in certain circumstances, an owner may elect to allow such failure if lightweight debris is determined to be an acceptable hazard to building occupants.

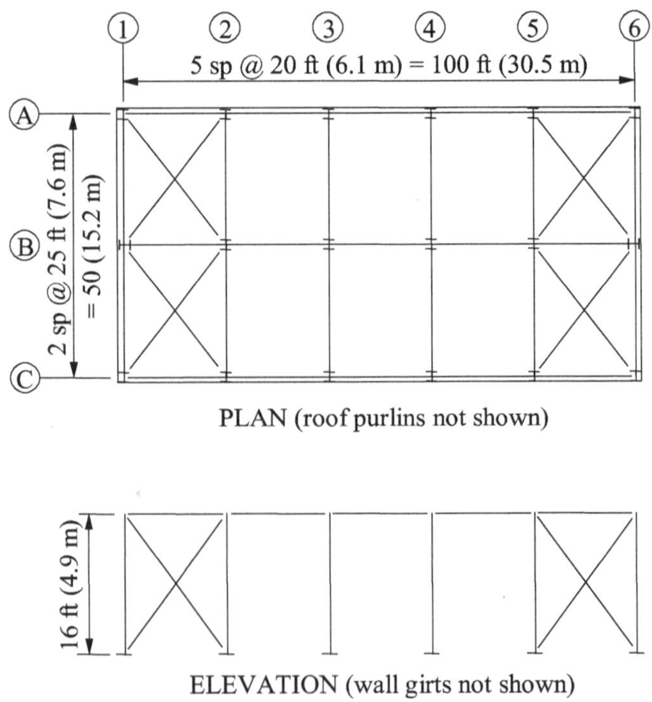

Figure 11-1. Framing plan and elevation.

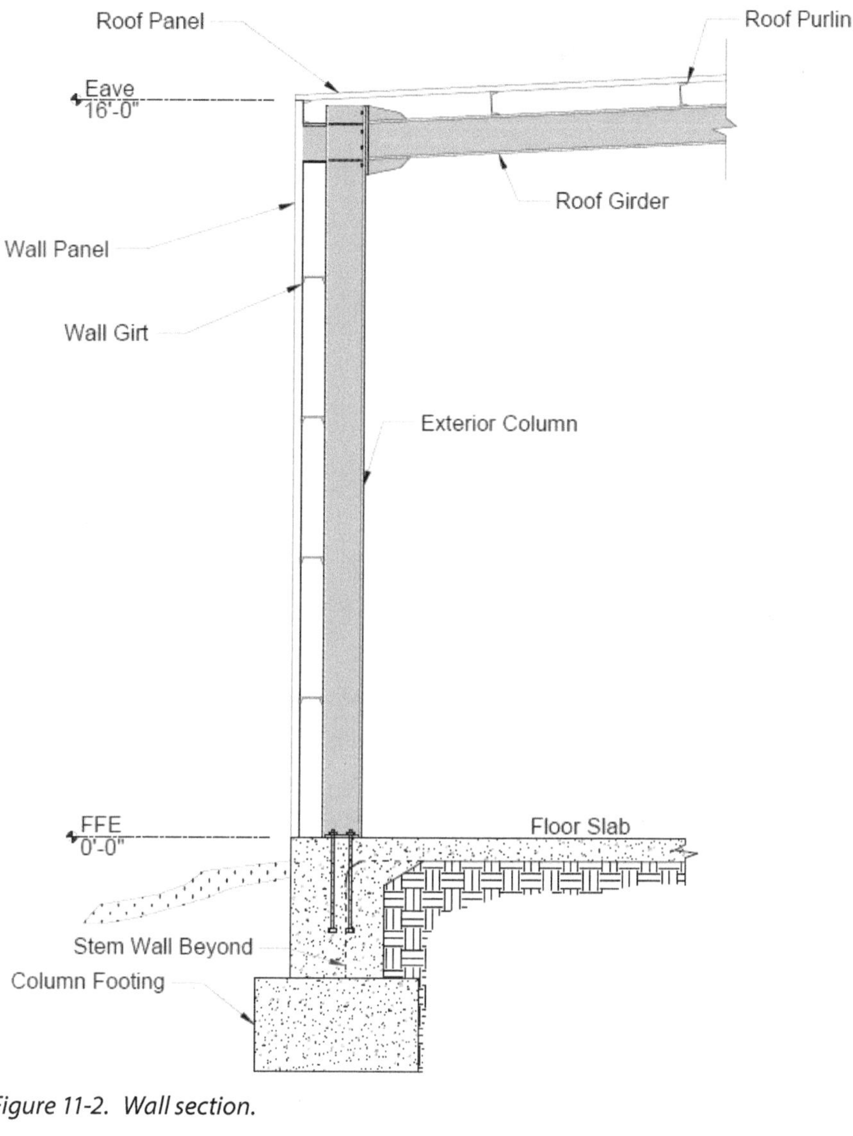

Figure 11-2. Wall section.

Metal panels and girts along the long sides will load the main frames of the building. Loads on the back wall will be ignored to maximize sidesway. Reactions from these members will be transferred to the frame. Loads on the side walls will be resisted by braced frames in the end bays.

A preliminary design for each member will be accomplished using required resistance formulas and then checked for response to time-dependent loads assuming that the connections are able to adequately transfer the blast loads/developed reactions. Final design would require the evaluation of connections, bracing, and other items that could potentially prevent the members from reaching their plastic capacity.

Determine required member sizes for

1. Roof deck,
2. Wall panels (facing blast),
3. Purlins,
4. Girts (facing blast),
5. Rigid frame (facing blast), and
6. Braced frame.

11.2.4 General Solution Procedure

1. Determine dynamic material properties.
2. Select trial sizes.
3. Compute section properties.
4. Compute SDOF properties (if applicable).
5. Compute the response and compare it to deformation limits.
6. Revise section as required.
7. Check secondary failure modes (shear, buckling, etc.).
8. Design connections for controlling reactions (not included for brevity).

Note: This general solution procedure is not rigid. It varies in some regard from the member design process steps presented in Chapter 7, Section 7.3, as well as the steps used in the Chapter 10 calculations.

11.3 DESIGN DATA

11.3.1 Material Properties

Metal decking: ASTM A653/A653M-20 (ASTM 2022), $F_y = 50$ ksi (345 MPa)

Structural steel (channels): ASTM A36/A36M-19 (ASTM 2019), $F_y = 36$ ksi (248 MPa)

Structural steel (W shapes): ASTM A992/A992M-01 (ASTM 2017), $F_y = 50$ ksi (345 MPa)

Strength increase factors (SIFs) and dynamic increase factors (DIFs) are taken from Chapter 5, Tables 5A-1 and A-3, respectively. These values are summarized in Table 11-1.

Because of low permissible dynamic response, per Chapter 5, Table 5A-5, $F_{ds} = F_{dy} =$ (SIF)(DIF) F_y. The modulus of elasticity, E_s, for all steel grades can be taken as 29,000 ksi (200,000 MPa).

Table 11-1. Dynamic Material Properties.

Material	F_y	*SIF*	*DIF*
Metal decking	50 ksi (345 MPa)	1.21	1.1
Structural steel (channels)	36 ksi (248 MPa)	1.1	1.29*
Structural steel (W shapes)	50 ksi (345 MPa)	1.1	1.19*

*For tension and compression checks, DIF = 1.19 (Grade 36) or 1.12 (Grade 50).

11.3.2 Building Performance Requirements: Deformation Limits

A medium response range is selected. The corresponding response limits for the members to be designed are taken from Chapter 5, Appendix 5B. These values are summarized in Table 11-2.

Maximum sidesway at eave: (Chapter 5, Table 5B-3)

$$X_a = B_H/35 = (16 \text{ ft} \times 12 \text{ in./ft})/35 = 5.49 \text{ in. } (139 \text{ mm})$$

11.3.3 Design Loads

Dead load:

Roof mechanical = 5 psf (239 Pa)

The following are initial estimates:

Roof deck = 3 psf (144 Pa) Wall siding = 2 psf (96 Pa)
Roof framing = 3 psf (144 Pa) Wall framing = 4 psf (192 Pa)

Blast load:

The design blast direction is parallel to the main frames. The long (front) wall receives a reflected load. All other walls and the roof receive a side-on (free air) load. Free-field blast wave parameters are assumed to have been provided by

Table 11-2. Response Limits.

Element	*Ductility ratio, μ_a*	*Support rotation, θ_a (degrees)*
Roof deck	3	2
Wall panels	3	2
Hot-rolled purlins*	10	6
Hot-rolled girts*	10	6
Rigid frames	2	1.5
Braced frames	2	1.5

*The frame components are assumed to have limited compression (i.e., the axial compressive load is less than 20% of the dynamic axial capacity of the component). If the calculated axial compression forces in the frame components are more than 20% of the dynamic axial capacity of the component, then lower response limits would need to be used.

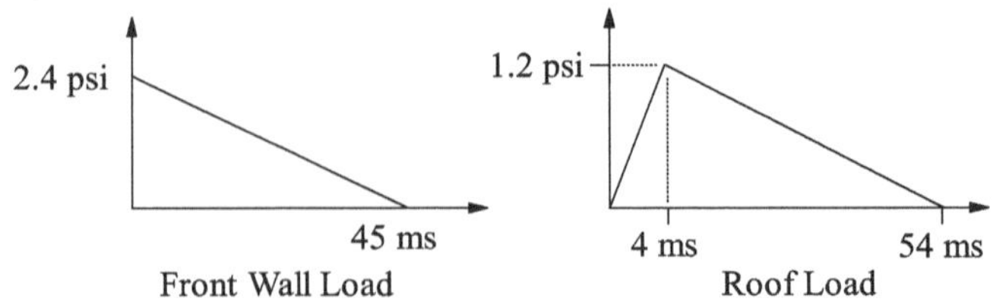

Figure 11-3. Design blast loads.

others. Calculations for blast pressures on the building surfaces are omitted for brevity (such a calculation is provided in Chapter 3, Appendix 3). The resulting blast loads indicated in Figure 11-3 represent a far range (low pressure) load.

Notes: The reflected blast load on the front wall can also be calculated by following Chapter 3, Section 3.5.1, to include the effects of clearing. As a rule of thumb, this typically provides a lower design blast load when the duration in units of milliseconds exceeds four times the building height in feet.

11.4 ROOF DECK

The roof deck spans continuously over the roof purlins. The worst-case span for deflection calculations is for the end bays with fixed-pinned boundary conditions. To add the effects of dead load to the SDOF calculations, each pressure-time pair will be increased by the magnitude of the dead load and the initial displacement will be set equal to the dead load deflection. This will create a balanced condition at the start of the SDOF response calculation (refer to the preload discussion in Chapter 7, Section 7.2.5).

Notes: The worst-case span for reaction calculations is for the typical fixed-fixed case outside of the end bays. See Section 11.5 for a follow-on discussion.

Treat the roof deck as a 1 in. (25.4 mm) wide, one-way strip.

Response limits: $\mu_a = 3$, $\theta_a = 2$ degrees

Dead load (initial guess):

$DL = 3$ psf or 0.02 psi (0.14 kPa)

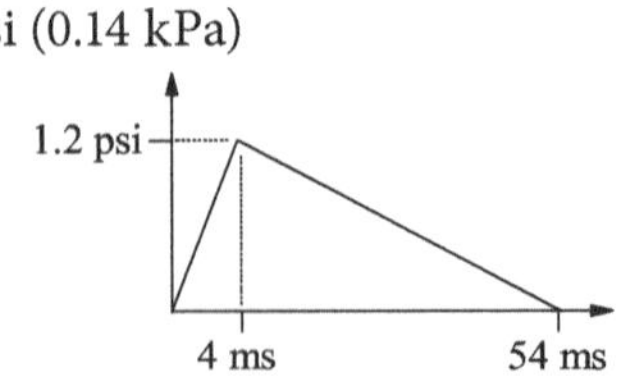

Blast load:

$BL = 1.2$ psi (8.3 kPa)

Impulse:

$$I_o = BL \times t_d/2 = (1.2 \text{ psi})(54 \text{ ms})/2$$
$$= 32.4 \text{ psi ms } (223 \text{ kPa ms})$$

11.4.1 Dynamic Material Properties

For dynamic bending and shear: (Chapter 5, Appendix 5A)

$$F_{ds} = F_{dy} = (\text{SIF})(\text{DIF})\ F_y = (1.21)(1.1)(50\ \text{ksi}) = 66.55\ \text{ksi}\ (459\ \text{MPa})$$

11.4.2 Calculate a Trial Size

Try 4 ft purlin spacing, $L = 4$ ft or 48 in. (1,220 mm).

Let the ductility demand equal the limiting value, $\mu_d = \mu_a = 3$.

As an initial guess, let $\tau = t_d/t_n = 3$ (in the dynamic range of response).

Load/resistance ratio: [Chapter 6, Equation (6-11)]

$$Fo/R_m = \frac{\sqrt{(2\mu_d - 1)}}{\pi\tau} + \frac{(2\mu_d - 1)\tau}{2\mu_d(\tau + 0.7)} + \frac{\sqrt{(2\times3-1)}}{\pi\times3} + \frac{(2\times3-1)\times3}{2\times3\times(3+0.7)} = 0.913$$

Peak load:

$$F_o = BL + DL = 1.2\ \text{psi} + 0.02\ \text{psi} = 1.22\ \text{psi}\ (8.42\ \text{kPa})$$

Resistance:

$$R_m = F_o/0.913 = (1.22\ \text{psi})/0.913 = 1.34\ \text{psi}\ (9.22\ \text{kPa})$$

Effective ultimate moment, from $R_b = 4\ (M_{ps} + 2\ M_{pc})/L$: (Chapter 6, Table 6-3)

$$M_p = R_m L/12 = (1.34\ \text{psi} \times 48\ \text{in.} \times 1\ \text{in.})(48\ \text{in.})/12 = 257\ \text{in.. lb}\ (29.0\ \text{N m})$$

Ultimate moment: (Chapter 5, Section 5.4.4)

$$M_u = M_p/0.9 = (257\ \text{in. lb})/0.9 = 285\ \text{in. lb}\ (32.2\ \text{N m})$$

Note: The ductile plastic moment capacity, M_p, for cold-formed components is assumed equal to 90% of the ultimate moment capacity, M_u, as discussed in Chapter 5, Section 5.4.4.

Section modulus:

$$S_x = M_u/F_{ds} = (285\ \text{in. lb})/(66{,}550\ \text{psi}) = 0.00429\ \text{in.}^3\ (70.2\ \text{mm}^3)$$

Notes: The elastic section modulus, S_x, is used to compute the moment capacity instead of the plastic section modulus, Z_x, mainly because elastic section modulus values are readily available. The difference is minor because of the relatively low response and due to capacity reductions from buckling of the thin web.

Select the panel type and thickness from the vendor catalog:

R panel, thickness = 24 gauge (t_p = 0.0258 in. or 0.655 mm)

Weight = 1.25 psf = 0.00868 psi (0.060 kPa)

$I_x = 0.0548\ \text{in.}^4/\text{ft} = 0.00457\ \text{in.}^4/\text{in.}\ (74.8\ \text{mm}^4/\text{mm})$

$S_x = 0.0573\ \text{in.}^3/\text{ft} = 0.00478\ \text{in.}^3/\text{in.}\ (3.08\ \text{mm}^3/\text{mm})$

$A_v = 0.0645\ \text{in.}^2/\text{ft} = 0.00538\ \text{in.}^2/\text{in.}\ (0.137\ \text{mm}^2/\text{mm})$

h (rib height) = 1.3 in. (33 mm)

Perform a detailed check of this section.

11.4.3 Compute Section Properties

Effective moment capacity: (Chapter 5, Section 5.4.4)

$$M_p = 0.9\, S_x\, F_{ds} = 0.9\ (0.00478\ \text{in.}^3)(66{,}550\ \text{psi}) = 286\ \text{in. lb}\ (32.3\ \text{N m})$$

Shear capacity:

Shear in webs of cold-formed members with $F_{ds} = 66$ ksi may be limited by shear buckling. Because $h/t = 1.3$ in./0.0258 in. $= 50$, use $F_{dv} = 15.75$ ksi (109 MPa) per Chapter 5, Table 5-6b, in UFC 3-340-02 (USACE/NAVFAC/AFCESA 2014) for the shear capacity of the panel at the support with combined bending and shear.

$$V_n = A_v\, F_{dv} = (0.00538\ \text{in.}^2)(15{,}750\ \text{psi}) = 84.7\ \text{lb}\ (377\ \text{N})$$

11.4.4 Compute SDOF Properties

Ultimate resistance, from $R_b = 4\ (M_{ps} + 2\ M_{pc})/L$: (Chapter 6, Table 6-3)

$$R_u = 12\ M_p/L = 12\ (286\ \text{in. lb})/(48\ \text{in.}) = 71.5\ \text{lb, or } \mathbf{1.49\ psi}\ (10.3\ \text{kPa})$$

Compute the effective "bilinear" elastic stiffness and deflection to determine the natural period and ductility ratios.

Effective "bilinear" elastic stiffness: (Chapter 6, Table 6-3)

$$\begin{aligned} K_e &= 160\ E_s\, I_x/L^3 \\ &= 160\ (29{,}000{,}000\ \text{psi})(0.00457\ \text{in.}^4)/(48\ \text{in.})^3 \\ &= 192\ \text{lb/in., or } \mathbf{3.99\ psi/in.}\ (1.08\ \text{kPa/mm}) \end{aligned}$$

Effective elastic yield deflection:

$$y_e = R_u/K_e = (1.49\ \text{psi})/(3.99\ \text{psi/in.}) = \mathbf{0.373\ in.}\ (9.49\ \text{mm})$$

Mass:

weight = (1.25 psf)(0.083 ft width)(4 ft length) = 0.417 lb, or 0.00868 psi (0.060 kPa)

$$\begin{aligned} M &= \text{weight}/g \\ &= (0.417\ \text{lb})/(386\ \text{in./s}^2) \\ &= 0.00108\ \text{lb s}^2\text{/in., or } \mathbf{22.5\ psi\ ms^2/in.}\ (6.10\ \text{kg/m}^2) \end{aligned}$$

Load-mass transformation factors: (Chapter 6, Table 6-3)

elastic, $K_{LM} = K_M/K_L = 0.45/0.58 = \mathbf{0.78}$
plastic, $K_{LM} = K_M/K_L = 0.33/0.50 = \mathbf{0.66}$

Assuming some plastic deformation, use an average value for the load-mass factor when calculating the natural period:

$$K_{LM} = (0.78 + 0.66)/2 = \mathbf{0.72}$$

Equivalent mass:

$$M_e = K_{LM}\, M = 0.72\ (22.5\ \text{psi·ms}^2\text{/in.}) = 16.2\ \text{psi·ms}^2\text{/in.}\ (4.40\ \text{kg/m}^2)$$

Natural period: [Chapter 6, Equation (6-8)]

$$t_n = 2\pi\sqrt{\frac{M_e}{K_e}} = 2\pi\sqrt{\frac{(16.2\text{ psi-ms}^2/\text{in.})}{(3.99\text{ psi/in.})}} = \mathbf{12.7\ ms}$$

Notes: With fixed-pinned boundary conditions, a trilinear resistance-deflection curve could alternatively be developed by using elastic and elastic-plastic stiffness and resistance values from Chapter 6, Table 6-3. In this case, elastic-plastic K_{LM} would equal 0.78.

11.4.5 Compute Response and Compare It to Deformation Limits (Chart Solution)

Note: Both charts and numerical integration need not be used but are presented in this sample design to illustrate implementation.

To use Chapter 6, Figure 6-6, an instantaneous load rise must be assumed.

t_d/t_n = (54 ms)/(12.7 ms) = 4.3 (3.0 was originally assumed)
R_u/F_o = (1.49 psi)/(1.2 psi) = 1.24

Using these values, $\mu_d \approx 2 < \mu_a$, **OK** (Chapter 6, Figure 6-6)

Maximum deflection:

$$y_m = \mu_d\, y_e = 2\ (0.373\text{ in.}) = 0.75\text{ in. } (19\text{ mm})$$

Support rotation: (Chapter 5, Figure 5-9)

$$\theta_d = \tan^{-1}(y_m/0.5\,L) = \tan^{-1}[(0.75\text{ in.})/(0.5)(48\text{ in.})] = 1.8° < \theta_a,\ \mathbf{OK}$$

The response is adequate. A revision is not necessary.

Note: 0.5 L is used in the support rotation calculation even for nonsymmetric boundary conditions.

11.4.6 Compute Response and Compare It to Deformation Limits (Numerical Integration Solution)

Static (dead load) deflection:

$$y_{st} = \text{weight}/K_e = (0.00868\text{ psi})/(3.99\text{ psi/in.}) = 0.0022\text{ in. } (0.055\text{ mm})$$

Note: Ordinarily, such a low static deflection would be insignificant; however, this deflection will be included to illustrate implementation.

Time increment:

$$t_n/10 = 12.7\text{ ms}/10 \approx 1.3\text{ ms}$$

However, it is recommended to use at most a 0.4 ms time step to obtain at least 10 increments of loading in the smallest, first section of the pressure-time history. Consider a time range of 100 ms to capture multiple deflection cycles and component behavior following load termination.

Table 11-3. Roof Deck Numerical Integration.

Example:	Section 11.4 Roof Deck						
time increment:	0.4 ms	**Blast Load**					
mass:	22.5 psi-ms²/in	t0:	0 ms	P0:	0 psi		
stiffness:	3.99 psi/in	t1:	4 ms	P1:	1.2 psi		
damping:	0 psi-ms/in	t2:	54 ms	P2:	0 psi		
positive resistance:	1.49 psi	**Dynamic Reaction Factors**			BC: F-S		
rebound resistance:	-1.49 psi	Crel:	0.43	Cfel:	0.19		
static load:	0.00868 psi	Crpl:	0.46	Cfpl:	0.12		

KEY = 0 when elastic
KEY = +1 when positive plastic
KEY = -1 when rebound plastic
Z = 1 when continued elastic
Z = 2 at start of positive plastic deformation
Z = 3 when increasing plastic deformation
Z = 4 at end of positive plastic deformation
Z = 6 at start of rebound plastic deformation
Z = 8 at end of rebound plastic deformation

time (ms)	load (psi)	y (in)	v (in/ms)	a (in/ms²)	ye+ (in)	ye- (in)	R (psi)	KEY -	KLM	V (psi)	eff K (psi/in)	incr F (psi)	incr y (in)	incr v (in/ms)	Z -
0.000	0.000	0.002	0.000	0.000	0.373	-0.373	0.000	0	0.78	0.000	662	0.1	0.000	0.001	
0.400	0.120	0.002	0.001	0.007	0.373	-0.373	0.009	0	0.78	0.027	662	0.8	0.001	0.004	1
0.800	0.240	0.004	0.005	0.013	0.373	-0.373	0.014	0	0.78	0.052	662	2.2	0.003	0.007	1
1.200	0.360	0.007	0.012	0.019	0.373	-0.373	0.028	0	0.78	0.080	662	4.3	0.006	0.009	1
1.600	0.480	0.013	0.021	0.025	0.373	-0.373	0.054	0	0.78	0.114	662	6.9	0.010	0.011	1
2.000	0.600	0.024	0.032	0.029	0.373	-0.373	0.095	0	0.78	0.155	662	10.0	0.015	0.012	1
2.400	0.720	0.039	0.044	0.033	0.373	-0.373	0.155	0	0.78	0.204	662	13.4	0.020	0.014	1
2.800	0.840	0.059	0.057	0.035	0.373	-0.373	0.236	0	0.78	0.261	662	17.1	0.026	0.014	1
3.200	0.960	0.085	0.072	0.036	0.373	-0.373	0.339	0	0.78	0.328	662	20.9	0.032	0.014	1
3.600	1.080	0.117	0.086	0.036	0.373	-0.373	0.465	0	0.78	0.405	662	24.6	0.037	0.014	1
4.000	1.200	0.154	0.100	0.034	0.373	-0.373	0.613	0	0.78	0.492	662	28.1	0.042	0.012	1
4.400	1.190	0.196	0.111	0.024	0.373	-0.373	0.782	0	0.78	0.563	662	30.6	0.046	0.007	1
4.800	1.181	0.242	0.119	0.013	0.373	-0.373	0.966	0	0.78	0.640	662	31.9	0.048	0.003	1
5.200	1.171	0.290	0.121	0.001	0.373	-0.373	1.159	0	0.78	0.721	662	32.0	0.048	-0.002	1
5.600	1.162	0.339	0.120	-0.010	0.373	-0.373	1.352	0	0.78	0.802	662	30.9	0.047	-0.006	1
6.000	1.152	0.385	0.113	-0.022	0.373	-0.373	1.490	1	0.66	0.824	557	24.2	0.044	-0.009	2
6.400	1.142	0.429	0.104	-0.023	0.373	-0.373	1.490	1	0.66	0.822	557	22.2	0.040	-0.009	3
6.800	1.133	0.469	0.095	-0.023	0.373	-0.373	1.490	1	0.66	0.821	557	20.1	0.036	-0.010	3
7.200	1.123	0.505	0.085	-0.024	0.373	-0.373	1.490	1	0.66	0.820	557	18.0	0.032	-0.010	3
7.600	1.114	0.537	0.076	-0.025	0.373	-0.373	1.490	1	0.66	0.819	557	15.7	0.028	-0.010	3
8.000	1.104	0.565	0.066	-0.025	0.373	-0.373	1.490	1	0.66	0.818	557	13.5	0.024	-0.010	3
8.400	1.094	0.590	0.055	-0.026	0.373	-0.373	1.490	1	0.66	0.817	557	11.2	0.020	-0.011	3
8.800	1.085	0.610	0.045	-0.027	0.373	-0.373	1.490	1	0.66	0.816	557	8.8	0.016	-0.011	3
9.200	1.075	0.625	0.034	-0.027	0.373	-0.373	1.490	1	0.66	0.814	557	6.3	0.011	-0.011	3
9.600	1.066	0.637	0.023	-0.028	0.373	-0.373	1.490	1	0.66	0.813	557	3.8	0.007	-0.011	3
10.000	1.056	0.644	0.012	-0.029	0.373	-0.373	1.490	1	0.66	0.812	557	1.3	0.002	-0.012	3
10.400	1.046	0.646	0.000	-0.029	0.373	-0.373	1.490	1	0.66	0.811	557	-1.3	-0.002	-0.012	3
10.800	1.037	0.644	-0.012	-0.025	0.646	-0.101	1.481	0	0.78	0.834	662	-4.4	-0.007	-0.010	4
11.200	1.027	0.637	-0.022	-0.024	0.646	-0.101	1.454	0	0.78	0.820	662	-6.9	-0.010	-0.009	1
11.600	1.018	0.627	-0.031	-0.022	0.646	-0.101	1.412	0	0.78	0.801	662	-9.3	-0.014	-0.008	1
12.000	1.008	0.613	-0.039	-0.019	0.646	-0.101	1.356	0	0.78	0.775	662	-11.3	-0.017	-0.007	1
12.400	0.998	0.595	-0.046	-0.016	0.646	-0.101	1.288	0	0.78	0.744	662	-13.0	-0.020	-0.006	1
12.800	0.989	0.576	-0.052	-0.012	0.646	-0.101	1.210	0	0.78	0.708	662	-14.3	-0.022	-0.004	1
13.200	0.979	0.554	-0.056	-0.008	0.646	-0.101	1.124	0	0.78	0.669	662	-15.1	-0.023	-0.002	1
13.600	0.970	0.532	-0.058	-0.003	0.646	-0.101	1.033	0	0.78	0.629	662	-15.4	-0.023	0.000	1
14.000	0.960	0.508	-0.058	0.002	0.646	-0.101	0.941	0	0.78	0.587	662	-15.2	-0.023	0.002	1
14.400	0.950	0.485	-0.057	0.006	0.646	-0.101	0.849	0	0.78	0.546	662	-14.6	-0.022	0.003	1
14.800	0.941	0.463	-0.053	0.011	0.646	-0.101	0.761	0	0.78	0.506	662	-13.4	-0.020	0.005	1
15.200	0.931	0.443	-0.048	0.015	0.646	-0.101	0.680	0	0.78	0.469	662	-11.9	-0.018	0.007	1
15.600	0.922	0.425	-0.041	0.018	0.646	-0.101	0.608	0	0.78	0.437	662	-9.9	-0.015	0.008	1
16.000	0.912	0.410	-0.034	0.021	0.646	-0.101	0.548	0	0.78	0.409	662	-7.7	-0.012	0.009	1
16.400	0.902	0.398	-0.025	0.023	0.646	-0.101	0.502	0	0.78	0.387	662	-5.3	-0.008	0.010	1
16.800	0.893	0.390	-0.015	0.025	0.646	-0.101	0.470	0	0.78	0.372	662	-2.7	-0.004	0.010	1
17.200	0.883	0.386	-0.005	0.025	0.646	-0.101	0.454	0	0.78	0.363	662	0.0	0.000	0.010	1
17.600	0.874	0.386	0.005	0.024	0.646	-0.101	0.454	0	0.78	0.361	662	2.5	0.004	0.009	1
18.000	0.864	0.390	0.014	0.023	0.646	-0.101	0.469	0	0.78	0.366	662	4.9	0.007	0.009	1

Pinned end reaction:	Fixed end reaction:
elastic, $V_p = 0.26R + 0.12F$	elastic, $V_f = 0.43R + 0.19F$
plastic, $V_p = 0.38R_u + 0.12F - M_p/L$	plastic, $V_f = 0.38R_u + 0.12F + M_p/L$

Note: $M_p/L = R_u/12$. Therefore, in the plastic range, $V_p = 0.30R_u + 0.12F$ and $V_f = 0.46R_u + 0.12F$.

Numerical integration follows the approach outlined in Chapter 6, Appendix 6. For the numerical integration through the time of initial rebound, refer to Table 11-3. For the corresponding output plots extending multiple cycles, see Figure 11-4.

Peak positive deflection: $y_m = +0.646$ in. (16.4 mm) at $t = 10.4$ ms
Peak rebound deflection: $y_{m-} = +0.153$ in. (3.88 mm) at $t = 56.8$ ms

Peak positive dynamic reaction from each span length (at the fixed end):

$$V_m = (+0.834 \text{ psi})(48 \text{ in.}) = +40.0 \text{ lb/in.} \ (7.01 \text{ N/mm}) \text{ at } t = 10.8 \text{ ms}$$

Peak rebound dynamic reaction from each span length (at the fixed end):

$$V_{m-} = (-0.206 \text{ psi})(48 \text{ in.}) = -9.87 \text{ lb/in.} \ (-1.73 \text{ N/mm}) \text{ at } t = 56.8 \text{ ms}$$

Notes: The peak rebound deflection and rebound dynamic reaction occur after numerous cycles and after the blast load disappears. The minimum deflection achieved is the initial static deflection of 0.0022 in. (0.055 mm). Without accounting for damping, the panel SDOF model oscillates about a permanent plastic deflection of 0.27 in. (7.0 mm).

Notes: Performing some sensitivity trials with the numerical integration method outlined in Appendix 6, it is found that a time step of 0.4 ms is sufficiently small to provide a converged solution and smooth output. In most cases, a time step of 0.1 ms provides adequate output. Larger time steps are considered here, however, because of space limitations of the tabulated output results.

Ductility:

$$\mu_d = y_m/y_e = (0.646 \text{ in.})/(0.373 \text{ in}) = 1.73 < \mu_a, \textbf{ OK}$$

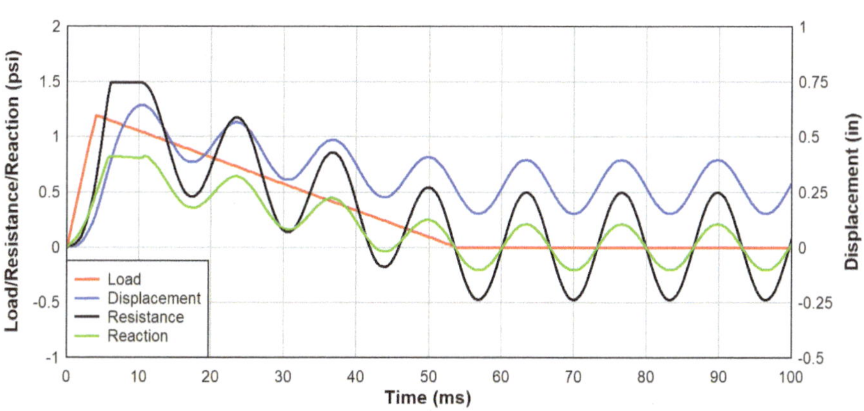

Figure 11-4. Output plots from roof deck numerical integration.

Support rotation: (Chapter 5, Figure 5-9)

$\theta_d = \tan^{-1}(y_m/0.5\,L) = \tan^{-1}[(0.646 \text{ in.})/(0.5)(48 \text{ in.})] = 1.54° < \theta_a$, **OK**

The response is adequate. A revision is not necessary.

11.4.7 Check Secondary Failure Modes (In This Case, Shear)

Reaction at the ultimate resistance (at the fixed end):

$V_u = 5\,R_u/8 = 5\,(71.5 \text{ lb})/8 = 44.7 \text{ lb } (199 \text{ N}) < V_n$, **OK**

The shear capacity is adequate.

Notes: The reaction at the ultimate resistance at the pinned end of the fixed-pinned panel would be calculated as 3 $R_u/8$. For the more typical fixed-fixed case, the reaction would be based on a higher ultimate resistance. See Section 11.5 for an example calculation.

11.5 WALL PANELS

The wall panels span continuously over the wall girts. The worst-case span for deflection calculations is at the top or bottom of the wall with fixed-pinned boundary conditions. Treat the wall panels as a 1 in. (25.4 mm) wide, one-way strip.

Response limits: $\mu_a = 3$, $\theta_a = 2$ degrees
Dead load: $DL = 0$

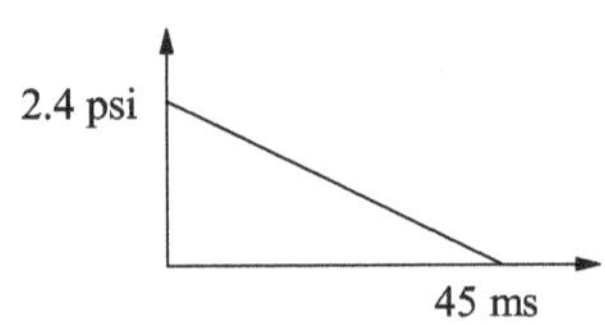

Blast load:

$BL = 2.4$ psi (16.5 kPa)

Impulse:

$$I_o = BL \times t_d/2 = (2.4 \text{ psi})(45 \text{ ms})/2 = 54 \text{ psi ms } (372 \text{ kPa ms})$$

11.5.1 Dynamic Material Properties

For dynamic bending or shear: (Chapter 5, Appendix 5A)

$F_{ds} = F_{dy} = (\text{SIF})(\text{DIF})\,F_y = (1.21)(1.1)(50 \text{ ksi}) = 66.55 \text{ ksi } (459 \text{ MPa})$

11.5.2 Calculate a Trial Size

Try 3 ft girt spacing, $L = 3$ ft or 36 in. (914 mm).

Let the ductility demand equal the limiting value, $\mu_d = \mu_a = 3$.
As an initial guess, let $\tau = t_d/t_n = 3$ (in the dynamic range of response).
Load/resistance ratio: [Chapter 6, Equation (6-11)]

$$Fo / R_m = \frac{\sqrt{(2\mu_d - 1)}}{\pi\tau} + \frac{(2\mu_d - 1)\tau}{2\mu_d(\tau + 0.7)} + \frac{\sqrt{(2\times3-1)}}{\pi\times3} + \frac{(2\times3-1)\times3}{2\times3\times(3+0.7)} = 0.913$$

Peak load:

$F_o = BL = 2.4$ psi (16.5 kPa)

Resistance:

$R_m = F_o/0.913 = (2.4 \text{ psi})/0.913 = 2.63$ psi (18.1 kPa)

Effective ultimate moment, from $R_b = 4\,(M_{ps} + 2\,M_{pc})/L$: (Chapter 6, Table 6-3)

$M_p = R_m\,L/12 = (2.63 \text{ psi} \times 36 \text{ in.} \times 1 \text{ in.})(36 \text{ in.})/12 = 284$ in. lb (32.1 N m)

Ultimate moment: (Chapter 5, Section 5.4.4)

$M_u = M_p/0.9 = (284 \text{ in. lb})/0.9 = 315$ in. lb (35.6 N m)

Section modulus:

$S_x = M_u/F_{ds} = (315 \text{ in. lb})/(66{,}550 \text{ psi}) = 0.00474 \text{ in.}^3$ (77.7 mm^3)

Notes: The section modulus, S_x, is used to compute the moment capacity instead of the plastic section modulus, Z_x, mainly because section modulus values are readily available. The difference is minor due to the relatively low response and owing to capacity reductions from buckling of the thin web.

Select the panel type and thickness consistent with the roof:

R panel, thickness = 24 gauge ($t_p = 0.0258$ in. or 0.655 mm)

Weight = 1.25 psf = 0.00868 psi (0.060 kPa)

$I_x = 0.0548 \text{ in.}^4/\text{ft} = 0.00457 \text{ in.}^4/\text{in.}$ (74.8 mm^4/mm)

$S_x = 0.0573 \text{ in.}^3/\text{ft} = 0.00478 \text{ in.}^3/\text{in.}$ (3.08 mm^3/mm)

$A_v = 0.0645 \text{ in.}^2/\text{ft} = 0.00538 \text{ in.}^2/\text{in.}$ (0.137 mm^2/mm)

h (rib height) = 1.3 in. (33 mm)

Perform a detailed check of this section.

11.5.3 Compute Section Properties

Effective moment capacity: (Section 5.4.4)

$M_p = 0.9\,S_x\,F_{ds} = 0.9\,(0.00478 \text{ in.}^3)(66{,}550 \text{ psi}) = 286$ in.·lb (32.3 N·m)

Shear capacity:

Shear in the webs of cold-formed members with $F_{ds} = 66$ ksi may be limited by shear buckling. Because $h/t = 1.3 \text{ in.}/0.0258 \text{ in.} = 50$, use $F_{dv} = 15.75$ ksi

(109 MPa) per Chapter 5, Table 5-6b, in UFC 3-340-02 for the shear capacity of the panel at the support with combined bending and shear.

$V_n = A_v F_{dv} = (0.00538 \text{ in.}^2)(15{,}750 \text{ psi}) = 84.7 \text{ lb } (377 \text{ N})$

11.5.4 Compute SDOF Properties (Fixed-Pinned Panel)

Ultimate resistance, from $R_b = 4\,(M_{ps} + 2\,M_{pc})/L$: (Table 6-3)

$R_u = 12\,M_p/L = 12\,(286 \text{ in. lb})/(36 \text{ in.}) = 95.3 \text{ lb}$, or **2.65 psi** (18.3 kPa)

Compute the effective "bilinear" elastic stiffness and deflection to determine the natural period and ductility ratios.

Effective "bilinear" elastic stiffness: (Chapter 6, Table 6-3)

$$
\begin{aligned}
K_e &= 160\,E_s\,I_x/L^3 \\
&= 160\,(29{,}000{,}000 \text{ psi})(0.00457 \text{ in.}^4)/(36 \text{ in.})^3 \\
&= 454 \text{ lb/in., or } \mathbf{12.6 \text{ psi/in.}} \text{ (3.42 kPa/mm)}
\end{aligned}
$$

Effective elastic yield deflection:

$y_e = R_u/K_e = (2.65 \text{ psi})/(12.6 \text{ psi/in.}) = \mathbf{0.210 \text{ in.}}$ (5.34 mm)

Mass:

weight = 1.25 psf (0.083 ft width)(3 ft length) = 0.312 lb, or 0.00868 psi (0.060 kPa)

$$
\begin{aligned}
M &= \text{weight/g} \\
&= (0.312 \text{ lb})/(386 \text{ in./s}^2) \\
&= 0.000809 \text{ lb-s}^2\text{/in., or } \mathbf{22.5 \text{ psi·ms}^2\text{/in.}} \text{ (6.10 kg/m}^2)
\end{aligned}
$$

Load-mass transformation factors: (Chapter 6, Table 6-3)

elastic, $K_{LM} = K_M/K_L = 0.45/0.58 = \mathbf{0.78}$
plastic, $K_{LM} = K_M/K_L = 0.33/0.50 = \mathbf{0.66}$

Assuming some plastic deformation, use an average value for the load-mass factor when calculating the natural period:

$K_{LM} = (0.78 + 0.66)/2 = \mathbf{0.72}$

Equivalent mass:

$M_e = K_{LM}\,M = 0.72\,(22.5 \text{ psi·ms}^2\text{/in.}) = 16.2 \text{ psi·ms}^2\text{/in.}$ (4.40 kg/m²)

Natural period: [Chapter 6, Equation (6-8)]

$$
t_n = 2\pi\sqrt{\frac{M_e}{K_e}} = 2\pi\sqrt{\frac{(16.2 \text{ psi-ms}^2/\text{in.})}{(12.6 \text{ psi/in.})}} = \mathbf{7.1 \text{ ms}}
$$

Notes: With fixed-pinned boundary conditions, a trilinear resistance-deflection curve could alternatively be developed by using elastic and elastic-plastic stiffness and resistance values from Table 6-3. In this case, elastic-plastic K_{LM} would equal 0.78.

11.5.5 Compute Response and Compare It to Deformation Limits (Numerical Integration Solution for Fixed-Pinned Panel)

Time increment:

$$t_n/10 = 7.1 \text{ ms}/10 \approx 0.7 \text{ ms}$$

A smaller time step can be used to potentially obtain a more refined solution with smoother output plots. Consider a time range of 100 ms to capture multiple deflection cycles and component behavior following load termination.

Pinned end reaction:	Fixed end reaction:
elastic, $V_p = 0.26R + 0.12F$	elastic, $V_f = 0.43R + 0.19F$
plastic, $V_p = 0.38R_u + 0.12F - M_p/L$	plastic, $V_f = 0.38R_u + 0.12F + M_p/L$

Note: $M_p/L = R_u/12$. Therefore, in the plastic range, $V_p = 0.30R_u + 0.12F$ and $V_f = 0.46R_u + 0.12F$.

Performing some sensitivity trials, a time step of 0.1 ms appears to be required to provide a converged solution and smooth output. For output plots of key parameters, see Figure 11-5.

Peak positive deflection: $y_m = +0.613$ in. (15.6 mm) at $t = 6.8$ ms
Peak rebound deflection: $y_{m\text{-}} = +0.350$ in. (8.90 mm) at $t = 47.6$ ms

Peak positive dynamic reaction from each span length (at the fixed end):

$$V_m = (+1.52 \text{ psi})(36 \text{ in.}) = +54.9 \text{ lb/in. } (9.61 \text{ N/mm}) \text{ at } t = 6.9 \text{ ms}$$

Peak rebound dynamic reaction from each span length (at the fixed end):

$$V_{m\text{-}} = (-0.286 \text{ psi})(36 \text{ in.}) = -10.3 \text{ lb/in. } (-1.80 \text{ N/mm}) \text{ at } t = 47.6 \text{ ms}$$

Note: The peak rebound deflection and rebound dynamic reaction occur after numerous cycles and after the blast load disappears.

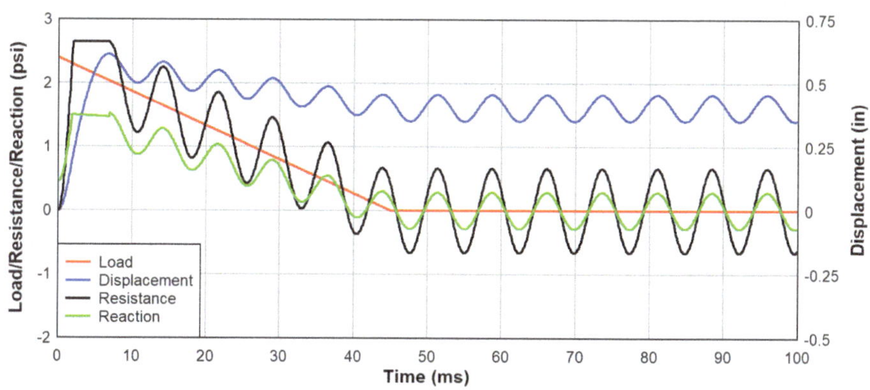

Figure 11-5. Output plots from fixed-pinned wall panel numerical integration.

Ductility:

$\mu_d = y_m/y_e = (0.613 \text{ in.})/(0.210 \text{ in.}) = 2.92 < \mu_a$, **OK**

Support rotation: (Chapter 5, Figure 5-9)

$\theta_d = \tan^{-1}(y_m/0.5\ L) = \tan^{-1}[(0.613 \text{ in.})/(0.5)(36 \text{ in.})] = 1.95° < \theta_a$, **OK**

The response is adequate. A revision is not necessary.

Notes: The achieved ductility and support rotation are close to the upper allowable limits. However, it should be recognized that a fixed-fixed panel is more common than a fixed-pinned condition, which occurs only at the top and bottom of the wall. The wall panel will be rerun in the following sections for comparison purposes.

11.5.6 Compute Single-Degree-of-Freedom Properties (Fixed-Fixed Panel)

Ultimate resistance, from $R_b = 8\ (M_{ps} + M_{pc})/L$: (Chapter 6, Table 6-2)

$R_u = 16\ M_p/L = 16\ (286 \text{ in.·lb})/(36 \text{ in.}) = 127$ lb, or **3.53 psi** (24.3 kPa)

Compute the effective "bilinear" elastic stiffness and deflection to determine the natural period and ductility ratios.

Effective "bilinear" elastic stiffness: (Chapter 6, Table 6-2)

$$K_e = 307\ E_s I_x/L^3$$
$$= 307\ (29{,}000{,}000 \text{ psi})(0.00457 \text{ in.}^4)/(36 \text{ in.})^3$$
$$= 871 \text{ lb/in., or } \mathbf{24.2\ psi/in.}\ (6.57 \text{ kPa/mm})$$

Effective elastic yield deflection:

$y_e = R_u/K_e = (3.53 \text{ psi})/(24.2 \text{ psi/in.}) =$ **0.146 in.** (3.71 mm)

Mass:

weight = 1.25 psf (0.083 ft width)(3 ft length) = 0.312 lb, or 0.00868 psi (0.060 kPa)

$$M = \text{weight}/g$$
$$= (0.312 \text{ lb})/(386 \text{ in./s}^2)$$
$$= 0.000809 \text{ lb-s}^2\text{/in., or } \mathbf{22.5\ psi{\cdot}ms^2/in.}\ (6.10 \text{ kg/m}^2)$$

Load-mass transformation factors: (Chapter 6, Table 6-2)

elastic, $K_{LM} = K_M/K_L = 0.41/0.53 = \mathbf{0.77}$
plastic, $K_{LM} = K_M/K_L = 0.33/0.50 = \mathbf{0.66}$

Assuming some plastic deformation, use an average value for the load-mass factor when calculating the natural period:

$K_{LM} = (0.77 + 0.66)/2 = \mathbf{0.715}$

Equivalent mass:

$M_e = K_{LM}\ M = 0.715\ (22.5 \text{ psi·ms}^2\text{/in.}) = 16.1 \text{ psi·ms}^2\text{/in.}\ (4.37 \text{ kg/m}^2)$

Natural period: (Equation 6.8)

$$t_n = 2\pi\sqrt{\frac{M_e}{K_e}} = 2\pi\sqrt{\frac{(16.1 \text{ psi} \cdot \text{ms}^2/\text{in.})}{(24.2 \text{ psi}/\text{in.})}} = \mathbf{5.1 \text{ ms}}$$

Notes: With fixed-fixed boundary conditions, a trilinear resistance-deflection curve could alternatively be developed by using elastic and elastic-plastic stiffness and resistance values from Chapter 6, Table 6-2. In this case, elastic-plastic K_{LM} would equal 0.78.

11.5.7 Compute Response and Compare It to Deformation Limits (Numerical Integration Solution for a Fixed-Fixed Panel)

Time increment:

$t_n/10 = 5.1$ ms/10 ≈ 0.5 ms

A smaller time step can be used to potentially obtain a more refined solution with smoother output plots. Consider a time range of 100 ms to capture multiple deflection cycles and component behavior following load termination.

End reaction:

elastic, $V = 0.36R + 0.14F$
plastic, $V = 0.38R_u + 0.12F$

For the numerical integration for the first 14 ms, refer to Table 11-4. For the corresponding output plots extending multiple cycles, see Figure 11-6.

Peak positive deflection: $y_m = +0.202$ in. (5.14 mm) at $t = 3.0$ ms
Peak rebound deflection: $y_{m\text{-}} = +0.001$ in. (0.02 mm) at $t = 54.0$ ms

Peak positive dynamic reaction from each span length:

$V_m = (+1.62$ psi$)(36$ in.$) = +58.2$ lb/in. (10.2 N/mm) at $t = 2.0$ ms

Peak rebound dynamic reaction from each span length:

$V_{m\text{-}} = (-0.484$ psi$)(36$ in.$) = -17.4$ lb/in. (-3.05 N/mm) at $t = 54.0$ ms

Note: The peak rebound deflection and rebound dynamic reaction occur after numerous cycles and after the blast load disappears.

Ductility:

$\mu_d = y_m/y_e = (0.202 \text{ in.})/(0.146 \text{ in.}) = 1.39 < \mu_a$, **OK**

Support rotation: (Chapter 5, Figure 5-9)

$\theta_d = \tan^{-1}(y_m/0.5\,L) = \tan^{-1}[(0.202 \text{ in.})/(0.5)(36 \text{ in.})] = 0.64° < \theta_a$, **OK**

The response is adequate. A revision is not necessary.

Note: The fixed-fixed panel results in a significantly lower maximum deflection than the fixed-pinned case, well under acceptable response limits.

Table 11-4. Fixed-Fixed Wall Panel Numerical Integration.

Example:	Section 11.5 Wall Panel (Fixed-Fixed)							KEY = 0 when elastic
								KEY = +1 when positive plastic
time increment:	0.5	ms	**Blast Load**					KEY = -1 when rebound plastic
mass:	22.5	psi-ms²/in	**t0:**	0 ms	**P0:**	2.4	psi	Z = 1 when continued elastic
stiffness:	24.2	psi/in	**t1:**	45 ms	**P1:**	0	psi	Z = 2 at start of positive plastic deformation
damping:	0	psi-ms/in	**t2:**	45 ms	**P2:**	0	psi	Z = 3 when increasing plastic deformation
positive resistance:	3.53	psi	**Dynamic Reaction Factors**		**BC: F-F**			Z = 4 at end of positive plastic deformation
rebound resistance:	-3.53	psi	**Crel:**	0.36	**Cfel:**	0.14		Z = 6 at start of rebound plastic deformation
static load:	0	psi	**Crpl:**	0.38	**Cfpl:**	0.12		Z = 8 at end of rebound plastic deformation

time (ms)	load (psi)	y (in)	v (in/ms)	a (in/ms²)	ye+ (in)	ye- (in)	R (psi)	KEY -	KLM	V (psi)	eff K (psi/in)	incr F (psi)	incr y (in)	incr v (in/ms)	Z -
0.000	2.400	0.000	0.000	0.139	0.146	-0.146	0.000	0	0.77	0.336	440	7.2	0.016	0.063	
0.500	2.373	0.016	0.063	0.114	0.146	-0.146	0.395	0	0.77	0.474	440	19.0	0.043	0.042	1
1.000	2.347	0.060	0.105	0.052	0.146	-0.146	1.442	0	0.77	0.848	440	24.5	0.056	0.006	1
1.500	2.320	0.115	0.111	-0.027	0.146	-0.146	2.788	0	0.77	1.329	440	21.7	0.049	-0.031	1
2.000	2.293	0.164	0.080	-0.083	0.146	-0.146	3.530	1	0.66	1.617	356	10.5	0.030	-0.042	2
2.500	2.267	0.194	0.038	-0.085	0.146	-0.146	3.530	1	0.66	1.613	356	2.9	0.008	-0.043	3
3.000	2.240	0.202	-0.005	-0.087	0.146	-0.146	3.530	1	0.66	1.610	356	-4.8	-0.013	-0.044	3
3.500	2.213	0.189	-0.049	-0.057	0.202	-0.090	3.204	0	0.77	1.463	440	-13.2	-0.030	-0.019	4
4.000	2.187	0.159	-0.067	-0.017	0.202	-0.090	2.479	0	0.77	1.199	440	-14.9	-0.034	0.003	1
4.500	2.160	0.125	-0.064	0.029	0.202	-0.090	1.658	0	0.77	0.899	440	-11.9	-0.027	0.024	1
5.000	2.133	0.098	-0.041	0.065	0.202	-0.090	1.002	0	0.77	0.660	440	-5.1	-0.012	0.036	1
5.500	2.107	0.086	-0.005	0.080	0.202	-0.090	0.720	0	0.77	0.554	440	3.2	0.007	0.037	1
6.000	2.080	0.093	0.033	0.068	0.202	-0.090	0.895	0	0.77	0.613	440	10.3	0.023	0.026	1
6.500	2.053	0.117	0.058	0.034	0.202	-0.090	1.461	0	0.77	0.814	440	13.8	0.031	0.006	1
7.000	2.027	0.148	0.064	-0.011	0.202	-0.090	2.223	0	0.77	1.084	440	12.7	0.029	-0.016	1
7.500	2.000	0.177	0.048	-0.053	0.202	-0.090	2.920	0	0.77	1.331	440	7.2	0.016	-0.033	1
8.000	1.973	0.193	0.015	-0.077	0.202	-0.090	3.313	0	0.77	1.469	440	-0.9	-0.002	-0.038	1
8.500	1.947	0.191	-0.023	-0.076	0.202	-0.090	3.264	0	0.77	1.448	440	-8.8	-0.020	-0.031	1
9.000	1.920	0.171	-0.055	-0.050	0.202	-0.090	2.781	0	0.77	1.270	440	-14.0	-0.032	-0.014	1
9.500	1.893	0.140	-0.069	-0.007	0.202	-0.090	2.013	0	0.77	0.990	440	-14.7	-0.033	0.008	1
10.000	1.867	0.106	-0.061	0.038	0.202	-0.090	1.206	0	0.77	0.696	440	-10.7	-0.024	0.027	1
10.500	1.840	0.082	-0.034	0.071	0.202	-0.090	0.617	0	0.77	0.480	440	-3.4	-0.008	0.038	1
11.000	1.813	0.074	0.004	0.080	0.202	-0.090	0.431	0	0.77	0.409	440	4.9	0.011	0.036	1
11.500	1.787	0.085	0.039	0.063	0.202	-0.090	0.702	0	0.77	0.503	440	11.4	0.026	0.022	1
12.000	1.760	0.111	0.061	0.025	0.202	-0.090	1.331	0	0.77	0.725	440	14.0	0.032	0.001	1
12.500	1.733	0.143	0.062	-0.021	0.202	-0.090	2.101	0	0.77	0.999	440	11.8	0.027	-0.020	1
13.000	1.707	0.170	0.042	-0.060	0.202	-0.090	2.750	0	0.77	1.229	440	5.5	0.013	-0.035	1
13.500	1.680	0.183	0.007	-0.079	0.202	-0.090	3.054	0	0.77	1.335	440	-2.7	-0.006	-0.038	1
14.000	1.653	0.176	-0.031	-0.072	0.202	-0.090	2.906	0	0.77	1.278	440	-10.2	-0.023	-0.028	1

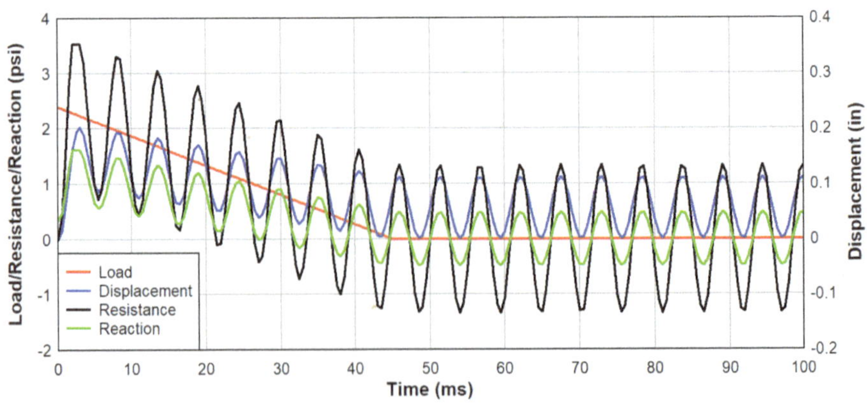

Figure 11-6. Output plots from fixed-fixed wall panel numerical integration.

11.5.8 Check Secondary Failure Modes (In This Case, Shear)

Reaction at the ultimate resistance (at the fixed end of a fixed-pinned panel):

$$V_u = 5\,R_u/8 = 5\,(95.3\text{ lb})/8 = 59.6\text{ lb }(265\text{ N}) < V_n,\ \textbf{OK}$$

Reaction at the ultimate resistance (for a fixed-fixed panel):

$$V_u = R_u/2 = (127\text{ lb})/2 = 63.6\text{ lb }(283\text{ N}) < V_n,\ \textbf{OK}$$

The shear capacity is adequate.

Note: With a higher ultimate resistance, the fixed-fixed panel achieves a higher reaction than at the fixed end of the fixed-pinned panel.

11.6 ROOF PURLINS

The purlins span continuously over the roof beams. The worst-case span for deflection calculations is for the end purlins with fixed-pinned boundary conditions. Assume A36 rolled shapes. The purlin is assumed to be laterally braced by the roof deck in tension.

To add the effects of dead load to the SDOF calculations, each pressure-time pair will be increased by the magnitude of the dead load and the initial displacement will be set equal to the dead load deflection. This will create a balanced condition at the start of the SDOF response calculation (refer to the preload discussion in Chapter 7, Section 7.2.5).

Span, $L = 20$ ft or 240 in. (6,100 mm)
Purlin spacing, $B = 4$ ft or 48 in. (1,220 mm)

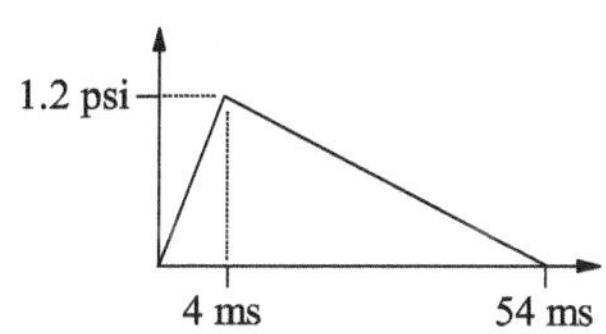

Response limits: $\mu_a = 10$, $\theta_a = 6$ degrees
Dead load (initial guess):

$$DL = 6\text{ psf, or }0.04\text{ psi }(0.29\text{ kPa})$$

Blast load:

$$BL = 1.2\text{ psi }(8.3\text{ kPa})$$

Impulse:

$$I_o = BL \times t_d/2 = (1.2\text{ psi})(54\text{ ms})/2 = 32.4\text{ psi·ms }(223\text{ kPa·ms})$$

The tributary width method will be used to apply the blast load acting on the roof deck to the purlin. Alternatively, a similar approach, as demonstrated in Section 11.7, can be used to apply the dynamic reaction from the roof deck to the purlin.

11.6.1 Dynamic Material Properties

For dynamic bending or shear: (Chapter 5, Appendix 5A)

$$F_{ds} = F_{dy} = \text{(SIF)(DIF)}\ F_y = (1.1)(1.29)(36\ \text{ksi}) = 51.1\ \text{ksi}\ (352\ \text{MPa})$$

11.6.2 Calculate a Trial Size

Tributary width, $B = 4$ ft or 48 in. (1,220 mm)

Let the ductility demand equal the limiting value, $\mu_d = \mu_a = 10$.

As an initial guess, let $\tau = t_d/t_n = 3$ (in the dynamic range of response).

Load/resistance ratio: [Chapter 6, Equation (6-11)]

$$Fo/R_m = \frac{\sqrt{(2\mu_d - 1)}}{\pi\tau} + \frac{(2\mu_d - 1)\tau}{2\mu_d(\tau + 0.7)} = \frac{\sqrt{(2\times 10 - 1)}}{\pi\times 3} + \frac{(2\times 10 - 1)\times 3}{2\times 10\times(3 + 0.7)} = 1.23$$

Peak load:

$$F_o = BL + DL = 1.2\ \text{psi} + 0.04\ \text{psi} = 1.24\ \text{psi}\ (8.56\ \text{kPa})$$

Resistance:

$$R_m = F_o/1.23 = (1.24\ \text{psi})/1.23 = 1.01\ \text{psi}\ (6.94\ \text{kPa})$$

Ultimate moment, from $R_b = 4\ (M_{ps} + 2\ M_{pc})/\text{L}$: (Chapter 6, Table 6-3)

$$\begin{aligned} M_p &= R_m\ L/12 \\ &= (1.01\ \text{psi} \times 240\ \text{in.} \times 48\ \text{in.})(240\ \text{in.})/12 \\ &= 232{,}000\ \text{in.·lb}\ (26{,}200\ \text{N·m}) \end{aligned}$$

Plastic section modulus:

$$Z_x = M_p/F_{ds} = (232{,}000\ \text{in.·lb})/(51{,}100\ \text{psi}) = 4.54\ \text{in.}^3\ (74{,}400\ \text{mm}^3)$$

The moment capacity is based on Z_x because the target $\mu > 3$.

Select C8 × 11.5: (AISC Manual)

$$\begin{aligned} I_x &= 32.5\ \text{in.}^4\ (13.5 \times 10^6\ \text{mm}^4) \\ Z_x &= 9.63\ \text{in.}^3\ (158{,}000\ \text{mm}^3) \\ A_v\ \text{(area of web)} &= 1.76\ \text{in.}^2\ (1{,}140\ \text{mm}^2) \end{aligned}$$

Note: A component with a significantly larger plastic section modulus than the trial value is selected to ensure that deflections to conventional loading are within acceptable limits per code.

Perform a detailed check of this section.

11.6.3 Compute Section Properties

Moment capacity: (AISC 360, Equation F2-1)

$$M_p = Z_x\ F_{ds} = (9.63\ \text{in.}^3)(51{,}100\ \text{psi}) = 492{,}000\ \text{in.·lb}\ (55{,}600\ \text{N·m})$$

Shear capacity: (AISC 360, Equation G2-1)

$$V_n = 0.6\ A_v\ F_{ds}\ C_v = 0.6\ (1.76\ \text{in.}^2)(51{,}100\ \text{psi})(1.0) = 53{,}900\ \text{lb}\ (240\ \text{kN})$$

11.6.4 Compute Single-Degree-of-Freedom Properties

Ultimate resistance, from $R_b = 4\ (M_{ps} + 2\ M_{pc})/L$: (Chapter 6, Table 6-3)

$$R_u = 12\ M_p/L = 12\ (492{,}000\ \text{in.-lb})/(240\ \text{in.}) = 24{,}600\ \text{lb, or } \mathbf{2.14\ psi}\ (14.7\ \text{kPa})$$

Compute the effective "bilinear" elastic stiffness and deflection to determine the natural period and ductility ratios.

Effective "bilinear" elastic stiffness: (Chapter 6, Table 6-3)

$$\begin{aligned} K_e &= 160\ E_s\ I_x/L^3 \\ &= 160\ (29{,}000{,}000\ \text{psi})(32.5\ \text{in.}^4)/(240\ \text{in.})^3 \\ &= 10{,}900\ \text{lb/in., or } \mathbf{0.947\ psi/in.}\ (0.257\ \text{kPa/mm}) \end{aligned}$$

Effective elastic yield deflection:

$$y_e = R_u/K_e = (2.14\ \text{psi})/(0.947\ \text{psi/in.}) = \mathbf{2.26\ in.}\ (57.4\ \text{mm})$$

Mass:

weight = (1.25 psf deck + 5 psf mechanical)(20 ft span)(4 ft tributary width) + (11.5 plf purlin)(20 ft span)
= 730 lb, or 0.0634 psi (0.437 kPa)

$$\begin{aligned} M &= \text{weight/g} \\ &= (730\ \text{lb})/(386\ \text{in./s}^2) \\ &= 1.89\ \text{lb-s}^2\text{/in., or } \mathbf{164\ psi{\cdot}ms^2/in.}\ (44.6\ \text{kg/m}^2) \end{aligned}$$

Static (dead load) deflection:

$$y_{st} = \text{weight}/K_e = (0.0634\ \text{psi})/(0.947\ \text{psi/in.}) = 0.067\ \text{in.}\ (1.70\ \text{mm})$$

Load-mass transformation factors: (Table 6-3)

elastic, $K_{LM} = K_M/K_L = 0.45/0.58 = \mathbf{0.78}$
plastic, $K_{LM} = K_M/K_L = 0.33/0.50 = \mathbf{0.66}$

Assuming some plastic deformation, use an average value for the load-mass factor when calculating the natural period:

$$K_{LM} = (0.78 + 0.66)/2 = \mathbf{0.72}$$

Equivalent mass:

$$M_e = K_{LM}\ M = 0.72\ (164\ \text{psi·ms}^2\text{/in.}) = 118\ \text{psi·ms}^2\text{/in.}\ (32.1\ \text{kg/m}^2)$$

Natural period: [Equation (6-8)]

$$t_n = 2\pi\sqrt{\frac{M_e}{K_e}} = 2\pi\sqrt{\frac{(118\ \text{psi-ms}^2\ /\ \text{in.})}{(0.947\ \text{psi}\ /\ \text{in.})}} = \mathbf{70.2\ ms}$$

Notes: The roof panel has a period, t_n, of 12.7 ms. Analyzing the roof panel and purlin separately should be adequate with this difference in periods (refer to Section 6.4.2).

Notes: With fixed-pinned boundary conditions, a trilinear resistance-deflection curve could alternatively be developed by using elastic and elastic-plastic stiffness and resistance values from Table 6-3. In this case, elastic-plastic K_{LM} would equal 0.78.

11.6.5 Compute Response and Compare It to Deformation Limits (Numerical Integration)

Time increment:

$$t_n/10 = 70.2 \text{ ms}/10 \approx 7 \text{ ms}$$

However, it is recommended to use at most a 0.4-ms time step to obtain at least 10 increments of loading in the smallest, first section of the pressure-time history. A smaller time step can be used to potentially obtain a more refined solution with smoother output plots. Consider a time range of 200 ms to capture multiple deflection cycles and component behavior following load termination.

Pinned end reaction:	Fixed end reaction:
elastic, $V_p = 0.26R + 0.12F$	elastic, $V_f = 0.43R + 0.19F$
plastic, $V_p = 0.38R_u + 0.12F - M_p/L$	plastic, $V_f = 0.38R_u + 0.12F + M_p/L$

Note: $M_p/L = R_u/12$. Therefore, in the plastic range, $V_p = 0.30R_u + 0.12F$ and $V_f = 0.46R_u + 0.12F$.

For output plots of key parameters, see Figure 11-7.

Peak positive deflection: $y_m = +1.90$ in. (48.2 mm) at $t = 33.6$ ms
Peak rebound deflection: $y_{m-} = -1.57$ in. (−39.9 mm) at $t = 72.4$ ms

Peak positive dynamic reaction (at the fixed end):

$$V_m = (+0.870 \text{ psi})(240 \text{ in.})(48 \text{ in.}) = +10{,}000 \text{ lb } (44.6 \text{ kN}) \text{ at } t = 32.4 \text{ ms}$$

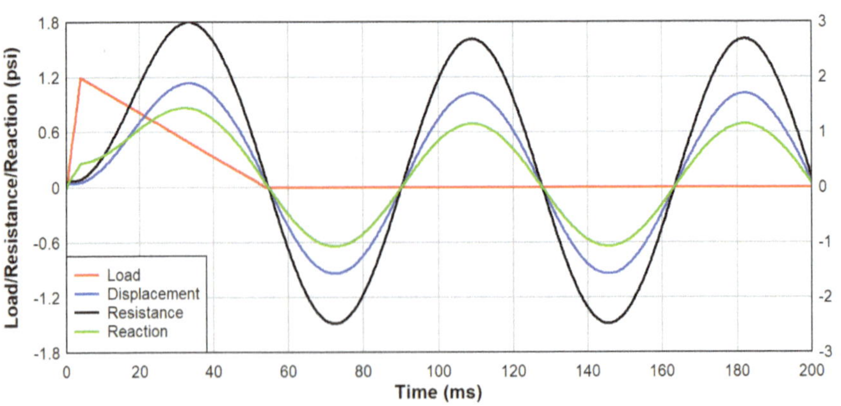

Figure 11-7. Output plots from roof purlin numerical integration.

Peak rebound dynamic reaction (at the fixed end):

$V_{m-} = (-0.640 \text{ psi})(240 \text{ in.})(48 \text{ in.}) = -7{,}370 \text{ lb } (-32.8 \text{ kN}) \text{ at } t = 72.4 \text{ ms}$

Ductility:

$\mu_d = y_m/y_e = (1.90 \text{ in.})/(2.26 \text{ in.}) = 0.84 < \mu_a$, **OK**

Support rotation: (Chapter 5, Figure 5-9)

$\theta_d = \tan^{-1}(y_m/0.5\,L) = \tan^{-1}[(1.90 \text{ in.})/(0.5)(240 \text{ in.})] = 0.91° < \theta_a$, **OK**

The response is adequate. A revision is not necessary.

11.6.6 Check Secondary Failure Modes (In This Case, Shear)

Reaction at the ultimate resistance (at the fixed end):

$V_u = 5\,R_u/8 = 5\,(24{,}600 \text{ lb})/8 = 15{,}400 \text{ lb } (68.4 \text{ kN}) < V_n$, **OK**

The shear capacity is adequate.

11.7 WALL GIRTS

The girts are flush with the columns and are simply supported. Assume A36 rolled shapes. The girts are assumed to be laterally braced by the wall panels in tension.

Span, $L = 20$ ft or 240 in. (6,100 mm)
Girt spacing, $B = 3$ ft or 36 in. (914 mm)

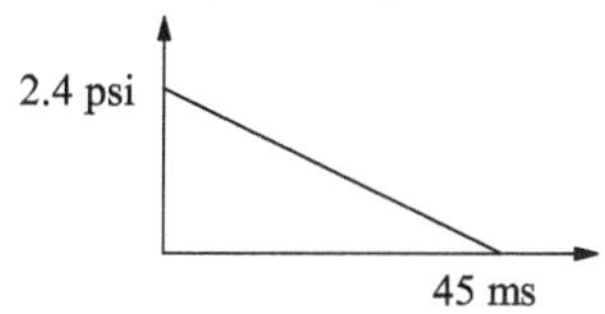

Response limits: $\mu_a = 10$, $\theta_a = 6$ degrees
Dead load: $DL = 0$

Blast load:
$BL = 2.4$ psi (16.5 kPa)
Impulse:

$I_o = BL \times t_d/2 = (2.4 \text{ psi})(45 \text{ ms})/2 = 54 \text{ psi·ms } (372 \text{ kPa·ms})$

Two methods for applying blast loads will be used. The first is the tributary width method (as was carried out for the purlin in Section 11.6), which applies the wall panel pressure-time history to the loaded area of the girt. The second method will use the dynamic reactions of the wall panel as the load applied to the girt. The girt load for the second method is determined at each time step as follows in Equation (11-1):

$$\text{Load, psi} = (2 \text{ sides})\,[\text{panel reaction, lb/in.}]/[\text{member spacing, in.}] \qquad (11\text{-}1)$$

11.7.1 Dynamic Material Properties

For dynamic bending or shear: (Chapter 5, Appendix 5A)

$$F_{ds} = F_{dy} = (\text{SIF})(\text{DIF})\ F_y = (1.1)(1.29)(36 \text{ ksi}) = 51.1 \text{ ksi } (352 \text{ MPa})$$

11.7.2 Calculate a Trial Size

Use the tributary width method for applying load to the girt in initial sizing.

Tributary width, $B = 3$ ft or 36 in. (914 mm)

Let the ductility demand equal the limiting value, $\mu_d = \mu_a = 10$.

As an initial guess, let $\tau = t_d/t_n = 3$ (in the dynamic range of response).

Load/resistance ratio: [Chapter 6, Equation (6-11)]

$$Fo/R_m = \frac{\sqrt{(2\mu_d - 1)}}{\pi\tau} + \frac{(2\mu_d - 1)\tau}{2\mu_d(\tau + 0.7)} = \frac{\sqrt{(2\times 10 - 1)}}{\pi \times 3} + \frac{(2\times 10 - 1)\times 3}{2\times 10\times(3 + 0.7)} = 1.23$$

Peak load: $F_o = BL = 2.4$ psi (16.5 kPa)

Resistance:

$$R_m = F_o/1.23 = (2.4 \text{ psi})/1.23 = 1.95 \text{ psi } (13.4 \text{ kPa})$$

Ultimate moment, from $R_b = 8\ M_{pc}/L$: (Chapter 6, Table 6-1)

$$\begin{aligned} M_p &= R_m\ L/8 \\ &= (1.95 \text{ psi} \times 240 \text{ in.} \times 36 \text{ in.})(240 \text{ in.})/8 \\ &= 505{,}000 \text{ in.·lb } (57{,}000 \text{ N·m}) \end{aligned}$$

Plastic section modulus:

$$Z_x = M_p/F_{ds} = (505{,}000 \text{ in.·lb})/(51{,}100 \text{ psi}) = 9.88 \text{ in.}^3 \ (162{,}000 \text{ mm}^3)$$

The moment capacity is based on Z_x because the target $\mu > 3$.

Notes: The value of Z_x determined previously is not a minimum requirement, but rather an initial estimate. Based on the results of the roof purlin calculation, a similar trial size will be selected.

Select C8 × 11.5: (AISC Manual)

$I_x = 32.5$ in.^{4} (13.5×10^6 mm^4)
$Z_x = 9.63$ in.^{3} (158,000 mm^3)
A_v (area of web) $= 1.76$ in.^{2} (1,140 mm^2)

Perform a detailed check of this section.

11.7.3 Compute Section Properties

Moment capacity: (AISC 360, Equation F2-1)

$$M_p = Z_x\ F_{ds} = (9.63 \text{ in.}^3)(51{,}100 \text{ psi}) = 492{,}000 \text{ in.·lb } (55{,}600 \text{ N·m})$$

Shear capacity: (AISC 360, Equation G2-1)

$$V_n = 0.6\ A_v\ F_{ds}\ C_v = 0.6\ (1.76 \text{ in.}^2)(51{,}100 \text{ psi})(1.0) = 53{,}900 \text{ lb } (240 \text{ kN})$$

11.7.4 Compute SDOF Properties

Ultimate resistance, from $R_b = 8\ M_{pc}/L$: (Chapter 6, Table 6-1)

$$R_u = 8\ M_p/L = 8\ (492{,}000\ \text{in.·lb})/(240\ \text{in.}) = 16{,}400\ \text{lb, or}\ \mathbf{1.90\ psi}\ (13.1\ \text{kPa})$$

Effective stiffness:

$$\begin{aligned} K_e &= 384\ E_s\ I_x/5\ L^3\ \text{(Chapter 6, Table 6-1)} \\ &= 384\ (29{,}000{,}000\ \text{psi})(32.5\ \text{in.}^4)/(5)(240\ \text{in.})^3 \\ &= 5{,}240\ \text{lb/in., or}\ \mathbf{0.606\ psi/in.}\ (0.165\ \text{kPa/mm}) \end{aligned}$$

Effective elastic yield deflection:

$$y_e = R_u/K_e = (1.90\ \text{psi})/(0.606\ \text{psi/in.}) = \mathbf{3.14\ in.}\ (79.6\ \text{mm})$$

Mass:

weight = (1.25 psf siding)(20 ft span)(3 ft tributary width) + (11.5 plf girt) (20 ft span)
= 305 lb, or 0.0353 psi (0.243 kPa)

$$\begin{aligned} M &= \text{weight}/g \\ &= (305\ \text{lb})/(386\ \text{in./s}^2) \\ &= 0.790\ \text{lb-s}^2\text{/in., or}\ \mathbf{91.4\ psi{\cdot}ms^2/in.}\ (24.8\ \text{kg/m}^2) \end{aligned}$$

Load-mass transformation factors: (Chapter 6, Table 6-1)

elastic, $K_{LM} = K_M/K_L = 0.50/0.64 = \mathbf{0.78}$
plastic, $K_{LM} = K_M/K_L = 0.33/0.50 = \mathbf{0.66}$

Assuming some plastic deformation, use an average value for the load-mass factor when calculating the natural period:

$$K_{LM} = (0.78 + 0.66)/2 = \mathbf{0.72}$$

Equivalent mass:

$$M_e = K_{LM}\ M = 0.72\ (91.4\ \text{psi·ms}^2\text{/in.}) = 65.8\ \text{psi·ms}^2\text{/in.}\ (17.9\ \text{kg/m}^2)$$

Natural period: [Chapter 6, Equation (6-8)]

$$t_n = 2\pi\sqrt{\frac{M_e}{K_e}} = 2\pi\sqrt{\frac{(65.8\ \text{psi-ms}^2/\text{in.})}{(0.606\ \text{psi}/\text{in.})}} = \mathbf{65.5\ ms}$$

Notes: The wall panel has a period, t_n, of 7.1 ms. Analyzing the wall panel and girt separately should be adequate with this difference in periods (refer to Chapter 6, Section 6.2.3).

11.7.5 Compute Response and Compare It to Deformation Limits (Numerical Integration of the Tributary Area Load)

Refer to Section 11.7.2 for the tributary area load.

Time increment:

$t_n/10 = 65.5\ \text{ms}/10 \approx 6.5\ \text{ms}$

A smaller time step can be used to potentially obtain a more refined solution with smoother output plots. Consider a time range of 200 ms to capture multiple deflection cycles and component behavior following load termination

End reaction:

elastic, $V = 0.39R + 0.11F$
plastic, $V = 0.38R_u + 0.12F$

Performing some sensitivity trials, it is found that a time step of 0.5 ms appears to be sufficiently small to provide a converged solution and smooth output. For the numerical integration through the time of peak positive deflection, refer to Table 11-5. For output plots of key parameters, see Figure 11-8.

Peak positive deflection: $y_m = +6.46$ in. (164 mm) at $t = 35.5$ ms
Peak rebound deflection: $y_{m\text{-}} = +0.270$ in. (6.85 mm) at $t = 71.0$ ms

Peak positive dynamic reaction:

$V_m = (+0.908\ \text{psi})(240\ \text{in.})(36\ \text{in.}) = +7{,}850\ \text{lb}\ (34.9\ \text{kN})$ at $t = 16.0$ ms

Peak rebound dynamic reaction:

$V_{m\text{-}} = (-0.721\ \text{psi})(240\ \text{in.})(36\ \text{in.}) = -6{,}230\ \text{lb}\ (-27.7\ \text{kN})$ at $t = 71.0$ ms

Ductility:

$\mu_d = y_m/y_e = (6.46\ \text{in.})/(3.14\ \text{in.}) = 2.06 < \mu_a$, **OK**

Support rotation: (Chapter 5, Figure 5-9)

$\theta_d = \tan^{-1}(y_m/0.5\,L) = \tan^{-1}[(6.46\ \text{in.})/(0.5)(240\ \text{in.})] = 3.08° < \theta_a$, **OK**

The response is adequate. A revision is not necessary.

11.7.6 Compute Response and Compare It to Deformation Limits (Numerical Integration of the Dynamic Reaction)

Refer to Table 11-4 and Figure 11-6 for the dynamic reaction load from the wall panel. On the basis of Equation (11-1), the dynamic reaction pressure load obtained from Table 11-4 and Figure 11-6 needs to be doubled prior to being applied to the girt. However, the dynamic reaction load is truncated after one full positive and negative phase of response as recommended in Chapter 7, Section 7.3.1, although this occurs after the peak deflection is reached and the original blast load terminates at $t = 45$ ms.

For the numerical integration through the time of peak positive deflection, refer to Table 11-6. For the corresponding output plots and truncated dynamic reaction load, see Figure 11-9.

Peak positive deflection: $y_m = +6.44$ in. (164 mm) at $t = 36.0$ ms
Peak rebound deflection: $y_{m\text{-}} = +0.154$ in. (3.92 mm) at $t = 70.5$ ms

Table 11-5. Wall Girt Numerical Integration (Tributary Width Method).

Example:	Section 11.7 Wall Girt (Tributary Width)								KEY = 0 when elastic
									KEY = +1 when positive plastic
time increment:	0.5	ms	**Blast Load**						KEY = -1 when rebound plastic
mass:	91.4	psi-ms^2/in	**t0:**	0	ms	**P0:**	2.4	psi	Z = 1 when continued elastic
stiffness:	0.606	psi/in	**t1:**	45	ms	**P1:**	0	psi	Z = 2 at start of positive plastic deformation
damping:	0	psi-ms/in	**t2:**	45	ms	**P2:**	0	psi	Z = 3 when increasing plastic deformation
positive resistance:	1.90	psi	**Dynamic Reaction Factors**			**BC: S-S**			Z = 4 at end of positive plastic deformation
rebound resistance:	-1.90	psi	**Crel:**	0.39		**Cfel:**	0.11		Z = 6 at start of rebound plastic deformation
static load:	0	psi	**Crpl:**	0.38		**Cfpl:**	0.12		Z = 8 at end of rebound plastic deformation

time (ms)	load (psi)	y (in)	v (in/ms)	a (in/ms^2)	ye+ (in)	ye- (in)	R (psi)	KEY -	KLM	V (psi)	eff K (psi/in)	incr F (psi)	incr y (in)	incr v (in/ms)	Z -
0.000	2.400	0.000	0.000	0.034	3.135	-3.135	0.000	0	0.78	0.264	1712	7.2	0.004	0.017	
0.500	2.373	0.004	0.017	0.033	3.135	-3.135	0.003	0	0.78	0.262	1712	21.4	0.013	0.017	1
1.000	2.347	0.017	0.033	0.033	3.135	-3.135	0.010	0	0.78	0.262	1712	35.4	0.021	0.016	1
1.500	2.320	0.037	0.049	0.032	3.135	-3.135	0.023	0	0.78	0.264	1712	49.2	0.029	0.016	1
2.000	2.293	0.066	0.065	0.032	3.135	-3.135	0.040	0	0.78	0.268	1712	62.7	0.037	0.016	1
2.500	2.267	0.103	0.081	0.031	3.135	-3.135	0.062	0	0.78	0.274	1712	75.9	0.044	0.015	1
3.000	2.240	0.147	0.096	0.030	3.135	-3.135	0.089	0	0.78	0.281	1712	88.9	0.052	0.015	1
3.500	2.213	0.199	0.111	0.029	3.135	-3.135	0.121	0	0.78	0.291	1712	101.4	0.059	0.014	1
4.000	2.187	0.258	0.126	0.028	3.135	-3.135	0.157	0	0.78	0.302	1712	113.6	0.066	0.014	1
4.500	2.160	0.325	0.140	0.028	3.135	-3.135	0.197	0	0.78	0.314	1712	125.4	0.073	0.014	1
5.000	2.133	0.398	0.153	0.027	3.135	-3.135	0.241	0	0.78	0.329	1712	136.7	0.080	0.013	1
5.500	2.107	0.478	0.166	0.025	3.135	-3.135	0.290	0	0.78	0.345	1712	147.6	0.086	0.012	1
6.000	2.080	0.564	0.179	0.024	3.135	-3.135	0.342	0	0.78	0.362	1712	158.1	0.092	0.012	1
6.500	2.053	0.656	0.191	0.023	3.135	-3.135	0.398	0	0.78	0.381	1712	168.0	0.098	0.011	1
7.000	2.027	0.755	0.202	0.022	3.135	-3.135	0.457	0	0.78	0.401	1712	177.4	0.104	0.011	1
7.500	2.000	0.858	0.213	0.021	3.135	-3.135	0.520	0	0.78	0.423	1712	186.3	0.109	0.010	1
8.000	1.973	0.967	0.223	0.019	3.135	-3.135	0.586	0	0.78	0.446	1712	194.6	0.114	0.009	1
8.500	1.947	1.081	0.232	0.018	3.135	-3.135	0.655	0	0.78	0.470	1712	202.4	0.118	0.009	1
9.000	1.920	1.199	0.241	0.017	3.135	-3.135	0.727	0	0.78	0.495	1712	209.5	0.122	0.008	1
9.500	1.893	1.321	0.249	0.015	3.135	-3.135	0.801	0	0.78	0.521	1712	216.1	0.126	0.007	1
10.000	1.867	1.448	0.256	0.014	3.135	-3.135	0.877	0	0.78	0.547	1712	222.0	0.130	0.007	1
10.500	1.840	1.577	0.263	0.012	3.135	-3.135	0.956	0	0.78	0.575	1712	227.3	0.133	0.006	1
11.000	1.813	1.710	0.268	0.011	3.135	-3.135	1.036	0	0.78	0.604	1712	232.0	0.136	0.005	1
11.500	1.787	1.846	0.274	0.009	3.135	-3.135	1.118	0	0.78	0.633	1712	236.0	0.138	0.004	1
12.000	1.760	1.984	0.278	0.008	3.135	-3.135	1.202	0	0.78	0.662	1712	239.3	0.140	0.004	1
12.500	1.733	2.123	0.281	0.006	3.135	-3.135	1.287	0	0.78	0.692	1712	242.0	0.141	0.003	1
13.000	1.707	2.265	0.284	0.005	3.135	-3.135	1.372	0	0.78	0.723	1712	244.0	0.143	0.002	1
13.500	1.680	2.407	0.286	0.003	3.135	-3.135	1.459	0	0.78	0.754	1712	245.3	0.143	0.001	1
14.000	1.653	2.551	0.287	0.002	3.135	-3.135	1.546	0	0.78	0.785	1712	246.0	0.144	0.000	1
14.500	1.627	2.694	0.288	0.000	3.135	-3.135	1.633	0	0.78	0.816	1712	246.0	0.144	0.000	1
15.000	1.600	2.838	0.287	-0.002	3.135	-3.135	1.720	0	0.78	0.847	1712	245.2	0.143	-0.001	1
15.500	1.573	2.981	0.286	-0.003	3.135	-3.135	1.807	0	0.78	0.878	1712	243.8	0.142	-0.002	1
16.000	1.547	3.124	0.284	-0.005	3.135	-3.135	1.893	0	0.78	0.908	1712	241.8	0.141	-0.003	1
16.500	1.520	3.265	0.281	-0.006	3.135	-3.135	1.900	1	0.66	0.904	1448	202.3	0.140	-0.003	2
17.000	1.493	3.405	0.278	-0.007	3.135	-3.135	1.900	1	0.66	0.901	1448	199.8	0.138	-0.003	3
17.500	1.467	3.543	0.274	-0.007	3.135	-3.135	1.900	1	0.66	0.898	1448	197.2	0.136	-0.004	3
18.000	1.440	3.679	0.271	-0.008	3.135	-3.135	1.900	1	0.66	0.895	1448	194.5	0.134	-0.004	3
18.500	1.413	3.813	0.267	-0.008	3.135	-3.135	1.900	1	0.66	0.892	1448	191.5	0.132	-0.004	3
19.000	1.387	3.946	0.263	-0.009	3.135	-3.135	1.900	1	0.66	0.888	1448	188.5	0.130	-0.004	3
19.500	1.360	4.076	0.258	-0.009	3.135	-3.135	1.900	1	0.66	0.885	1448	185.2	0.128	-0.005	3
20.000	1.333	4.204	0.254	-0.009	3.135	-3.135	1.900	1	0.66	0.882	1448	181.8	0.126	-0.005	3
20.500	1.307	4.329	0.249	-0.010	3.135	-3.135	1.900	1	0.66	0.879	1448	178.3	0.123	-0.005	3
21.000	1.280	4.452	0.244	-0.010	3.135	-3.135	1.900	1	0.66	0.876	1448	174.5	0.121	-0.005	3
21.500	1.253	4.573	0.238	-0.011	3.135	-3.135	1.900	1	0.66	0.872	1448	170.7	0.118	-0.005	3
22.000	1.227	4.691	0.233	-0.011	3.135	-3.135	1.900	1	0.66	0.869	1448	166.6	0.115	-0.006	3
22.500	1.200	4.806	0.227	-0.012	3.135	-3.135	1.900	1	0.66	0.866	1448	162.4	0.112	-0.006	3
23.000	1.173	4.918	0.221	-0.012	3.135	-3.135	1.900	1	0.66	0.863	1448	158.1	0.109	-0.006	3
23.500	1.147	5.027	0.215	-0.012	3.135	-3.135	1.900	1	0.66	0.860	1448	153.5	0.106	-0.006	3
24.000	1.120	5.133	0.209	-0.013	3.135	-3.135	1.900	1	0.66	0.856	1448	148.9	0.103	-0.007	3
24.500	1.093	5.236	0.202	-0.013	3.135	-3.135	1.900	1	0.66	0.853	1448	144.0	0.099	-0.007	3
25.000	1.067	5.336	0.196	-0.014	3.135	-3.135	1.900	1	0.66	0.850	1448	139.0	0.096	-0.007	3

(Continued)

Table 11-5. Wall Girt Numerical Integration (Tributary Width Method). (Continued)

time (ms)	load (psi)	y (in)	v (in/ms)	a (in/ms²)	ye+ (in)	ye- (in)	R (psi)	KEY -	KLM	V (psi)	eff K (psi/in)	incr F (psi)	incr y (in)	incr v (in/ms)	Z -
25.500	1.040	5.432	0.189	-0.014	3.135	-3.135	1.900	1	0.66	0.847	1448	133.9	0.092	-0.007	3
26.000	1.013	5.524	0.181	-0.015	3.135	-3.135	1.900	1	0.66	0.844	1448	128.5	0.089	-0.007	3
26.500	0.987	5.613	0.174	-0.015	3.135	-3.135	1.900	1	0.66	0.840	1448	123.1	0.085	-0.008	3
27.000	0.960	5.698	0.166	-0.016	3.135	-3.135	1.900	1	0.66	0.837	1448	117.4	0.081	-0.008	3
27.500	0.933	5.779	0.158	-0.016	3.135	-3.135	1.900	1	0.66	0.834	1448	111.6	0.077	-0.008	3
28.000	0.907	5.856	0.150	-0.016	3.135	-3.135	1.900	1	0.66	0.831	1448	105.7	0.073	-0.008	3
28.500	0.880	5.929	0.142	-0.017	3.135	-3.135	1.900	1	0.66	0.828	1448	99.5	0.069	-0.009	3
29.000	0.853	5.998	0.133	-0.017	3.135	-3.135	1.900	1	0.66	0.824	1448	93.3	0.064	-0.009	3
29.500	0.827	6.062	0.124	-0.018	3.135	-3.135	1.900	1	0.66	0.821	1448	86.8	0.060	-0.009	3
30.000	0.800	6.122	0.115	-0.018	3.135	-3.135	1.900	1	0.66	0.818	1448	80.2	0.055	-0.009	3
30.500	0.773	6.178	0.106	-0.019	3.135	-3.135	1.900	1	0.66	0.815	1448	73.5	0.051	-0.009	3
31.000	0.747	6.228	0.097	-0.019	3.135	-3.135	1.900	1	0.66	0.812	1448	66.5	0.046	-0.010	3
31.500	0.720	6.274	0.087	-0.020	3.135	-3.135	1.900	1	0.66	0.808	1448	59.5	0.041	-0.010	3
32.000	0.693	6.315	0.077	-0.020	3.135	-3.135	1.900	1	0.66	0.805	1448	52.2	0.036	-0.010	3
32.500	0.667	6.351	0.067	-0.020	3.135	-3.135	1.900	1	0.66	0.802	1448	44.8	0.031	-0.010	3
33.000	0.640	6.382	0.057	-0.021	3.135	-3.135	1.900	1	0.66	0.799	1448	37.3	0.026	-0.011	3
33.500	0.613	6.408	0.046	-0.021	3.135	-3.135	1.900	1	0.66	0.796	1448	29.5	0.020	-0.011	3
34.000	0.587	6.429	0.035	-0.022	3.135	-3.135	1.900	1	0.66	0.792	1448	21.7	0.015	-0.011	3
34.500	0.560	6.443	0.024	-0.022	3.135	-3.135	1.900	1	0.66	0.789	1448	13.6	0.009	-0.011	3
35.000	0.533	6.453	0.013	-0.023	3.135	-3.135	1.900	1	0.66	0.786	1448	5.4	0.004	-0.011	3
35.500	0.507	6.457	0.002	-0.023	3.135	-3.135	1.900	1	0.66	0.783	1448	-2.9	-0.002	-0.012	3
36.000	0.480	6.455	-0.010	-0.020	6.457	0.186	1.899	0	0.78	0.793	1712	-12.8	-0.007	-0.010	4

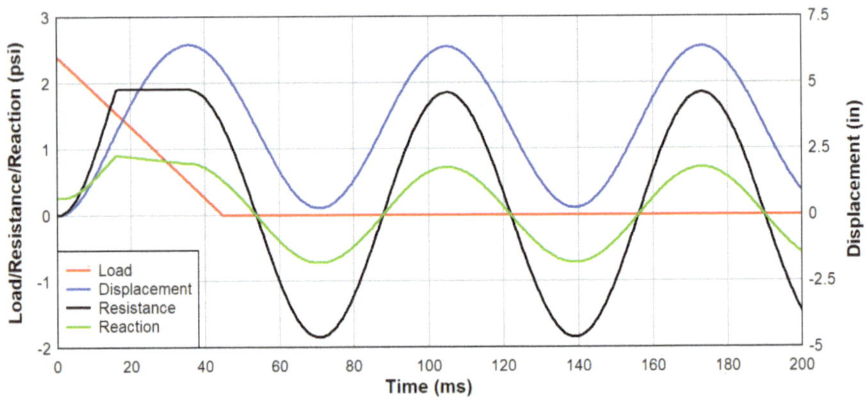

Figure 11-8. Output plots from wall girt numerical integration (tributary width method).

Ductility:

$\mu_d = y_m/y_e = (6.44 \text{ in.})/(3.14 \text{ in.}) = 2.05 < \mu_a$, **OK**

Support rotation: (Chapter 5, Figure 5-9)

$\theta_d = \tan^{-1}(y_m/0.5\,L) = \tan^{-1}[(6.44 \text{ in.})/(0.5)(240 \text{ in.})] = 3.07° < \theta_a$, **OK**

The response is adequate. A revision is not necessary.

Notes: The peak positive deflection achieved from the dynamic reaction load is approximately equal to that produced by applying the blast load by tributary area. Comparing Figures 11-8 and 11-9, it is found that the traces of deflection and resistance obtained from the two methods are nearly identical until their peak values are reached, after which an approximately 4-ms time lag develops. Dynamic reactions are not calculated in Table 11-6 or shown in Figure 11-9, as their use should be limited to a single-supported component, as discussed in Chapter 7, Section 7.3.1.

11.7.7 Check Secondary Failure Modes (In This Case, Shear)

Reaction at the ultimate resistance:

$$V_u = R_u/2 = (16{,}400\ \text{lb})/2 = 8{,}200\ \text{lb}\ (36.5\ \text{kN}) < V_n,\ \textbf{OK}$$

The shear capacity is adequate.

11.8 RIGID FRAMES

For a schematic frame elevation with acting loads from the purlins and girts, refer to Figure 11-10.

Frame spacing: B = 20 ft or 240 in. (6,100 mm)

Column bases are pinned.

Table 11-6. Wall Girt Numerical Integration (Dynamic Reaction Method).

Example:	Section 11.7 Wall Girt (Dynamic Reaction)								KEY = 0 when elastic
									KEY = +1 when positive plastic
time increment:	0.5	ms	**Blast Load**						KEY = -1 when rebound plastic
mass:	91.4	psi-ms²/in	**t0:**	0	ms	**P0:**	2.4	psi	Z = 1 when continued elastic
stiffness:	0.606	psi/in	**t1:**	45	ms	**P1:**	0	psi	Z = 2 at start of positive plastic deformation
damping:	0	psi-ms/in	**t2:**	45	ms	**P2:**	0	psi	Z = 3 when increasing plastic deformation
positive resistance:	1.90	psi	**Dynamic Reaction Factors**			**BC: S-S**			Z = 4 at end of positive plastic deformation
rebound resistance:	-1.90	psi	**Crel:**	0.39		**Cfel:**	0.11		Z = 6 at start of rebound plastic deformation
static load:	0	psi	**Crpl:**	0.38		**Cfpl:**	0.12		Z = 8 at end of rebound plastic deformation

time (ms)	load (psi)	y (in)	v (in/ms)	a (in/ms²)	ye+ (in)	ye- (in)	R (psi)	KEY -	KLM	V (psi)	eff K (psi/in)	incr F (psi)	incr y (in)	incr v (in/ms)	Z -
0.000	**0.672**	0.000	0.000	0.009	3.135	-3.135	0.000	0	0.78	--	1712	2.3	0.001	0.006	
0.500	**0.949**	0.001	0.006	0.013	3.135	-3.135	0.001	0	0.78	--	1712	8.4	0.005	0.009	1
1.000	**1.695**	0.006	0.015	0.024	3.135	-3.135	0.004	0	0.78	--	1712	18.8	0.011	0.015	1
1.500	**2.657**	0.017	0.030	0.037	3.135	-3.135	0.010	0	0.78	--	1712	34.3	0.020	0.021	1
2.000	**3.233**	0.037	0.051	0.045	3.135	-3.135	0.023	0	0.78	--	1712	53.0	0.031	0.022	1
2.500	**3.227**	0.068	0.073	0.045	3.135	-3.135	0.041	0	0.78	--	1712	72.1	0.042	0.022	1
3.000	**3.220**	0.110	0.095	0.044	3.135	-3.135	0.067	0	0.78	--	1712	90.7	0.053	0.021	1
3.500	**2.927**	0.163	0.116	0.040	3.135	-3.135	0.099	0	0.78	--	1712	107.5	0.063	0.018	1
4.000	**2.397**	0.226	0.134	0.032	3.135	-3.135	0.137	0	0.78	--	1712	121.0	0.071	0.014	1
4.500	**1.799**	0.297	0.148	0.023	3.135	-3.135	0.180	0	0.78	--	1712	130.8	0.076	0.010	1
5.000	**1.319**	0.373	0.157	0.015	3.135	-3.135	0.226	0	0.78	--	1712	137.6	0.080	0.007	1
5.500	**1.108**	0.454	0.164	0.012	3.135	-3.135	0.275	0	0.78	--	1712	142.9	0.084	0.006	1
6.000	**1.227**	0.537	0.170	0.013	3.135	-3.135	0.326	0	0.78	--	1712	148.6	0.087	0.008	1
6.500	**1.627**	0.624	0.178	0.018	3.135	-3.135	0.378	0	0.78	--	1712	156.3	0.091	0.010	1
7.000	**2.168**	0.715	0.188	0.024	3.135	-3.135	0.433	0	0.78	--	1712	166.6	0.097	0.014	1
7.500	**2.662**	0.813	0.202	0.030	3.135	-3.135	0.492	0	0.78	--	1712	179.4	0.105	0.016	1
8.000	**2.938**	0.918	0.218	0.033	3.135	-3.135	0.556	0	0.78	--	1712	193.4	0.113	0.016	1
8.500	**2.895**	1.031	0.234	0.032	3.135	-3.135	0.624	0	0.78	--	1712	206.7	0.121	0.014	1
9.000	**2.540**	1.151	0.249	0.026	3.135	-3.135	0.698	0	0.78	--	1712	217.6	0.127	0.011	1
9.500	**1.980**	1.278	0.259	0.017	3.135	-3.135	0.775	0	0.78	--	1712	224.8	0.131	0.006	1
10.000	**1.391**	1.410	0.265	0.008	3.135	-3.135	0.854	0	0.78	--	1712	228.1	0.133	0.002	1

(Continued)

Table 11-6. Wall Girt Numerical Integration (Dynamic Reaction Method). (Continued)

time	load	y	v	a	ye+	ye-	R	KEY	KLM	V	eff K	incr F	incr y	incr v	Z
(ms)	(psi)	(in)	(in/ms)	(in/ms^2)	(in)	(in)	(psi)	-		(psi)	(psi/in)	(psi)	(in)	(in/ms)	-
10.500	**0.959**	1.543	0.267	0.000	3.135	-3.135	0.935	0	0.78	--	1712	228.6	0.134	-0.001	1
11.000	**0.818**	1.677	0.267	-0.003	3.135	-3.135	1.016	0	0.78	--	1712	227.7	0.133	-0.001	1
11.500	**1.006**	1.810	0.266	-0.001	3.135	-3.135	1.097	0	0.78	--	1712	227.4	0.133	0.001	1
12.000	**1.451**	1.942	0.266	0.004	3.135	-3.135	1.177	0	0.78	--	1712	229.2	0.134	0.004	1
12.500	**1.998**	2.076	0.270	0.010	3.135	-3.135	1.258	0	0.78	--	1712	233.5	0.136	0.007	1
13.000	**2.458**	2.213	0.276	0.016	3.135	-3.135	1.341	0	0.78	--	1712	240.0	0.140	0.008	1
13.500	**2.670**	2.353	0.285	0.017	3.135	-3.135	1.426	0	0.78	--	1712	247.1	0.144	0.008	1
14.000	**2.555**	2.497	0.293	0.015	3.135	-3.135	1.513	0	0.78	--	1712	253.1	0.148	0.006	1
14.500	**2.143**	2.645	0.298	0.008	3.135	-3.135	1.603	0	0.78	--	1712	256.1	0.150	0.001	1
15.000	**1.560**	2.795	0.300	-0.002	3.135	-3.135	1.694	0	0.78	--	1712	255.3	0.149	-0.003	1
15.500	**0.991**	2.944	0.296	-0.011	3.135	-3.135	1.784	0	0.78	--	1712	250.8	0.147	-0.007	1
16.000	**0.614**	3.091	0.289	-0.018	3.135	-3.135	1.873	0	0.78	--	1712	243.5	0.142	-0.009	1
16.500	**0.544**	3.233	0.280	-0.022	3.135	-3.135	1.900	1	0.66	--	1448	198.7	0.137	-0.010	2
17.000	**0.797**	3.370	0.270	-0.018	3.135	-3.135	1.900	1	0.66	--	1448	192.3	0.133	-0.007	3
17.500	**1.279**	3.503	0.262	-0.010	3.135	-3.135	1.900	1	0.66	--	1448	188.7	0.130	-0.003	3
18.000	**1.824**	3.633	0.260	-0.001	3.135	-3.135	1.900	1	0.66	--	1448	188.1	0.130	0.001	3
18.500	**2.241**	3.763	0.261	0.006	3.135	-3.135	1.900	1	0.66	--	1448	189.9	0.131	0.003	3
19.000	**2.386**	3.894	0.264	0.008	3.135	-3.135	1.900	1	0.66	--	1448	192.4	0.133	0.003	3
19.500	**2.200**	4.027	0.267	0.005	3.135	-3.135	1.900	1	0.66	--	1448	194.0	0.134	0.001	3
20.000	**1.738**	4.161	0.268	-0.003	3.135	-3.135	1.900	1	0.66	--	1448	192.9	0.133	-0.004	3
20.500	**1.142**	4.294	0.264	-0.013	3.135	-3.135	1.900	1	0.66	--	1448	188.4	0.130	-0.009	3
21.000	**0.600**	4.425	0.256	-0.022	3.135	-3.135	1.900	1	0.66	--	1448	180.8	0.125	-0.012	3
21.500	**0.283**	4.549	0.243	-0.027	3.135	-3.135	1.900	1	0.66	--	1448	171.4	0.118	-0.013	3
22.000	**0.286**	4.668	0.230	-0.027	3.135	-3.135	1.900	1	0.66	--	1448	162.0	0.112	-0.012	3
22.500	**0.599**	4.780	0.218	-0.022	3.135	-3.135	1.900	1	0.66	--	1448	154.4	0.107	-0.009	3
23.000	**1.111**	4.886	0.209	-0.013	3.135	-3.135	1.900	1	0.66	--	1448	149.7	0.103	-0.004	3
23.500	**1.643**	4.990	0.205	-0.004	3.135	-3.135	1.900	1	0.66	--	1448	148.0	0.102	-0.001	3
24.000	**2.012**	5.092	0.204	0.002	3.135	-3.135	1.900	1	0.66	--	1448	148.4	0.102	0.001	3
24.500	**2.086**	5.195	0.206	0.003	3.135	-3.135	1.900	1	0.66	--	1448	149.2	0.103	0.000	3
25.000	**1.832**	5.298	0.206	-0.001	3.135	-3.135	1.900	1	0.66	--	1448	148.5	0.103	-0.003	3
25.500	**1.326**	5.400	0.203	-0.010	3.135	-3.135	1.900	1	0.66	--	1448	145.0	0.100	-0.007	3
26.000	**0.726**	5.500	0.196	-0.019	3.135	-3.135	1.900	1	0.66	--	1448	138.0	0.095	-0.012	3
26.500	**0.220**	5.596	0.184	-0.028	3.135	-3.135	1.900	1	0.66	--	1448	128.2	0.089	-0.015	3
27.000	**-0.033**	5.684	0.169	-0.032	3.135	-3.135	1.900	1	0.66	--	1448	116.9	0.081	-0.016	3
27.500	**0.042**	5.765	0.154	-0.031	3.135	-3.135	1.900	1	0.66	--	1448	106.1	0.073	-0.014	3
28.000	**0.411**	5.838	0.140	-0.025	3.135	-3.135	1.900	1	0.66	--	1448	97.3	0.067	-0.010	3
28.500	**0.944**	5.905	0.130	-0.016	3.135	-3.135	1.900	1	0.66	--	1448	91.6	0.063	-0.006	3
29.000	**1.455**	5.969	0.124	-0.007	3.135	-3.135	1.900	1	0.66	--	1448	88.7	0.061	-0.002	3
29.500	**1.768**	6.030	0.122	-0.002	3.135	-3.135	1.900	1	0.66	--	1448	87.6	0.060	-0.001	3
30.000	**1.770**	6.090	0.120	-0.002	3.135	-3.135	1.900	1	0.66	--	1448	86.5	0.060	-0.002	3
30.500	**1.452**	6.150	0.118	-0.007	3.135	-3.135	1.900	1	0.66	--	1448	83.6	0.058	-0.006	3
31.000	**0.910**	6.208	0.112	-0.016	3.135	-3.135	1.900	1	0.66	--	1448	77.6	0.054	-0.011	3
31.500	**0.314**	6.261	0.101	-0.026	3.135	-3.135	1.900	1	0.66	--	1448	68.2	0.047	-0.015	3
32.000	**-0.148**	6.309	0.086	-0.034	3.135	-3.135	1.900	1	0.66	--	1448	56.2	0.039	-0.018	3
32.500	**-0.332**	6.347	0.069	-0.037	3.135	-3.135	1.900	1	0.66	--	1448	43.1	0.030	-0.018	3
33.000	**-0.187**	6.377	0.051	-0.035	3.135	-3.135	1.900	1	0.66	--	1448	30.9	0.021	-0.016	3
33.500	**0.231**	6.398	0.035	-0.028	3.135	-3.135	1.900	1	0.66	--	1448	21.0	0.014	-0.012	3
34.000	**0.775**	6.413	0.024	-0.019	3.135	-3.135	1.900	1	0.66	--	1448	14.2	0.010	-0.007	3
34.500	**1.257**	6.423	0.016	-0.011	3.135	-3.135	1.900	1	0.66	--	1448	10.1	0.007	-0.004	3
35.000	**1.509**	6.430	0.012	-0.006	3.135	-3.135	1.900	1	0.66	--	1448	7.4	0.005	-0.004	3
35.500	**1.439**	6.435	0.008	-0.008	3.135	-3.135	1.900	1	0.66	--	1448	4	0.003	-0.005	3
36.000	**1.061**	6.438	0.003	-0.014	3.135	-3.135	1.900	1	0.66	--	1448	-1	-0.001	-0.009	3
36.500	**0.492**	6.437	-0.006	-0.020	6.438	0.167	1.900	0	0.78	--	1712	-10	-0.006	-0.012	4
37.000	**-0.091**	6.431	-0.018	-0.028	6.438	0.167	1.896	0	0.78	--	1712	-22	-0.013	-0.015	1

Allowable sidesway: $X_a = 5.49$ in. (139 mm) (Section 11.3.4)

Initial member sizes for the rigid frame will be estimated using an SDOF approximation of the frame. These estimated sizes will be used in the more detailed MDOF frame analysis to verify adequacy. Maximum deflection of individual members, as well as frame sidesway, will be used to evaluate the adequacy of the selected members.

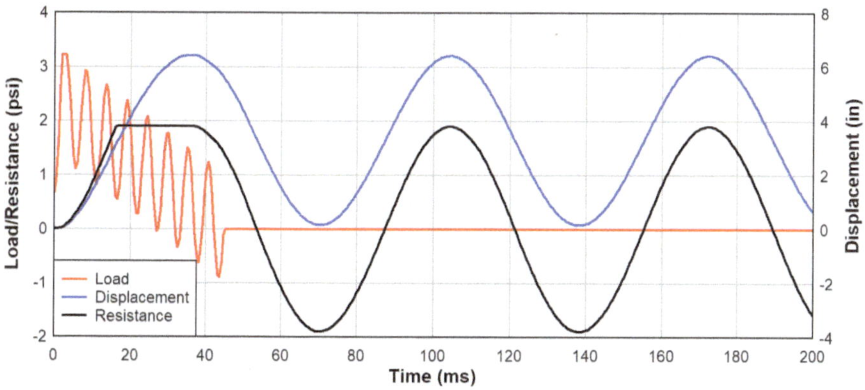

Figure 11-9. Output plots from wall girt numerical integration (dynamic reaction method).

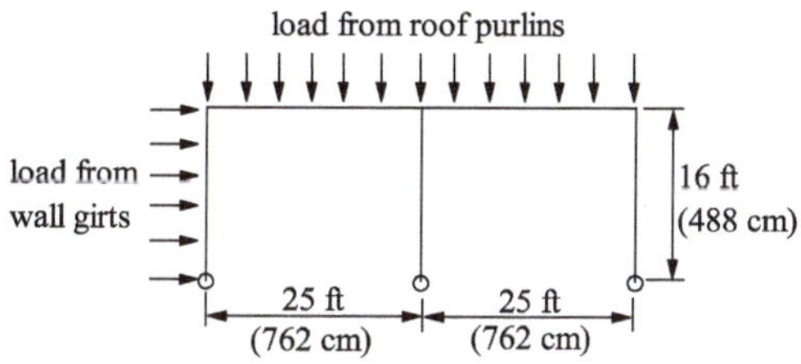

Figure 11-10. Frame elevation.

11.8.1 Dynamic Material Properties

For dynamic bending and shear: (Chapter 5, Appendix 5A)

$$F_{ds} = F_{dy} = (\text{SIF})(\text{DIF})\ F_y = (1.1)(1.19)(50\ \text{ksi}) = 65.45\ \text{ksi}\ (451\ \text{MPa})$$

11.8.2 Trial Sizing (General)

Use the tributary width method for applying the load to the frame for initial sizing (refer to Chapter 6, Section 6.2.3). The tributary width will be equal to the frame spacing of 20 ft (6,100 mm). Blast loads will consist of the externally applied roof and wall loads (refer to Section 11.3.3).

The natural periods of the frame and girts/purlins are sufficiently different to make this simplification valid. Girt response would not be significantly affected by including interaction, but column flexural response would probably be less. Dynamic reactions from the girts and purlins will have a lower pressure than the blast load but may have a longer duration. Because of the relatively slow global response of the frame, this mode is assumed to be out of phase with the dynamic response of the roof beams causing a compression load in the adjoining columns, and it is assumed that sidesway will be controlled by impulse and will not be significantly affected by load shape.

In addition, only the reflected blast load on the blastward side of the building is included in this example problem. A net blast lateral blast load including a "wrap-around" positive pressure on the leeward side of the building could be considered because the relatively long response time of the building in sidesway may cause the peak sidesway response to occur after the onset of the wrap-around pressure on the leeward side of the building.

11.8.3 Trial Sizing (Column)

To get an initial column size, treat it as a fixed-pinned member supported at the floor and eave.

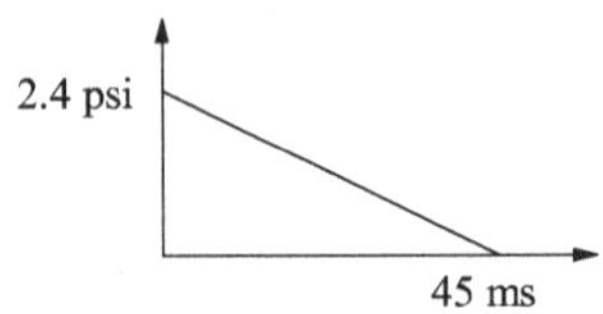

Response limits: $\mu_a = 2$, $\theta_a = 1.5$ degrees, $X_a = 5.49$ in. (Chapter 5, Table 5B-3)

Dead load: $DL = 0$

Blast load (front wall load):

$BL = 2.4$ psi (16.5 kPa)

Let the ductility demand equal the limiting value, $\mu_d = \mu_a = 2$.

As an initial guess, let $\tau = t_d/t_n = 1$.

Load/resistance ratio: [Chapter 6, Equation (6-11)]

$$Fo / R_m = \frac{\sqrt{(2\mu_d - 1)}}{\pi\tau} + \frac{(2\mu_d - 1)\tau}{2\mu_d\,(\tau + 0.7)} = \frac{\sqrt{(2\times 2 - 1)}}{\pi \times 1} + \frac{(2\times 2 - 1)\times 1}{2\times 2\times(1 + 0.7)} = 0.993$$

Peak load: $F_o = BL = 2.4$ psi (16.5 kPa)

Resistance:

$R_m = F_o/0.993 = (2.4\text{ psi})/0.993 = 2.42$ psi (16.7 kPa)

Effective ultimate moment, from $R_b = 4\,(M_{ps} + 2\,M_{pc})/L$: (Chapter 6, Table 6-3)

$M_p = R_m\,L/12 = (2.42\text{ psi} \times 240\text{ in.} \times 192\text{ in.})(192\text{ in.})/12$
$= 1{,}783{,}000$ in.·lb (201,400 N·m)

Section modulus (for selection purposes):

$Z_x = M_p/F_{ds} = (1{,}783{,}000\text{ in.·lb})/(65{,}450\text{ psi}) = 27.2\text{ in.}^3$ ($446{,}000\text{ mm}^3$)

Select W10 × 30: (AISC Manual)

$I_x = 170\text{ in.}^4$ ($70.8 \times 10^6\text{ mm}^4$) $I_y = 16.7\text{ in.}^4$ ($6.95 \times 10^6\text{ mm}^4$)
$S_x = 32.4\text{ in.}^3$ ($531 \times 10^3\text{ mm}^3$) $S_y = 5.75\text{ in.}^3$ ($94.2 \times 10^3\text{ mm}^3$)
$Z_x = 36.6\text{ in.}^3$ ($600 \times 10^3\text{ mm}^3$) $Z_y = 8.84\text{ in.}^3$ ($145 \times 10^3\text{ mm}^3$)

$A_s = 8.84 \text{ in.}^2$ (5,700 mm²)
A_v (area of web) $= dt_w =$ (10.5 in.)(0.3 in.) = 3.15 in.² (2,030 mm²)

Note: Although a smaller W10 section can be specified for the columns to better match the target Z_x value, a W10 × 30 is used to match the section required for the beams (as calculated in Section 11.8.4).

Notes: Columns and girders of rigid frames are considered primary structural members. To achieve higher levels of ductility, frame sections should be selected in accordance with AISC seismic requirements. Refer to Section 10.10.4 for an example.

Moment capacity: (AISC 360, Equation F2-1)

$$M_p = Z_x F_{ds} = (36.6 \text{ in.}^3)(65{,}450 \text{ psi}) = 2{,}395{,}000 \text{ in.·lb } (270{,}700 \text{ N·m})$$

Check the sidesway response:

The columns are simply supported at the base and continuous at the beam-column connection. This configuration can be modeled as three cantilevered columns acting in parallel with a concentrated load and mass at the tip, as shown schematically in Figure 11-11.

The tributary area for the sidesway load is equal to ½ the wall base to eave height. Use the wall blast load impulse for the forcing function.

Impulse:

$$I_o = (2.4 \text{ psi})(192 \text{ in.}/2)(240 \text{ in.})(45 \text{ ms})/2 = 1{,}240{,}000 \text{ lb-ms } (5{,}530 \text{ kN·ms})$$

Single column resistance:

$$R_u = M_p/L = (2{,}395{,}000 \text{ in.·lb})/(192 \text{ in.}) = 12{,}500 \text{ lb } (55.5 \text{ kN})$$

Total resistance = (3 columns)(12,500 lb) = 37,400 lb (166 kN)
Single column stiffness: (AISC Manual, Beam Diagrams)

$$\begin{aligned} K_e &= 3\, E_s\, I_x/L^3 \\ &= 3\,(29{,}000{,}000 \text{ psi})(170 \text{ in.}^3)/(192 \text{ in.})^3 \\ &= 2{,}090 \text{ lb/in. } (366 \text{ kN/m}) \end{aligned}$$

Total stiffness = (3 columns)(2,090 lb/in.) = 6,270 lb/in. (1,100 kN/m)
Sidesway elastic deflection:

$$X_e = R_u/K_e = (37{,}400 \text{ lb})/(6{,}270 \text{ lb/in.}) = 5.97 \text{ in. } (152 \text{ mm})$$

Mass:

Use 1/3 mass of walls + mass of roof.
weight = [(2 walls)(16 ft height/3) + (50 ft roof span)] (20 ft width)(say 6 psf)
= 7,280 lb (32.4 kN)

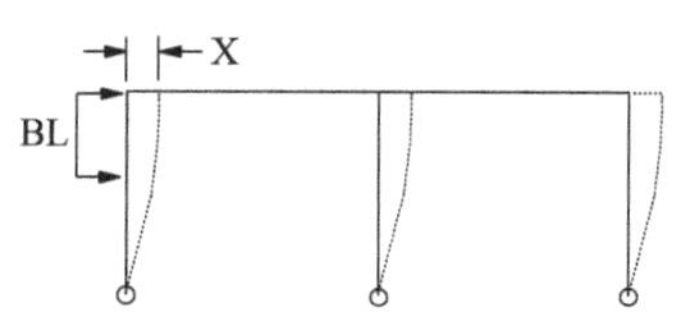

Figure 11-11. Schematic of three columns working in parallel.

M = weight/g
= (7,280 lb)/(386 in./s^2)
= 18.9 lb-s^2/in., or 18,900,000 lb-ms^2/in. (3,300 kg/m^2)

Because the load can be treated as an impulse, use an energy balance method to determine the response.

Kinetic energy of the load: (Biggs, Chapter 5, Section 5.5)

$KE = (I_o)^2/2\ M_e$
= (1,240,000 lb-ms)^2/(2)(18,900,000 lb-ms^2/in.)
= 41,000 lb-in. (4,640 N·m)

Strain energy of the frame, where X is the sidesway at the top of the frame:

$SE = R_u\ (X_e)/2 + R_u\ (X_m - X_e)$

Rearranging and substituting KE for SE,

$X_m = [KE + 0.5\ R_u\ X_e]/R_u$
= [(41,000 lb-in.) + (0.5)(37,400 lb)(5.97 in.)]/(37,400 lb)
= 4.08 in. (104 mm) $< X_a$, **OK**

11.8.4 Trial Sizing (Beam)

To obtain an initial beam size, treat it as a fixed-pinned member supported at the eave and center support.

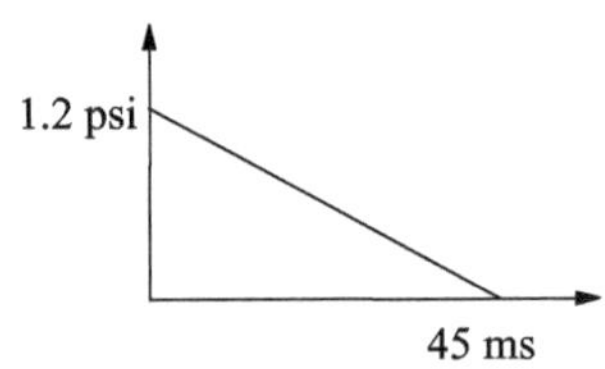

Response limits: $\mu_a = 2$, $\theta_a = 1.5$ degrees

Dead load:

DL = 11 psf, or 0.08 psi (0.53 kPa)

Blast load:

BL = 1.2 psi (8.27 kPa)

Impulse: $I_o = BL \times t_d/2$ = (1.2 psi)(45 ms)/2 = 27 psi·ms (186 kPa·ms)

Let the ductility demand equal the limiting value, $\mu_d = \mu_a = 2$.

As an initial guess, let $\tau = t_d/t_n = 1$.

Load/resistance ratio: [Equation (6-11)]

$$Fo / R_m = \frac{\sqrt{(2\mu_d - 1)}}{\pi\tau} + \frac{(2\mu_d - 1)\tau}{2\mu_d(\tau + 0.7)} = \frac{\sqrt{(2\times2-1)}}{\pi\times1} + \frac{(2\times2-1)\times1}{2\times2\times(1+0.7)} = 0.993$$

Peak load: $F_o = DL + BL$ = 0.08 psi + 1.2 psi = 1.28 psi (8.80 kPa)

Resistance:

$R_m = F_o/0.993 = (1.28 \text{ psi})/0.993 = 1.29 \text{ psi } (8.87 \text{ kPa})$

Effective ultimate moment, from $R_b = 4\,(M_{ps} + 2\,M_{pc})/L$: (Table 6-3)

$$\begin{aligned} M_p &= R_m\,L/12 \\ &= (1.29 \text{ psi} \times 240 \text{ in.} \times 300 \text{ in.})(300 \text{ in.})/12 \\ &= 2{,}315{,}000 \text{ in.·lb } (261{,}500 \text{ N·m}) \end{aligned}$$

Section modulus (for selection purposes):

$Z_x = M_p/F_{ds} = (2{,}315{,}000 \text{ in.·lb})/(65{,}450 \text{ psi}) = 35.4 \text{ in.}^3\ (580{,}000 \text{ mm}^3)$
Select W10 × 30: (AISC Manual)

$I_x = 170 \text{ in.}^4\ (70.8 \times 10^6 \text{ cm}^4)$
$S_x = 32.4 \text{ in.}^3\ (531 \times 10^3 \text{ mm}^3)$
$Z_x = 36.6 \text{ in.}^3\ (600 \times 10^3 \text{ mm}^3)$
$A_s = 8.84 \text{ in.}^2\ (5{,}700 \text{ mm}^2)$
A_v (area of web) $= 3.15 \text{ in.}^2\ (2{,}030 \text{ mm}^2)$

It is advisable to perform a dynamic analysis of the columns and beams as isolated components to verify that flexural responses are acceptable before analyzing the entire frame because these SDOF runs are much quicker to perform than a frame analysis. These SDOF steps are not shown in this solution. Frame response will include the effects of axial loads that reduce the flexural capacity of members. This will increase the response with respect to component analysis.

11.8.5 Multi-Degree-of-Freedom Analysis (Model)

A multi-degree-of-freedom (MDOF) plane frame analysis program is used to determine the response of the frame to the blast loads applied over a tributary area. The structure is discretized into elements, and loads are applied to nodes. Input includes nodal coordinates, nodal mass, member connectivity, member properties, supports, and solution control parameters (e.g., time step, duration of numerical integration). Dynamic material properties are input into the program by defining the elastic and ultimate yield strengths and associated strain or by defining member capacities. The latter approach will be used for this problem. Nonlinear response will be included in the analysis through the tracking of plastic hinge formation and reformulation of the stiffness matrix at each time point.

This analysis is similar to a conventional static analysis, with the exception of nonlinear member properties and pressure-time loadings. Member adequacy is judged by maximum deflection, support rotation, and/or the amount of strain achieved rather than the member stress criteria used in static design.

Output includes node displacements, member end forces, and support reactions. A three-dimensional model would produce more accurate results, but a two-dimensional analysis is normally sufficient for this type of structure. Members will be subjected to loads from both the long and the short walls. The member capacity used in the model or the allowable deformation must be limited to account

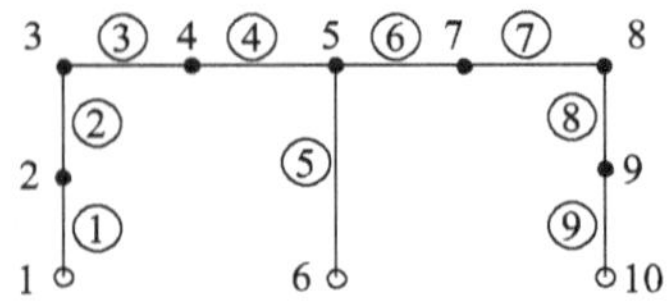

Figure 11-12. MDOF frame model.

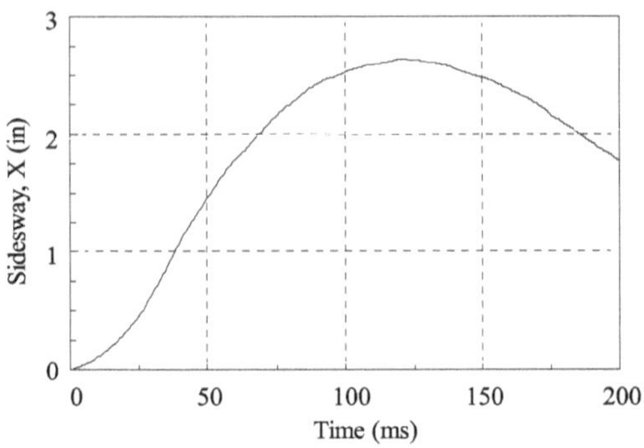

Figure 11-13. Plot of Node 3 sidesway response.

for the fact that the members will be subjected to simultaneous biaxial loading but checked for each direction separately. A typical capacity reduction factor is 25%. This factor reflects the fact that peak stresses from each direction rarely occur at the same time. This is a simpler alternative to combining the time histories in two directions.

The MDOF frame model with nodes and members (elements) labeled is shown in Figure 11-12.

Columns: W10 × 30 Beams: W10 × 30

A single load case was run, wherein the blast loads were applied over a tributary area. The results of the analysis are indicated in the following section. For the corresponding output plot, see Figure 11-13.

11.8.6 Multi-Degree-of-Freedom Analysis (Tributary Area Loading)

Maximum sidesway (Node 3):

$X_m = 2.64$ in. (67.1 mm) $< X_a$, **OK**

A dynamic analysis of the columns and beams as isolated components was performed (not included for the sake of brevity), and it showed adequate response with relatively low ductility values and support rotations for all members with the selected section size.

Peak support reactions:

Horizontal reaction = 18.8 kips (83.6 kN) at Node 1
Vertical reaction = 96.4 kips (429 kN) at Node 6

Shear capacity: (AISC 360, Equation G2-1)

$V_n = 0.6\ A_v\ F_{ds}\ C_v = 0.6\ (3.15\ \text{in.}^2)(65.45\ \text{ksi})(1.0) = 124\ \text{kips}\ (550\ \text{kN})$

Reaction at the ultimate resistance:

$V_u = 18.8$ kips (83.6 kN) $< V_n$, **OK**

The shear capacity is adequate.

Notes: The frame can alternatively be analyzed by using dynamic reactions. With the low ductility and support rotation values achieved considering tributary loading, the use of less-robust sections can be investigated.

11.8.7 Additional Considerations

Connections—Design to develop the ultimate strength of the members being connected to ensure ductile behavior in the member as assumed in the analysis. Peak member end forces from the frame analysis can be used; however, this can be extremely conservative because the peak force occurs for only a short time.

Bracing—Members should be braced in accordance with AISC specification requirements to develop the full moment capacity of each member. Both flanges should be braced to accommodate rebound forces.

11.9 BRACED FRAMES

The lateral load–resisting system for blast loads along the short side of the building is shown schematically in Figure 11-14.

Loads applied to panels on the short wall will be resisted by the three end columns. The roof panels will act as a diaphragm to distribute the loads, but they must also resist vertical blast loads in bending, which reduces in-plane capacity. To avoid this problem, the top of the center column will be supported by a truss in the roof spanning between the end bays. This truss will utilize the rigid frame beams as chord members with additional angles added to form the struts. Braced

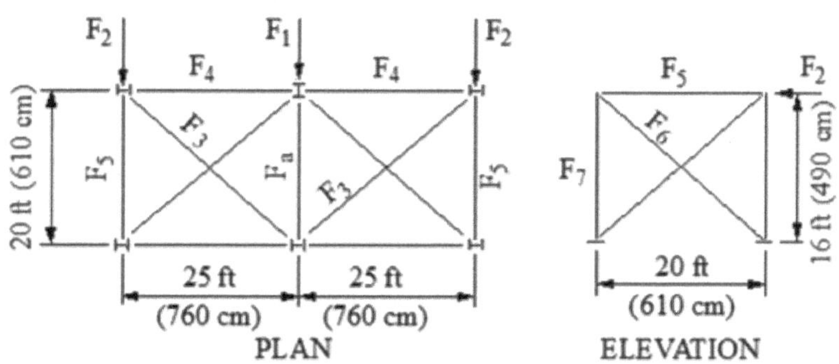

Figure 11-14. Braced frame layout.

frames in the end bay wall will resist the support reaction for the roof truss as well as the load from the corner columns. The end bay braced frame will consist of the rigid frame columns and x-bracing. Because the columns must resist loads from both directions, the axial capacity in each direction is artificially reduced for the analysis.

The braced frame will be designed using a static design process based on the capacity of supported members. Bracing provides a stiff system that responds to pressure without absorbing much energy.

11.9.1 Dynamic Material Properties

For dynamic bending and shear: (Chapter 5, Appendix 5A)

Grade 36: $F_{ds} = F_{dy}$ = (SIF)(DIF) F_y = (1.1)(1.29)(36 ksi) = 51.1 ksi (352 MPa)
Grade 50: $F_{ds} = F_{dy}$ = (SIF)(DIF) F_y = (1.1)(1.19)(50 ksi) = 65.45 ksi (451 MPa)

For dynamic tension and compression: (Chapter 5, Appendix 5A)

Grade 36: $F_{ds} = F_{dy}$ = (SIF)(DIF) F_y = (1.1)(1.19)(36 ksi) = 47.1 ksi (325 MPa)
Grade 50: $F_{ds} = F_{dy}$ = (SIF)(DIF) F_y = (1.1)(1.12)(50 ksi) = 61.6 ksi (425 MPa)

11.9.2 Determine Blast Load

The braced frame must develop the ultimate capacity of the members that it supports, namely, the girts and end wall columns. The force applied to the top of the column is equal to the tributary area times the resistance as a static load. Each braced frame will be designed to resist the entire load, even though there will be a frame at each end of the building. This will provide redundancy and will eliminate large axial forces in the top perimeter beams at the interior frames.

1. Load based on center column capacity

   Column ultimate moment (strong axis bending): (Section 11.8.3)
   M_p = 2,395,000 in.·lb (270,200 N·m)

   Column resistance: (Chapter 6, Table 6-3)

   $R_b = 4\,(M_{ps} + 2\,M_{pc})/L = 12\,M_p/L$
   = 12 (2,395,000 in.·lb)/(192 in.)
   = 150,000 lb (666 kN)

   Column unit resistance:

   $R_u = R_b/(25$ ft width)(16 ft height)
   = (150,000 lb)/(300 in.)(192 in.)
   = 2.60 psi (17.9 kPa)

2. Load based on corner column capacity

   Column ultimate moment (weak axis bending): (AISC 360, Equation F2-1)

   $M_p = Z_x\,F_{ds}$ = (8.84 in.^{3})(65,450 psi) = 578,600 in.·lb (65,370 N·m)

Column resistance: (Chapter 6, Table 6-3)

$$\begin{aligned} R_b &= 4\,(M_{ps} + 2\,M_{pc})/L = 12\,M_p/L \\ &= 12\,(578{,}600\text{ in.·lb})/(192\text{ in.}) \\ &= 36{,}200\text{ lb }(161\text{ kN}) \end{aligned}$$

Column unit resistance:

$$\begin{aligned} R_u &= R_b/(25\text{ ft}/2\text{ width})(16\text{ ft height}) \\ &= (36{,}200\text{ lb})/(300\text{ in.})(192\text{ in.}) \\ &= 1.26\text{ psi }(8.66\text{ kPa}) \end{aligned}$$

3. Load based on girt capacity

Girt ultimate moment: (Section 11.7.3)

$$M_p = 492{,}000\text{ in.·lb }(55{,}600\text{ N·m})$$

Girt resistance: (Chapter 6, Table 6-1)

$$\begin{aligned} R_b &= 8\,M_{pc}/L = 8\,M_p/L \\ &= 8\,(492{,}000\text{ in.·lb})/(300\text{ in.}) \\ &= 13{,}100\text{ lb }(58.4\text{ kN}) \end{aligned}$$

Girt unit resistance:

$$\begin{aligned} R_u &= R_b/(3\text{ ft width})(25\text{ ft length}) \\ &= (13{,}100\text{ lb})/(36\text{ in.})(300\text{ in.}) \\ &= 1.21\text{ psi }(8.37\text{ kPa}) \end{aligned}$$

The smaller girt controls the capacity, as it will fail first. Design bracing members by controlling the failure element capacity with an increase factor of 1.25 as per common capacity design practice. Therefore, use **$R_u = 1.25 \times 1.21$ psi = 1.52 psi (10.5 kPa)**.

11.9.3 Braced Frame Forces

Because of differences in stiffness, compression forces will be neglected in cross members.

Load on top of the center column:

$$\begin{aligned} F_1 &= R_u\,(25\text{ ft width})(16\text{ ft height}/2) \\ &= (1.52\text{ psi})(300\text{ in.})(96\text{ in.}) \\ &= 43{,}700\text{ lb }(195\text{ kN}) \end{aligned}$$

Load at the top of the corner columns:

$$\begin{aligned} F_2 &= R_u\,(25\text{ ft width}/2)(16\text{ ft height}/2) \\ &= (1.52\text{ psi})(150\text{ in.})(96\text{ in.}) \\ &= 21{,}900\text{ lb }(97.3\text{ kN}) \end{aligned}$$

Angle of the roof brace:

$$\alpha_1 = \tan^{-1}(25\text{ ft}/20\text{ ft}) = 51.3\text{ degrees}$$

Tension force in the roof brace:

$$F_3 = \tfrac{1}{2} F_1/\cos(\alpha_1)$$
$$= 1/2\ (43{,}700\ \text{lb})/\cos(51.3\ \text{degrees})$$
$$= 35{,}000\ \text{lb}\ (156\ \text{kN})$$

Axial force in the roof truss chord member:

$$F_4 = F_3 \sin(\alpha_1)$$
$$= (46{,}800\ \text{lb}) \sin(51.3\ \text{degrees})$$
$$= 27{,}300\ \text{lb}\ (122\ \text{kN})$$

Load in the top perimeter member along the long wall:

$$F_5 = F_2 + 1/2\ F_1$$
$$= (21{,}900\ \text{lb}) + \tfrac{1}{2}\ (43{,}700\ \text{lb})$$
$$= 43{,}700\ \text{lb}\ (195\ \text{kN})$$

Angle of the vertical brace:

$$\alpha_2 = \tan^{-1}(16\ \text{ft}/20\ \text{ft}) = 38.7\ \text{degrees}$$

Load in the vertical brace will include contribution from the roof truss:

$$F_6 = F_5/\cos(\alpha_2)$$
$$= (43{,}700\ \text{lb})/\cos(38.7\ \text{degrees})$$
$$= 56{,}000\ \text{lb}\ (249\ \text{kN})$$

Load in the column owing to braced frame response:

$$F_7 = F_6 \sin(\alpha_2)$$
$$= (56{,}000\ \text{lb}) \sin(38.7\ \text{degrees})$$
$$= 35{,}000\ \text{lb}\ (156\ \text{kN})$$

11.9.4 Design Top Perimeter Members

Controlling applied load: (Section 11.9.3)

$$P_u = F_5 = 43.7\ \text{kips}\ (195\ \text{kN})$$

Try a W10 × 19: (AISC Manual)

$I_x = 96.3$ in.^{4} (40.1×10^6 mm^4) $I_y = 4.29$ in.^{4} (1.79×10^6 mm^4)
$r_x = 4.14$ in. (105 mm) $r_y = 0.874$ in. (22.2 mm)
$A_s = 5.62$ in.^{2} (3,630 mm^2)

For compression, capacity is determined by buckling.

A brace will be provided at midspan for the weak axis. Therefore, the unbraced lengths are as follows:

$L_x = 20$ ft or 240 in. (6,100 mm)
$L_y = 20$ ft/2 or 120 in. (3,050 mm)

Use effective length factor, $K = 1.0$.

$K_x L_x/r_x = (1.0)(240 \text{ in.})/(4.14 \text{ in.}) = 58$
$K_y L_y/r_y = (1.0)(120 \text{ in.})/(0.874 \text{ in.}) = \mathbf{137}$ **[CONTROLS]**

Column slenderness factor: (AISC 360, Section E3)

$$\lambda_c = 4.71\sqrt{\frac{E_s}{F_{dy}}} = 4.71\sqrt{\frac{(29{,}000 \text{ ksi})}{(61.6 \text{ ksi})}} = 102 < 137$$

Elastic critical buckling stress: (AISC 360, Equation E3-4)

$F_e' = \pi^2 E_s/(K_y L_y/r_y)^2 = \pi^2 (29{,}000 \text{ ksi})/(137)^2 = 15.2 \text{ ksi } (105 \text{ MPa})$

Flexural buckling stress: (AISC 360, Equation E3-3)

$F_{cr} = 0.877 \ = 0.877\,(15.2 \text{ ksi}) = 13.3 \text{ ksi } (91.8 \text{ MPa})$

Required area: (AISC 360, Equation E3-1)

$A_n = P_u/F_{cr}$
$= (43.7 \text{ kips})/(13.3 \text{ ksi})$
$= 3.28 \text{ in.}^2 \ (2{,}120 \text{ mm}^2) < A_s$, **OK**

USE W10 × 19.

11.9.5 Design Cross Bracing

Controlling applied load: (Section 11.9.3)

$P_u = F_6 = 56.0 \text{ kips } (249 \text{ kN})$

Required area for tension: (AISC 360, Equation D2-1)

$A_n = P_n/F_y = P_u/F_{ds}$
$= (56.0 \text{ kips})/(47.1 \text{ ksi})$
$= 1.19 \text{ in.}^2 \ (767 \text{ mm}^2)$

For L3 × 3 × 1/4, $A_s = 1.44 \text{ in.}^2 \ (929 \text{ mm}^2) > A_n$, **OK**
USE L3 × 3 × 1/4.

11.10 FOUNDATION

As noted in Chapter 7, foundation design based on conventional loading will usually result in adequate performance during a blast event.

References

AISC (American Institute of Steel Construction). 2017. *Steel construction manual.* 15th ed. Chicago: AISC.

ANSI/AISC (American National Standards Institute/American Institute of Steel Construction). 2019. *Specification for structural steel buildings*, ANSI/AISC 360-16. Revised 2019. Chicago: AISC.

ASTM International. 2017. *Standard specification for steel for structural shapes for use in building framing*, ASTM A992/A992M-01. West Conshohocken, PA: ASTM.

ASTM. 2019. *Standard specification for carbon structural steel*, ASTM A36/A36M-19. West Conshohocken, PA: ASTM.

ASTM. 2022. *Standard specification for steel sheet, zinc-coated (galvanized) or zinc-iron alloy-coated (galvannealed) by the hot-dip process*, ASTM A653/A653M-20. West Conshohocken, PA: ASTM.

Biggs, J. M. 1964. *Introduction to structural dynamics*. New York: McGraw Hill.

USACE/NAVFAC/AFCESA (US Army Corp of Engineers/Naval Facilities Engineering Command/Air Force Civil Engineer Support Agency). 2014. *Structures to resist the effects of accidental explosions*, UFC 3-340-02, Change 2. Washington, DC: National Institute of Building Sciences.

CHAPTER 12

Masonry Retrofit Design Example

12.1 INTRODUCTION

This chapter provides an example of the evaluation and retrofit of the masonry infill walls of an existing reinforced concrete framed building using the principles outlined in Chapter 9, "Evaluation and Upgrade of Existing Buildings." The explosion magnitude and front wall blast load are determined by others. Only the analysis of the exterior walls and upgrade options for those walls are presented in detail in this example. Checks for secondary items (e.g., shear walls, window openings) are usually only noted but not presented in detail.

Structural code provisions, as applicable, are taken from TMS 402/602-22 (TMS 2022), *Building Code Requirements and Specifications for Masonry Structures*, and ACI 318-19(22) (ACI 2022a), *Building Code Requirements for Structural Concrete and Commentary*.

Additional references

- NCMA TEK 14-01B (NCMA 2007), *Section Properties of Concrete Masonry Walls*
- ACI PRC-440.7-22 (ACI 2022b), *Externally Bonded Fiber-Reinforced Polymer Systems Design and Construction for Strengthening Masonry Structures—Guide*
- PDC-TR 06-08 Rev 1 (USACE 2008), *Single Degree of Freedom Structural Response Limits for Antiterrorism Design*
- UFC 3-340-02 (USACE/NAVFAC/AFCESA 2014), *Structures to Resist the Effects of Accidental Explosions*

12.2 STRUCTURAL SYSTEM

The structure in this example uses unreinforced masonry wall construction as described in Section 4.3.1.

Note: Although unreinforced masonry is not recommended for blast design because of a lack of ductility, it is often encountered in existing buildings.

12.2.1 Description of the Structure

- One-story reinforced concrete frame structure with 8 in. (203 mm) nominal unreinforced masonry infill walls.
- Plan dimensions are 60 ft (18.3 m) by 80 ft (24.4 m).
- Wall height is 10 ft (3.0 m).
- Roof slab consists of a reinforced concrete section (dimensions not important in the context of this example).

12.2.2 Framing Plan

For the general framing plan and elevation, refer to Figure 12-1.

12.3 DESIGN DATA

12.3.1 Material Properties

Concrete: $f_c' = 4{,}000$ psi (27.6 MPa)

Reinforcing steel: ASTM A615, grade 60, $F_y = 60$ ksi (414 MPa)

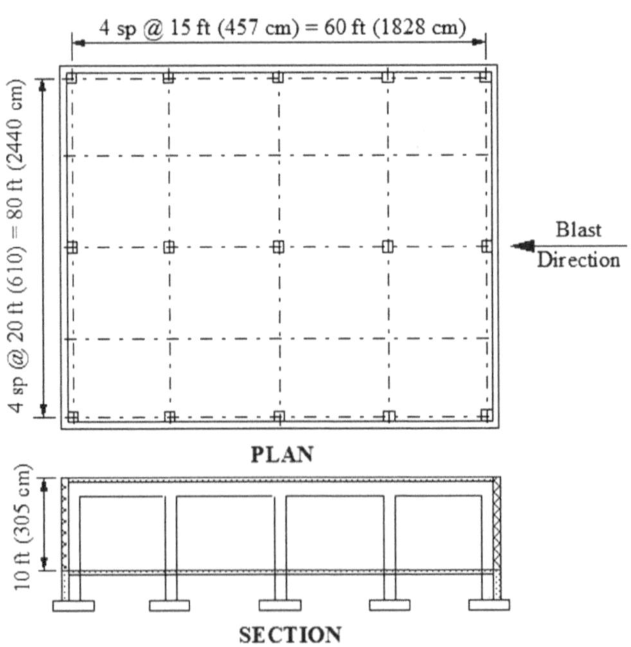

Figure 12-1. Framing plan and elevation.

Table 12-1. Dynamic Material Properties.

Material	F_y *or* f'_c	*SIF*	*DIF*	*Modulus of elasticity,* E_s *or* E_c
Concrete	4 ksi (27.6 MPa)	1.0	1.19	3,605 ksi (24,900 MPa)
Masonry	1.5 ksi (10.3 MPa)	1.0*	1.19	1,607 ksi (11,100 MPa)
Reinforcing steel	60 ksi (414 MPa)	1.1	1.22	29,000 ksi (200,000 MPa)

*SIF for masonry is not presented in this book. Value is defaulted to 1.0.

Masonry (hollow units, running bond): $f'_m = 1{,}500$ psi (10.3 MPa)

Medium weight masonry: 115 pcf (1,840 kg/m^3)

Mortar: Type N, 120 pcf (1,920 kg/m^3)

Strength increase factors (SIFs) and dynamic increase factors (DIFs) are taken from Chapter 5, Tables 5A-1 through 5A-3. These values are summarized in Table 12-1.

12.3.2 Design Loads

The following loads are computed from free-field blast wave parameters. Refer to Chapter 3, "Determination of Loads," for a general load determination procedure.

Peak reflected overpressure: $P_r = 3.4$ psi (23 kPa)

Effective duration: $t_d = 90$ ms

Figure 12-2 shows a pressure–time diagram of the design load.

12.3.3 Building Performance Requirements: Deformation Limits

For unreinforced masonry, the failure mode is often based on tensile cracking. To avoid the resulting catastrophic failure, the existing wall in this example is to remain elastic. Thus, $\mu_a = 1.0$.

Notes: Other response limits for unreinforced masonry are presented in PDC-TR 06-08 Rev 1 (USACE 2008) and UFC 3-340-02 (USACE/NAVFAC/AFCESA 2014) based on a brittle material model with non-elastic, perfectly plastic resistance function. The reader is referred to these and related documents for a more detailed coverage on analyzing unreinforced masonry elements to blast loads.

For upgrade options, a medium response is selected in accordance with Chapter 5, Appendix 5B.

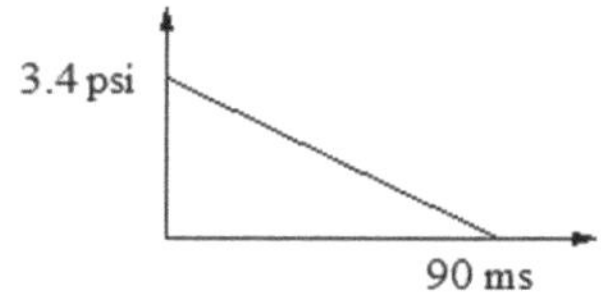

Figure 12-2. Design blast load.

12.4 EXISTING WALL EVALUATION

The exterior masonry walls facing the blast are 10 ft (3,048 mm) high and 40 ft (12,190 mm) wide between supports, a 4 to 1 ratio. Therefore, the wall will be analyzed as a one-way simply supported beam spanning vertically between the grade beam and the roof beam, as shown in Fig. 12.2.

Span, $L = 10$ ft or 120 in. (3,048 mm)

For an 8 in. (203 mm) nominal concrete masonry unit (CMU) wall with ungrouted cells, the following section properties are based on a 1 in. (25.4 mm) width of the wall:

Actual thickness: $t_w = 7.625$ in. (194 mm)

Area: $A_n = 3.46$ in.^{2} (2,230 mm^2) (NCMA, Table 3)

Moment of inertia: $I_g = 27.8$ in.^{4} (11,570,000 mm^4) (NCMA, Table 3)

12.4.1 Compute Required Resistance

For dynamic bending: (Chapter 5, Appendix 5A)

$$f'_{dm} = (\text{SIF})(\text{DIF})\ f'_m = (1.0)(1.19)(1{,}500\text{ psi}) = 1{,}785\text{ psi (12.3 MPa)}$$

Modulus of elasticity (TMS 402, Table 4.2.2):

$$E_m = 900\ f'_{dm} = 900\ (1{,}785\text{ psi}) = 1{,}607{,}000\text{ psi (11,100 MPa)}$$

Elastic stiffness: (Chapter 6, Table 6-1)

$$K_e = 384\ E_m\ I_g/5\ L^3$$

$$= 384\ (1{,}607{,}000\text{ psi})(27.8\text{ in.}^4)/5\ (120\text{ in.})^3$$

$$= 1{,}985\text{ lb/in., or } \mathbf{16.5\ psi/in.}\ (4.49\text{ kPa/mm})$$

Mass:

Weight = (115 pcf)(3.46 in.^{2}/144)(10 ft) = 27.6 lb, or 0.23 psi (1.59 kPa)

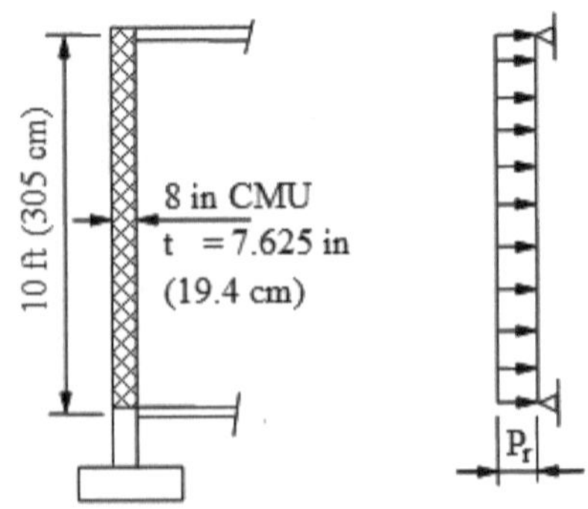

Figure 12-3. Wall section.

$$M = \text{weight/g}$$
$$= (27.6 \text{ lb})/(386 \text{ in./s}^2)$$
$$= 0.0716 \text{ lb·s}^2\text{/in., or } \mathbf{596\ psi{\cdot}ms^2/in.}\ (162 \text{ kg/m}^2)$$

Load-mass transformation factors: (Chapter 6, Table 6-1)

$$\text{elastic, } K_{LM} = K_M/K_L = 0.50/0.64 = \mathbf{0.78}$$

Equivalent mass:

$$M_e = K_{LM}\, M = 0.78\ (596 \text{ psi·ms}^2\text{/in.}) = 465 \text{ psi·ms}^2\text{/in.}\ (126 \text{ kg/m}^2)$$

Natural period [Equation (6-8)]:

$$t_n = 2\pi\sqrt{\frac{M_e}{K_e}} = 2\pi\sqrt{\frac{(465 \text{ psi-ms}^2\text{/in.})}{(16.5 \text{ psi/in.})}} = \mathbf{33.4\ ms}$$

Duration-period ratio:

$$\tau = t_d/t_n = (90 \text{ ms})/(33.4 \text{ ms}) = 2.7$$

Let the ductility demand equal the limiting value, $\mu_d = \mu_a = 1$.
Load/resistance ratio [Equation (6-11)]:

$$F_o/R_m = \frac{\sqrt{(2\mu_d - 1)}}{\pi\tau} + \frac{(2\mu_d - 1)\tau}{2\mu_d(\tau + 0.7)} = \frac{\sqrt{(2\times1-1)}}{\pi\times2.7} + \frac{(2\times1-1)\times2.7}{2\times1\times(2.7+0.7)} = 0.515$$

Required resistance with $F_o = P_r$:

$$R_m = P_r/0.515 = (3.4 \text{ psi})/0.515 = 6.6 \text{ psi (46 kPa)}$$

Note: Because of the symmetry of the wall system, the rebound resistance case will not be the controlling factor and is therefore not included in this example.

12.4.2 Available Flexural Capacity

For unreinforced masonry, flexure is based on the cracking strength of the masonry.

Notes: It is conservatively assumed that the unreinforced masonry wall fails upon reaching its cracking moment, thus neglecting the tensile strength of the mortar. Other approaches can be used to evaluate unreinforced masonry under blast loading, including SDOF analyses with the use of a resistance-deflection "rocking" behavior.

Modulus of rupture: $f_r = 64$ psi (441 kPa) (TMS 402, Table 9.1.9.1)
Weight of the wall above the midpoint:

$$\text{weight} = (0.23 \text{ psi})(120 \text{ in. height/2})(1 \text{ in. width}) = 13.8 \text{ lb (61.5 N)}$$

Cracking moment:

$$M_{cr} = (f_r + \text{weight}/A_n)\ I_o/0.5\ t_w$$
$$= [64 \text{ psi} + (13.8 \text{ lb})/(3.46 \text{ in.}^2)]\ (27.8 \text{ in.}^4)/0.5\ (7.625 \text{ in.})$$
$$= 496 \text{ lb·in. (56.0 N·m)}$$

Resistance at cracking moment: (Chapter 6, Table 6-1)

$$R_b = 8\ M_{cr}/L = 8\ (496\ \text{lb·in.})/(120\ \text{in.}) = 33.1\ \text{lb, or } 0.275\ \text{psi } (1.90\ \text{kPa})$$

12.4.3 Available Shear Capacity

For dynamic shear: (Chapter 5, Appendix 5A)

$$f'_{dm} = (\text{SIF})(\text{DIF})\ f'_m = (1.0)(1.0)(1{,}500\ \text{psi}) = 1{,}500\ \text{psi } (10.3\ \text{MPa})$$

Factored axial load:

$$P_u = 1.2\ (\text{weight}) = 1.2\ (13.8\ \text{lb}) = 16.6\ \text{lb } (73.7\ \text{N})$$

Nominal shear capacity: (TMS 402, Section 9.2.6.1)

$$V_{n1} = 3.8\ A_{nv}\sqrt{f'_m} = 3.8\ (3.46\ \text{in.}^2)\sqrt{1{,}500\ \text{psi}} = 509\ \text{lb } (2.27\ \text{kN})$$

$$V_{n2} = 300\ A_{nv} = 300\ (3.46\ \text{in.}^2) = 1{,}038\ \text{lb } (4.62\ \text{kN})$$

$$V_{n3} = 56\ A_{nv} + 0.45\ P_u = 56\ (3.46\ \text{in.}^2) + 0.45\ (16.6\ \text{lb}) = 201\ \text{lb } (895\ \text{N})$$

$$V_n = \min\ [V_{n1}, V_{n2}, V_{n3}] = 201\ \text{lb } (895\ \text{N})$$

The critical section for shear is located at a distance t_w from the support:

$$R_s = V_n\ L/(0.5\ L - t_w)$$

$$= (201\ \text{lb})(120\ \text{in.})/[0.5\ (120\ \text{in.}) - (7.625\ \text{in.})]$$

$$= 461\ \text{lb, or } 3.84\ \text{psi } (26.5\ \text{kPa})$$

12.4.4 Available Resistance

Because $R_b < R_s$, bending is the controlling behavior.

Positive and rebound resistance, $R_u = -R_{u-} = R_b =$ **0.275 psi** (1.90 kPa)

The existing wall only provides approximately 5% of the required resistance for the specified blast load. Therefore, the wall is not OK in its current configuration. For adequate resistance, the existing wall must either be strengthened or a new wall must be added next to the existing wall.

Multiple options are feasible. For this example, the first two options will be mainly explored with commentary included for two additional options:

Option 1: Fill wall cavities solid with concrete grout and add reinforcing steel.

Option 2: Add a new reinforced concrete wall exterior to the existing wall.

Option 3: Add steel posts to reduce the effective masonry wall span.

Option 4: Apply fiber-reinforced polymer (FRP) strips or mats to the existing wall.

12.5 OPTION 1: FULLY GROUTED AND REINFORCED EXISTING WALL

In this upgrade option, longitudinal #5 rebars are initially provided at 16 in. (406 mm) on the center and the wall cavities are filled solid with concrete grout.

For an 8 in. nominal (203 mm) CMU wall with fully grouted cells, the following section properties are based on a 1 in. (25.4 mm) width of wall:

Actual thickness: $t_w = 7.625$ in. (194 mm)

Area: $A_n = 7.625$ in.^{2} (4,920 mm^2) (NCMA, Table 3)

Moment of inertia: $I_g = 36.9$ in.^{4} (15,360,000 mm^4) (NCMA, Table 3)

#5 @ 16 in. (406 mm), vertical within the center of the cell

#5, diameter = 0.625 in. (15.9 mm); #5, area = 0.307 in.^{2} (198 mm^2)

12.5.1 Compute Bending Resistance

For dynamic bending: (Chapter 5, Appendix 5A)

$F_{ds} = F_{dy} =$ (SIF)(DIF) F_y = (1.1)(1.22)(60,000 psi) = 80,500 psi (555 MPa)

f'_{dm} = (SIF)(DIF) f'_m = (1.0)(1.19)(1,500 psi) = 1,785 psi (12.3 MPa)

For the nominal design width:

A_s = (0.307 in.^{2})/(16 in. bar spacing) = 0.0192 in.^{2} (12.4 mm^2)

For bending tension on either face:

$d = t_w/2$ = (7.625 in.)/2 = 3.81 in. (96.8 mm)

$d_{sb} = A_s F_{dy}/0.8 f'_{dm} b$ (TMS 402, Section 9.3.4)

= (0.0192 in.^{2})(80,500 psi)/(0.8)(1,785 psi)(1 in.)

= 1.08 in. (27.5 mm)

Note: In *ACI 318*, the parameter "d_{sb}" is defined as "a."

$M_p = A_s F_{ds} [d - d_{sb}/2]$ (TMS 402, Section 9.3.4)

= (0.0192 in.^{2})(80,500 psi) [(3.81 in.) – (1.08 in.)/2]

= 5,050 lb·in. (571 N·m)

$M_p > 1.3 M_{cr}$, **OK** (TMS 402, Section 9.3.3.2.2.1)

$R_b = 8 M_p/L$ = 8 (5,050 lb·in.)/(120 in.) = 337 lb, or 2.81 psi (19.3 kPa) (Chapter 6, Table 6-1)

12.5.2 Compute Shear Resistance

For dynamic shear: (Chapter 5, Appendix 5A)

$$f'_{dm} = (\text{SIF})(\text{DIF})\ f'_m = (1.0)(1.0)(1{,}500\text{ psi}) = 1{,}500\text{ psi } (10.3\text{ MPa})$$

$$V_n = 4\,A_n\sqrt{f'_{dm}}\quad [\text{TMS 402, Equation (9-17)}]$$

$$= 4\,(7.625\text{ in.}^2)\sqrt{1{,}500\text{ psi}} = 1{,}180\text{ lb } (5.25\text{ kN})$$

The critical section for shear is at a distance t_w from the support:

$$R_s = V_n\,L/(0.5\,L - t_w)$$

$$= (1{,}180\text{ lb})(120\text{ in.})/[0.5\,(120\text{ in.}) - (7.625\text{ in.})]$$

$$= 2{,}710\text{ lb, or } 22.6\text{ psi } (156\text{ kPa})$$

12.5.3 Resistance and Permissible Response

Because $R_b < R_s$, bending is the controlling factor.

$R_u = -R_{u-} = R_b =$ **2.81 psi** (19.3 kPa)

Response Limits: $\theta_a = 2$ degrees (Chapter 5, Table 5B-4)

12.5.4 Compute Other Single-Degree-of-Freedom Properties

Masonry modulus of elasticity: (TMS 402, Table 4.2.2)

$$E_m = 900\ f'_{dm} = 900\,(1{,}785\text{ psi}) = 1{,}607{,}000\text{ psi } (11{,}100\text{ MPa})$$

Rebar modulus of elasticity: (TMS 402, Table 4.2.2)

$$E_s = 29{,}000{,}000\text{ psi } (200{,}000\text{ MPa})$$

Modular ratio:

$$n = E_s/E_m = (29{,}000{,}000\text{ psi})/(1{,}607{,}000\text{ psi}) = 18.1$$

Transformed rebar area:

$$n\,A_s = (18.1)(0.0192\text{ in.}^2) = 0.346\text{ in.}^2\ (223\text{ mm}^2)$$

Location of a transformed neutral axis:

$$d_{na} = \frac{-nA_s + \sqrt{nA_s(nA_s + 2bd)}}{b}$$

$$= \frac{-0.346\text{ in.}^2 + \sqrt{0.346\text{ in.}^2(0.346\text{ in.}^2 + 2(1\text{ in.})(3.81\text{ in.}))}}{1\text{ in.}}$$

$$= 1.31\text{ in. } (33.4\text{ mm})$$

Cracked moment of inertia:

$$I_{cr} = b\, d_{na}^3/3 + n\, A_s\, (d - d_{na})^2$$
$$= (1 \text{ in.})(1.31 \text{ in.})^3/3 + (0.346 \text{ in.}^2)(3.81 \text{ in.} - 1.31 \text{ in.})^2$$
$$= 2.92 \text{ in.}^4\ (1{,}214{,}000 \text{ mm}^4)$$

Average moment of inertia:

$$I_a = (I_g + I_{cr})/2 = (36.9 \text{ in.}^4 + 2.92 \text{ in.}^4)/2 = 19.9 \text{ in.}^4\ (8{,}287{,}000 \text{ mm}^4)$$

Elastic stiffness:

$$K_e = 384\, E_m\, I_a/5\, L^3 \text{ (Chapter 6, Table 6-1)}$$
$$= 384\, (1{,}606{,}500 \text{ psi})(19.9 \text{ in.}^4)/5\, (120 \text{ in.})^3$$
$$= 1{,}421 \text{ lb/in., or } \mathbf{11.8 \text{ psi/in.}}\ (3.22 \text{ kPa/mm})$$

Elastic yield deflection: $y_e = R_u/K_e = (2.81 \text{ psi})/(11.8 \text{ psi/in.}) = \mathbf{0.238 \text{ in.}}$ (6.05 mm)

Mass:

$$\text{weight} = [(115 \text{ pcf})(3.46 \text{ in.}^2) + (120 \text{ pcf})(7.625 - 3.46 \text{ in.}^2)]/144 \times 10 \text{ ft}$$
$$= 62.3 \text{ lb, or } 0.52 \text{ psi } (3.58 \text{ kPa})$$

$$M = \text{weight/g}$$
$$= (62.3 \text{ lb})/(386 \text{ in./s}^2)$$
$$= 0.161 \text{ lb·s}^2\text{/in., or } \mathbf{1{,}345 \text{ psi·ms}^2\text{/in.}}\ (365 \text{ kg/m}^2)$$

Load-mass transformation factors: (Chapter 6, Table 6-1)

$$\text{elastic, } K_{LM} = K_M/K_L = 0.50/0.64 = \mathbf{0.78}$$
$$\text{plastic, } K_{LM} = K_M/K_L = 0.33/0.50 = \mathbf{0.66}$$

Assuming some plastic deformation, use an average value for the load-mass factor when calculating the natural period:

$$K_{LM} = (0.78 + 0.66)/2 = \mathbf{0.72}$$

Equivalent mass:

$$M_e = K_{LM}\, M$$
$$= 0.72\, (1{,}345 \text{ psi·ms}^2\text{/in.})$$
$$= 968 \text{ psi·ms}^2\text{/in. } (263 \text{ kg/m}^2)$$

Natural period: [Equation (6-8)]

$$t_n = 2\pi\sqrt{\frac{M_e}{K_e}} = 2\pi\sqrt{\frac{(968 \text{ psi-ms}^2\text{/in.})}{(11.8 \text{ psi/in.})}} = \mathbf{56.9 \text{ ms}}$$

12.5.5 Compute Response and Compare It to Deformation Limits (Chart Solution)

$t_d/t_n = (90 \text{ ms})/(56.9 \text{ ms}) = 1.58$

$R_u/F_o = (2.81 \text{ psi})/(3.4 \text{ psi}) = 0.83$

Using these values, $\mu_d \approx 4.5$. (Chapter 6, Figure 6-6)

Maximum deflection: $y_m = \mu_d\, y_e = 4.5\,(0.238 \text{ in.}) = 1.07 \text{ in.}\ (27.2 \text{ mm})$

Support rotation: (Chapter 5, Figure 5-9)

$\theta_d = \tan^{-1}(y_m/0.5\,L) = \tan^{-1}[(1.07 \text{ in.})/(0.5)(120 \text{ in.})] = 1.02 \text{ degrees} < \theta_a$, **OK**

The response is adequate. A revision is not necessary.

Notes: The reinforcement ratio should be checked to ensure that limits in ACI 318 and UFC 3-340-02 are not exceeded, as applicable. For blast design, the reinforcement ratio should not approach that of a balanced condition, as testing has shown that more ductility is available in reinforced masonry walls with lower reinforcement ratios. Placing rebars within every cell should be avoided. For thicker CMU walls (e.g., 12 in. nominal), one might consider using two rebars within a given grouted cell, one near the front and back block shell, to meet response requirements.

12.6 OPTION 2: NEW REINFORCED CONCRETE WALL

A new reinforced concrete wall has been determined to be a constructible solution to provide the required blast resistance. The new wall is simply supported at the top and bottom, similarly to the details shown in Chapter 9, Figure 9-14.

Span, $L = 10$ ft or 120 in. (3,048 mm) from the foundation to the base of the extended roof slab

Use a nominal design width, $b = 1$ ft or 12 in. (305 mm).

Try:

An 8 in thick concrete wall (203 mm)

#5 @ 16 in. (406 mm) at the center of the wall, vertical

#5, diameter = 0.625 in. (15.9 mm); #5, area = 0.307 in.^{2} (198 mm^2)

#4 @ 12 in. (305 mm) at the center of the wall, horizontal

#4, diameter = 0.5 in. (12.7 mm); #4, area = 0.196 in.^{2} (127 mm^2)

12.6.1 Compute Bending Resistance

For dynamic bending: (Chapter 5, Appendix 5A)

$F_{ds} = F_{dy} = (\text{SIF})(\text{DIF})\ F_y = (1.1)(1.22)(60{,}000 \text{ psi}) = 80{,}500 \text{ psi}\ (555 \text{ MPa})$

$f'_{dc} = (\text{SIF})(\text{DIF})\ f'_c = (1.0)(1.19)(4{,}000 \text{ psi}) = 4{,}760 \text{ psi}\ (32.8 \text{ MPa})$

For the nominal design width:

$A_s = (0.307 \text{ in.}^2)(12 \text{ in./ft})/(16 \text{ in. bar spacing}) = 0.23 \text{ in.}^2 \ (148 \text{ mm}^2)$

For bending tension on either face:

$$d = t_w/2 = 4.0 \text{ in. } (102 \text{ mm})$$

$$A_{s,\min} = 3\frac{\sqrt{f'_{dc}}}{F_{ds}}bd = 3\frac{\sqrt{4{,}760 \text{ psi}}}{80{,}500 \text{ psi}}(12 \text{ in.})(4.0 \text{ in.}) \text{ (ACI 318, Section 9.6.1.2)}$$

$$= 0.123 \text{ in.}^2 \ (79.6 \text{ mm}^2) < A_s, \quad \textbf{OK}$$

$$A_{s,\min} = 200\ bd/F_{dy} = (200)(12 \text{ in.})(4.0 \text{ in.})/(80{,}500 \text{ psi}) \text{ (ACI 318, Section 9.6.1.2)}$$

$$= 0.119 \text{ in.}^2 \ (76.9 \text{ mm}^2) < A_s, \quad \textbf{OK}$$

$$d_{sb} = A_s\ F_{ds}/0.85\ f'_{dc}\ b$$

$$= (0.23 \text{ in.}^2)(80{,}500 \text{ psi})/(0.85)(4{,}760 \text{ psi})(12 \text{ in.})$$

$$= 0.38 \text{ in. } (9.7 \text{ mm})$$

$$M_p - A_s\ F_{ds}\ [d - d_{sb}/2]$$

$$= (0.23 \text{ in.}^2)(80{,}500 \text{ psi})\ [(4.0 \text{ in.}) - (0.38 \text{ in.})/2]$$

$$= 70{,}600 \text{ lb·in. } (7{,}970 \text{ N·m})$$

Positive (inward) bending resistance based on pinned ends:

$$R_b = 8\ M_p/L = 8\ (70{,}600 \text{ lb·in.})/(120 \text{ in}). \text{ (Chapter 6, Table 6-1)}$$

$$= 4{,}705 \text{ lb, or } 3.27 \text{ psi } (22.5 \text{ kPa})$$

12.6.2 Compute Shear Resistance

For dynamic shear: (Chapter 5, Appendix 5A)

$$f'_{dc} = (\text{SIF})\ (\text{DIF})\ f'_c = (1.0)(1.0)(4{,}000 \text{ psi}) = 4{,}000 \text{ psi } (27.6 \text{ MPa})$$

$$V_n = 2\sqrt{f'_{dc}}\ bd \text{ (ACI 318, Table 22.5.5.1)}$$

$$= 2\sqrt{4{,}000 \text{ psi}}\ (12 \text{ in.})(4.0 \text{ in.}) = 6{,}070 \text{ lb } (27.0 \text{ kN})$$

The critical section for shear is at a distance d from the support.

$$R_s = V_n\ L/(0.5\ L - d)$$

$$= (6{,}070 \text{ lb})(120 \text{ in.})/[0.5\ (120 \text{ in.}) - (4.0 \text{ in.})]$$

$$= 13{,}010 \text{ lb, or } 9.04 \text{ psi } (62.3 \text{ kPa})$$

12.6.3 Resistance and Permissible Response

Because $R_b < R_s$, bending is the controlling factor.

$R_u = -R_{u-} = R_b =$ **3.27 psi** (22.5 kPa)

Response limit: $\theta_a = 2$ degrees (Chapter 5, Table 5B-4)

12.6.4 Compute Other Single-Degree-of-Freedom Properties

Gross moment of inertia:

$$I_g = bh^3/12 = (12\text{ in.})(8\text{ in.})^3/12 = 512\text{ in.}^4\ (213.1 \times 10^6\text{ mm}^4)$$

Concrete modulus of elasticity (based on flexure): (ACI 318, Equation 19.2.2.1.b)

$$E_c = 57{,}000\sqrt{f_c'} = 57{,}000\sqrt{4{,}000\text{ psi}} = 3{,}605{,}000\text{ psi } (24{,}900\text{ MPa})$$

Rebar modulus of elasticity: (ACI 318, Section 20.2.2.2)

$$E_s = 29{,}000{,}000\text{ psi } (200{,}000\text{ MPa})$$

Modular ratio:

$$n = E_s/E_c = (29{,}000{,}000\text{ psi})/(3{,}605{,}000\text{ psi}) = 8.04$$

Transformed rebar area:

$$n\,A_s = (8.04)(0.23\text{ in.}^2) = 1.85\text{ in.}^2\ (1{,}190\text{ mm}^2)$$

Location of a transformed neutral axis:

$$\begin{aligned} d_{na} &= \frac{-nA_s + \sqrt{nA_s(nA_s + 2bd)}}{b} \\ &= \frac{-1.85\text{ in.}^2 + \sqrt{1.85\text{ in.}^2(1.85\text{ in.}^2 + 2(12\text{ in.})(4.0\text{ in.}))}}{12\text{ in.}} \\ &= 0.967\text{ in. } (24.6\text{ mm}) \end{aligned}$$

Cracked moment of inertia:

$$\begin{aligned} I_{cr} &= bd_{na}^{\ 3}/3 + nA_s\,(d - d_{na})^2 \\ &= (12\text{ in.})(0.967\text{ in.})^3/3 + (1.85\text{ in.}^2)(4.0\text{ in.} - 0.967\text{ in.})^2 \\ &= 20.6\text{ in.}^4\ (8{,}593{,}000\text{ mm}^4) \end{aligned}$$

Average moment of inertia:

$$I_a = (I_g + I_{cr})/2 = (512\text{ in.}^4 + 20.6\text{ in.}^4)/2 = 266\text{ in.}^4\ (110.9 \times 10^6\text{ mm}^4)$$

Elastic stiffness:

$$\begin{aligned} K_e &= 384\,E_c\,I_a/5\,L^3 \text{ (Chapter 6, Table 6-1)} \\ &= 384\,(3{,}605{,}000\text{ psi})(266\text{ in.}^4)/5\,(120\text{ in.})^3 \\ &= 42{,}670\text{ lb/in., or } 29.6\text{ psi/in. } (8.04\text{ kPa/mm}) \end{aligned}$$

Elastic yield deflection: $y_e = R_u/K_e$ = (3.27 psi)/(29.6 psi/in.) = **0.11 in.** (2.8 mm)

Mass:

$$\text{weight} = (150\text{ pcf})(0.667\text{ ft thick})(1\text{ ft nominal width})(10\text{ ft span})$$
$$= 1{,}000\text{ lb, or } 0.694\text{ psi } (4.79\text{ kPa})$$

$$M = \text{weight}/g$$
$$= (1{,}000\text{ lb})/(386\text{ in./s}^2)$$
$$= 2.59\text{ lb·s}^2\text{/in., or } \mathbf{1{,}800\text{ psi·ms}^2\text{/in.}}\ (488\text{ kg/m}^2)$$

Load-mass transformation factors: (Chapter, Table 6-1)

$$\text{elastic, } K_{LM} = K_M/K_L = 0.50/0.64 = \mathbf{0.78}$$
$$\text{plastic, } K_{LM} = K_M/K_L = 0.33/0.50 = \mathbf{0.66}$$

Assuming some plastic deformation, use an average value for the load-mass factor when calculating the natural period:

$$K_{LM} = (0.78 + 0.66)/2 = \mathbf{0.72}$$

Equivalent mass:

$$M_e = K_{LM}\,M = (0.72)(1{,}800\text{ psi·ms}^2\text{/in.}) = 1{,}296\text{ psi·ms}^2\text{/in. } (352\text{ kg/m}^2)$$

Natural period: [Equation (6-8)]

$$t_n = 2\pi\sqrt{\frac{M_e}{K_e}} = 2\pi\sqrt{\frac{(1{,}296\text{ psi-ms}^2\text{/in.})}{(29.6\text{ psi/in.})}} = \mathbf{41.6\text{ ms}}$$

12.6.5 Compute the Response and Compare It to Deformation Limits (Chart Solution)

$$t_d/t_n = (90\text{ ms})/(41.6\text{ ms}) = 2.16$$
$$R_u/F_o = (3.27\text{ psi})/(3.4\text{ psi}) = 0.96$$

Using these values, $\mu_d \approx 3.5$ (Chapter 6, Figure 6-6).
Maximum deflection: $y_m = \mu_d\, y_e$ = 3.5 (0.11 in.) = 0.39 in. (9.8 mm)
Support rotation: (Chapter 5, Figure 5-9)

$$\theta_d = \tan^{-1}(y_m/0.5\,L) = \tan^{-1}[(0.39\text{ in.})/(0.5)(120\text{ in.})] = 0.37\text{ degrees} < \theta_a,\quad \textbf{OK}$$

The response is adequate. A revision is not necessary.

An illustration of this option is presented in Chapter 9, Figure 9-7.

Notes: The reinforced concrete wall nets a smaller deflection than the reinforced masonry upgrade of similar thickness and reinforcement layout calculated in Section 12.6. This is the result of the concrete wall having a 16% higher resistance and a 34% higher mass.

Notes: The reinforced concrete wall needs to be positioned sufficiently away from the existing masonry wall to not only avoid direct impact during its deflection but also avoid creating a significant gap pressure between the walls. The response of the existing wall needs be checked against the pressure generated as a result of the volume displacing between the walls. (The gap pressure decreases as the separation distance, and thereby the volume, between the walls increases.) This calculation is commonly performed using a time-stepping process similar to that illustrated in Chapters 10 and 11. The design of a new reinforced concrete wall should also consider any required check set forth by ACI 318 ACI (2017) for conventional reinforced concrete structures.

12.7 OPTION 3: STEEL POST UPGRADE

The steel post retrofit option presented in Chapter 9, Section 9.7.1, is briefly discussed here. The vertical post spacing would need to be such that the resulting reduced horizontal span of the masonry wall can adequately sustain the design blast load. From the calculation performed in Section 12.4.1, the unreinforced masonry wall of this example would need approximately a resistance of 6.6 psi to respond elastically to the 3.4 psi blast load. This translates to a required reduced span of about 2 to 3 ft, making the approach potentially viable.

The SDOF properties for flexural response would need be rechecked to confirm performance, particularly as a reduced span would not only increase the wall resistance but also increase its stiffness, thereby resulting in a net reduction in the elastic yield deflection and allowable deflection. Rotation-based response limits for unreinforced masonry such as those presented in PDC-TR 06-08 Rev 1 could potentially be referenced to determine a maximum permissible horizontal wall span. In the absence of empirical testing, the design of steel posts is usually limited to ensure that the peak deflection of the steel members is sustainable by the connected masonry.

Besides flexural checks for the masonry wall and posts, shear checks would need to be performed to determine the size and spacing of anchors supporting the steel posts (either tube sections with angles or wide flange sections). Unless confirmed by testing, anchors should be inserted into fully grouted cells only.

12.8 OPTION 4: FIBER-REINFORCED POLYMER WALL UPGRADE

The FRP retrofit option presented in Chapter 9, Section 9.7.2, is also briefly discussed here. The application would be as shown in Chapter 9, Figures 9-10 and 9-11. Several documents on the use of FRP for structural upgrades are published by ACI (e.g., ACI PRC-440.7) (ACI 2022b). A dedicated task team (ACI 370) is currently working on a recommended practice document for the use of FRP for blast applications.

The primary advantage of FRP systems is that they are relatively thin and do not require much space to be applied to a wall surface. A major drawback, however, is that FRP systems would at minimum need to be applied to the interior (i.e., tension) wall face, which may not be practical from an operations perspective. Fiber-reinforced polymer application to the exterior wall face may also be needed to prevent adverse rebound response, particularly for load-bearing walls. Wall grouting would also likely be needed to ensure that adequate shear capacity is provided. Additional considerations for using FRP systems in industrial facilities include the type of FRP material being specified within a particular chemical environment and the potential for hazardous material breakdown when subjected to fires.

12.9 CONCLUDING REMARKS

The analysis of the existing masonry wall revealed that the wall provides only a small percentage of the required resistance for the specified blast load.

A complete analysis of the walls will need to include the evaluation of the existing connection of the wall to the grade beam and roof. The evaluation may reveal that the connections are inadequate for the specified loads, and thus a corrective procedure should be specified. For Option 1 where reinforcement is added to the existing wall, dowels should be specified and embedded in the masonry and the concrete beams. Alternatively, continuous steel angles can be used to connect the walls to the roof and grade beam. This approach can also be applied to Option 2. However, if new concrete walls are added next to the existing wall, the wall can potentially be directly bolted to the grade beam and roof beam (refer to Chapter 9, Figure 9-14). In this case, the bolts should be checked for rebound pullout forces.

In this example, Option 1 may be the most cost-effective upgrade because of the minimal usage of new materials and formwork, with work being performed exclusively outside of the building and usually within the existing building footprint. Option 3 is also attractive construction-wise, with work being performed exclusively outside of the building and only a small distance away from the building to accommodate the added posts. These options might not be feasible, however, because of existing obstructions, or due to extensive alterations required to achieve the proposed reinforcement scheme. If Option 1 or Option 3 is not practical or feasible, one of the other options may be used, although each presents its own challenges. Option 2 requires a more significant expansion from the existing building footprint. Meanwhile, Option 4 inherently requires interior work to be performed for application of the FRP strips or mats on the tension side of the wall, which may result in interruption of the building/process operation.

It should be noted that the example presented addresses only the wall response, without a specific focus on how the upgraded wall blast reactions are delivered to the rest of the structure. A thorough blast assessment should take into consideration the response of the existing roof including out-of-plane (deck + joists) and in-plane (diaphragm) forces delivered by the walls under blast loading.

A complete blast assessment of an existing building would also take into consideration nonstructural elements that comprise the outer building envelope, such as doors, windows, louvers, and other openings. Doors, for example, can be designed to remain functional after a blast event or simply not to compromise the building envelope or become projectiles depending on the acceptable damage level. Wall areas near openings need to be capable of supporting reactions developed by nonstructural elements that are to stay in place, while having a reduced net area. Adequate reinforcement (or steel posts or FRP strips depending on the chosen option) should be provided around openings to account for load concentrations. Windows, if present, may be blocked in with new fully grouted CMU blocks adequately doweled into the surrounding wall or otherwise covered to limit glazing projectile hazards.

References

ACI (American Concrete Institute). 2017. *Guide for the design and construction of externally bonded FRP systems for strengthening concrete structures*, ACI PRC-440.2-17. Farmington Hills, MI: ACI.

ACI. 2022a. *Building code requirements for structural concrete and commentary*, ACI 318-19(22). Reaffirmed 2022. Farmington Hills, MI: ACI.

ACI. 2022b. *Externally Bonded fiber-reinforced polymer systems design and construction for strengthening masonry structures—Guide*, ACI PRC-440.7-22. Farmington Hills, MI: ACI.

NCMA (National Concrete Masonry Association). 2007. *Section properties of concrete masonry walls*, Guidance Document TEK 14-01B. Herndon, VA: NCMA.

TMS (The Masonry Society). 2022. *Building code requirements and specifications for masonry structures*, TMS 402/602-22. Longmont, CO: TMS.

USACE (US Army Corp of Engineers). 2008. *Single degree of freedom structural response limits for antiterrorism design*, PDC-TR 06-08 Rev 1. Washington, DC: USACE.

USACE/NAVFAC/AFCESA (US Army Corp of Engineers/Naval Facilities Engineering Command/Air Force Civil Engineer Support Agency). 2014. *Structures to resist the effects of accidental explosions*, UFC 3-340-02. Washington, DC: National Institute of Building Sciences.

Glossary

Angle of Incidence:	Angle between the direction of the blast wave movement and a flat surface.
Blast Wave:	A transient change in the gas density, pressure, and velocity of the air surrounding an explosion.
BLEVE:	Boiling Liquid Expanding Vapor Explosion
BRM:	Blast Resistant Modular
CFD:	Computational Fluid Dynamics
Conventional Loads:	Load normally considered in structural design such as dead loads, live loads, wind loads, and seismic loads.
Deflagration:	A propagating chemical reaction of a substance in which the reaction front advances into the unreacted substance rapidly but at less than sonic velocity.
Detonation:	A propagating chemical reaction of a substance in which the reaction front advances into the unreacted substance at or greater than sonic velocity.
Ductility Ratio:	A measure of the energy absorbing capacity of a structural member; ratio that is computed by dividing the element's maximum deformation by the yield deformation.
Duration:	Time from initial change in pressure to return to ambient pressure.
Dynamic Increase Factor (DIF):	Ratio of dynamic to static strength that is used to compute the effect of a rapidly applied load to the strength of a structural element.
Dynamic Reactions:	Support reaction of a structural component due to the dynamic blast loading, taking into account inertia effects.
Elastic Region:	Deformation range from zero up to the formation of the first plastic hinge.
Elasto-Plastic Region:	Deformation range from formation of the first plastic hinge up to formation of the final plastic hinge (i.e., ultimate capacity).
FEA:	Finite Element Analysis

Flammable Range: Range of mixture of fuel and air that will support flame propagation.

Free Field: Air or ground blast waves that are unimpeded by obstructions in the path of the wave.

Hinge Rotation: A measure of the energy absorbing capacity of a structural member; angle of deformation at a plastic hinge.

Impulse: Integrated area under the overpressure time curve.

Inelastic: Beyond the elastic response range.

Incident Side-On Overpressure: Initial peak pressure rise, above ambient, produced by a shock wave or pressure wave as felt by a flat surface oriented parallel to the direction of wave propagation.

Incipent Failure: Level of deformation where collapse can be expected to occur.

Linear: A response limited to the elastic range.

Lower Flammable Limit: Lowest mixture of fuel in air that will support flame propagation.

MDOF: Multi-Degree of Freedom

Negative Phase: Portion of the pressure-time history typically following the positive (overpressure) phase in which the pressure is below ambient pressure (suction).

Neutral Risk: Idea where a person inside a building should not be at greater risk of injury than another person just outside.

Nonlinear: A response that includes the elastic-plastic and/or plastic ranges.

OSHA: Occupational Safety and Health Administration (or Occupational Safety Hazards Act)

Overpressure: Pressure rise above ambient produced by a shock wave or pressure wave.

Plastic Region: Deformation range from ultimate capacity up to failure of the element.

Positive Phase: Portion of the pressure time history in which the pressure is above ambient pressure.

Pressure Wave: A blast wave that produces a gradual rise in pressure.

Rebound: Deformation in the direction opposing the initial blast pressure. This occurs after a component has reached a peak deformation and returns toward its initial position.

Reflected Overpressure: Rise in pressure produced by a shock wave or pressure wave as felt by a flat surface oriented perpendicular to the direction of wave propagation.

Resistance-Deflection Function: Value of the stress in a structural element as the deformation is increased from zero through the elastic range, the elastic-plastic range, ultimate capacity, and finally to failure of the element.

SDOF: Single Degree of Freedom.

Shockwave: A blast wave that produces a near instantaneous rise in pressure.

Side-on Pressure: Rise in pressure above ambient produced by a shock wave sweeping unimpeded across any surface (walls or roof) not facing the blast source.

Sidesway: Lateral movement of a structure due to vertical or horizontal loads.

Strain Energy: Energy stored within a structural element deformed due to the application of load. Value of strain energy is the area under the resistance-deflection function.

Strain Hardening: Observed increase in strength as a material is deformed well into the plastic range.

Strain Rate: Speed at which a material if deformed. The higher the strain rate, the higher the observed material strength.

Strength Increase Factor (SIF): Ratio of actual to nominal strength of a material. This factor takes into account conservatism in the manufacturing process.

Support Rotation: A measure of the blast absorbing capacity of a structural element. This is the same as hinge rotation except that the angle is computed at the member's support location.

TNT: Trinitrotoluene, a high explosive used as the basis for many charts describing blast effects.

TNT Equivalent: Amount of TNT which will produce similar effects as the actual amount of explosive material under consideration. An equivalence between two explosives can be determined by equating the quantity of energy released or by relating observed levels of damage.

Ultimate Capacity: Load applied to a structural element as the final plastic hinge, or collapse mechanism, is formed.

Ultimate Strength:	A method of design in which structural members are proportioned by total section capacities rather than by extreme fiber allowable stresses.
Upper Flammable Limit:	Maximum mixture of fuel in air that will support flame propagation.
Volatile:	Description of a substance that evaporates quickly or is unstable.
VCE:	Vapor Cloud Explosion

Index

Page numbers followed by *f* indicate a figure; page numbers followed by *t* indicate a table.

FSC
www.fsc.org
MIX
Papier | Fördert gute Waldnutzung
FSC® C083411

Kolibri 360 GmbH
Ferdinand-Jühlke-Straße 7
99095 Erfurt, Deutschland
produktsicherheit@kolibri360.de